BIODIVERSITY IN BRITISH COLUMBIA:

OUR CHANGING ENVIRONMENT

Editors:

Lee E. Harding
Emily McCullum

Environment
Canada
Canadian Wildlife Service

Environnement
Canada
Service Canadien de la Faune

CANADIAN CATALOGUING IN PUBLICATION DATA

Main entry under title:

Biodiversity in British Columbia:
Our Changing Environment

ISBN 0-662-20671-1
DSS cat. no. En40-463/1994E

1. Biological diversity — British Columbia.
2. Biological diversity conservation — British
Columbia. I. Harding, Lee. E. II. McCullum,
Emily. III. Canada. Environment Canada.
Canadian Wildlife Service. Pacific and Yukon
Region.

QH77.B74B56 1994 574.5 09711 C93-099586-4

Acknowledgements

Environment Canada wishes to acknowledge gratefully all those who generously contributed their knowledge and time to this project. In addition to the authors, whose names are listed in the Table of Contents, we would like to thank the resource people and the reviewers, whose names are given at the conclusion of each Chapter or section. Also deserving of much gratitude are those who helped with the mechanics of putting this book together: Christiane Coté, whose help with publication details was invaluable; Andrew Fabro, who kept the editors up to date on useful information sources and proofread the manuscript; Melissa Turnbull, who did a great deal of the typing and preparation work for electronic publication; Dick Boak, who produced many of the graphics; Pam Krannitz, who reviewed the entire manuscript; and David McCullum, who did the electronic publishing. Leslie Churchland is especially to be thanked for her advice and support throughout the preparation of this document.

Environment Canada also wishes to thank the agencies whose staff contributed to this project as authors, resource people or reviewers. They include the British Columbia Ministries of Environment, Lands and Parks; Forestry; and Agriculture, Fish and Food; the Royal British Columbia Museum; the Association of Professional Biologists; Forestry Canada; Fisheries and Oceans Canada; the University of British Columbia; the University of Victoria; and Simon Fraser University. Without their cooperation, this book would not have been possible.

Table of Contents

Part IV: Prospects for the Future

List of Figures

List of Tables

Preface

In June, 1992, at the United Nations Conference on Environment and Development (the "Earth Summit"), Canada and about 160 other countries agreed to a Global Convention on Biodiversity. Then Prime Minister of Canada, The Honourable Brian Mulroney, said, "The biodiversity accord is an extremely important aspect of sustainable development because it covers, quite simply, life on earth."[1] On December 4, 1992, in Delta, British Columbia, the Prime Minister affirmed his statement by signing the biodiversity convention on behalf of Canada. This convention provides a framework for conserving the diversity of the planet's living organisms and maintaining their habitats.

Environment Canada is a lead agency responsible for following through on Canada's commitment to maintaining biodiversity. This book, in which the existing knowledge of biodiversity in British Columbia is surveyed and recommendations for conservation of biodiversity are made, comprises a step toward that end. Information provided in this book was contributed by authorities in various fields who were encouraged to express their opinions and make recommendations based on their expertise. Their opinions and recommendations do not, therefore, necessarily reflect the position or policies of Environment Canada, nor of their own agencies. While the facts contained in this book are reliable, the accuracy of the predictions can only be proven by future events. Some will certainly be wrong--such is the nature of today's rapidly changing environment. But the changes in biodiversity that result from our use of the environment will have sufficient impact on current and future generations of British Columbians that we should not ignore the best guesses of our experts.

---

[1]Notes for an address by Prime Minister Brian Mulroney. Environment Week, Canadian Museum of Civilization, Hull, Quebec, June 1, 1992.

Sommaire

Le Canada et quelque 160 autres pays ont signé la Convention mondiale sur la biodiversité. Actuellement, le gouvernement fédéral, les provinces et les territoires élaborent des stratégies pour la mise en oeuvre de cette convention. Le présent rapport expose certaines des questions qui devront être prises en compte dans une stratégie de mise en oeuvre.

Partie I : Comprendre et catégoriser la biodiversité

Le chapitre 1 traite de l'importance de conserver les écosystèmes entiers, contrairement à la catégorisation et à la protection de chaque espèce individuellement. Dans la section intitulée «Terms of Endangerment», les auteurs donnent un aperçu des processus scientifiques et des mécanismes institutionnels utilisés pour la catégorisation en fonction de la rareté. La section se termine par un examen des outils scientifiques nécessaires pour déterminer le statut sur le plan de la conservation.

Partie II : La diversité des espèces

Treize chapitres passent brièvement en revue le statut actuel de groupes choisis d'espèces indigènes en Colombie-Britannique. On y trouve des chapitres sur les invertébrés terrestres, les papillons de jour et les papillons de nuit, les invertébrés marins, les bryophytes, les champignons, les plantes vasculaires, les algues marines, les reptiles et les amphibiens, les poissons, les mammifères et les oiseaux. Dans ces chapitres, on démontre que les connaissances sur certains groupes d'organismes, par exemple les plantes non vasculaires et les invertébrés, sont insuffisantes pour appuyer leur classement parmi les espèces rares ou en danger. Les auteurs de ces chapitres sont d'avis qu'au moins 50 espèces d'invertébrés, peut-être 15 espèces d'algues marines benthiques, cinq amphibiens, 26 poissons, 63 bryophytes, 124 plantes vasculaires et plusieurs espèces d'oiseaux sont menacés ou en danger de disparition, et qu'un nombre beaucoup plus grand d'espèces sont peut-être rares ou vulnérables. Au moins 23 espèces et sous-espèces (cinq mammifères, quatre oiseaux, un reptile, un poisson, cinq bryophytes, quatre plantes vasculaires et trois invertébrés) sont éteintes ou disparues en Colombie-Britannique, bien que trois d'entre elles ont été réintroduites depuis. En outre, 26 plantes qui ont été signalées par le passé n'ont pas été observées dans la province depuis 1950. Ces chiffres sont plus élevés que le nombre d'espèces officiellement désignées par les organismes fédéraux et provinciaux comme menacées, en danger de disparition ou disparues. La situation est due en partie au fait que les mécanismes officiels, bien qu'ils s'améliorent, ne peuvent encore traiter de l'éventail complet des espèces assurant la biodiversité, et en partie au fait que la science de la catégorisation ne suit pas le même rythme que les facteurs responsables de la rareté et de l'extinction.

Le déclin de la biodiversité des espèces indigènes est plus que compensé, quant au nombre, par les nombreuses espèces végétales et animales introduites. Les perturbations environnementales, par exemple la fragmentation, l'exploitation forestière et le pâturage, permettent à certaines espèces adventices de concurrencer avec succès les espèces indigènes. Certaines espèces adventices sont avantageuses pour la société, mais beaucoup d'entre elles causent des dommages à l'environnement et à l'économie.

Partie III : La diversité des écosystèmes

Aperçu de la diversité des écosystèmes

Étant donné la grande superficie de la Colombie-Britannique, son climat varié et sa physiographie, on trouve dans cette province les écosystèmes les plus diversifiés de tout le Canada. Nos écosystèmes sont façonnés par les changements naturels qui se sont produits, et qui se produisent encore, depuis le dernier âge glaciaire. Depuis que les Européens et les Asiatiques se sont établis dans la province, il y a une centaine d'années, le rythme du changement a augmenté. Environ la moitié des 124 écosystèmes classifiés ont été gravement fragmentés par des routes de même que par des exploitations industrielles et agricoles et par l'expansion urbaine.

Les forêts

La diversité des écosystèmes forestiers est déterminée par le large éventail de climats et de formes de relief ainsi que par la fréquence et le genre des perturbations. Les perturbations naturelles, par exemple les incendies et les infestations d'insectes, sont des fonctions essentielles des écosystèmes forestiers et on leur doit la diversité des habitats. Les feux sont plus fréquents à l'intérieur des terres que sur la côte, où l'absence prolongée d'incendie a permis la croissance de vastes peuplements regroupant certains des plus gros arbres du continent, c'est-à-dire la forêt ombrophile composée de «peuplements vieux». On constate une augmentation de la fréquence des incendies, peut-être à cause des étés chauds et secs des dernières années, mais la superficie détruite par le feu diminue grâce aux méthodes efficaces de lutte contre l'incendie. Au cours des années 1980, on a observé quelques vastes pullulations d'insectes forestiers; parmi les causes possibles, mentionnons les effets combinés de l'utilisation des terres (par exemple l'exploitation forestière) et les effets globaux, par exemple la situation météorologique changeante. Les taux de récolte du bois ont augmenté jusqu'en 1989 lorsque les coupes annuelles ont commencé à diminuer pour parvenir à des niveaux durables. En Colombie-Britannique, la coupe à blanc est utilisée pour plus de 90% du bois récolté, suivie par le reboisement ou le repeuplement naturel et la récolte des arbres de taille commercialisable. L'aménagement forestier à l'aide de ces méthodes impose de nouveaux processus de contrôle du développement des écosystèmes. Le résultat est une perte de la diversité. De nombreux changements positifs ont été apportés ou sont actuellement apportés aux pratiques d'aménagement forestier, mais il n'est pas certain que l'exploitation forestière intensive soit compatible avec la viabilité de l'écosystème forestier à tous les niveaux.

Les prairies

Les prairies de la Colombie-Britannique sont uniques en Amérique du Nord. On y trouve des espèces différentes de celles qui habitent la prairie canadienne ou d'autres prairies de l'Ouest, et une proportion beaucoup plus petite de ce territoire, comparativement à ce qui s'est produit ailleurs en Amérique du Nord, a été convertie à la production agricole. Pratiquement toutes les prairies ont été utilisées comme pâturage pour le bétail, et c'est pourquoi les écosystèmes des prairies intacts sont actuellement extrêmement rares. Des herbes graminiformes pérennes plus petites, des herbes graminiformes annuelles et des herbes non graminéennes, y compris de nombreuses espèces introduites, ont tendance à s'étendre dans les zones fortement utilisées comme pâturage, ce qui se traduit par des tendances à la baisse au niveau de la condition du parcours et par une diminution de la qualité de l'habitat convenant au bétail et à la faune. Par ailleurs, une utilisation modérée des prairies comme pâturage favorise la croissance des espèces fourragères souhaitables. Les prairies de la Colombie-Britannique se rétablissent actuellement de l'utilisation intensive dont elles ont fait l'objet, comme pâturage, au début de la colonisation par les Européens. Parmi les autres facteurs qui modifient les écosystèmes naturels des prairies, mentionnons l'utilisation pour les loisirs, l'empiètement des forêts, l'exploitation agricole et le développement urbain, les programmes de réensemencement pastoral, l'aménagement de routes et de sentiers, l'introduction d'espèces adventices, la lutte contre les prédateurs et la chasse. Dans la vallée de l'Okanagan, les secteurs consacrés au développement urbain et à l'exploitation agricole ont remplacé 90% des prairies relativement naturelles restantes. Dans les prairies du secteur Chilcotin-Cariboo, il est possible que la biodiversité soit surtout menacée par l'empiètement des forêts qui ont peut-être remplacé jusqu'à 30% des prairies depuis 1960.

La zone côtière

Rien ne permet de conclure à une diminution de la diversité des communautés pélagiques (phytoplancton/zooplancton) ou benthiques (endofaune) dans les principaux secteurs du détroit de Géorgie et du détroit Juan de Fuca. Les communautés benthiques sont saines sur le plan structural. Environ le tiers des stocks de saumon ont été réduits ou éliminés. Des éléments écologiques importants ont subi des changements, surtout à la suite de la pêche et de l'introduction d'espèces nouvelles. On ne constate aucun indice ferme d'instabilité écologique; cependant, l'augmentation apparente des proliférations d'algues toxiques, la contamination, la prévalence de la maladie et les changements possibles de la biomasse des principales espèces de poisson justifient la prise de mesures de surveillance. De plus, on possède certaines preuves que la situation météorologique changeante constatée depuis le milieu des années 1970 a modifié la structure et les mouvements des masses d'eau de mer, et ce phénomène peut avoir des répercussions sur le plan biologique.

Les milieux urbains

Les milieux urbains subissent des pertes énormes sur le plan de la biodiversité à cause de la disparition de l'habitat naturel. La ville de Victoria est située au coeur de l'un des écosystèmes les plus menacés en Amérique du Nord. Même si les habitats inchangés sont très rares, de vastes zones sont encore couvertes de communautés végétales indigènes représentatives, en périphérie. Les communautés de plantes indigènes restantes à Victoria sont peut-être suffisamment intactes pour que l'on cherche à les protéger et à les rétablir.

Dans le district régional de Vancouver, environ la moitié du territoire est considéré comme important pour le maintien des fonctions écologiques. Ce territoire comprend les terres boisées de la rive nord, l'estuaire du fleuve Fraser, d'autres vallées baignées par un cours d'eau, des terres humides, des forêts et des tourbières. On trouve 55 plantes vasculaires rares et 23 espèces fauniques menacées, en danger de disparition ou vulnérables dans le secteur Lower Mainland. Au moins 140 ravins en milieu urbain, d'une superficie totale de plus de 15 kilomètres carrés, abritent des milieux semi-naturels reliques habités par des espèces végétales et animales indigènes dont la diversité est considérable.

Partie IV : Transformation du globe

Déforestation tropicale

De nombreuses espèces d'oiseaux chanteurs de Colombie-Britannique se reproduisent au Canada et passent l'hiver sous les tropiques. Dans l'ensemble, ces espèces migratoires néotropicales diminuent en nombre partout en Amérique du Nord. Deux principales raisons expliquent la situation: la perte de l'habitat d'hivernage tropical provoquée par la déforestation tropicale et la fragmentation de l'habitat de reproduction nordique dans les forêts tempérées. On ne possède que très peu de données fiables sur les tendances des oiseaux forestiers en Colombie-Britannique, mais plusieurs se trouvent sur la liste des espèces menacées et l'on sait que les populations de d'autres espèces sont à la baisse. Certains des oiseaux forestiers qui sont touchés par la déforestation tropicale et par la fragmentation des habitats de reproduction jouent un rôle important dans la lutte contre les populations d'insectes forestiers.

Le changement climatique

La situation météorologique, en Colombie-Britannique, subit des changements depuis que l'on tient des registres fiables. À l'intérieur des terres, au nord et au sud de la Colombie-Britannique, on constate un réchauffement important de la température. Sur la côte, les températures sont demeurées constantes depuis 1938, mais les précipitations totales augmentent régulièrement. Parmi les autres tendances observées, mentionnons l'augmentation du manteau nival et du débit de l'eau dans les stations du nord et la diminution du manteau neigeux et du débit de l'eau dans les stations situées à l'intérieur des terres, au sud. Quelle que soit la cause de ces changements passés, ils sont similaires à ceux auxquels on peut s'attendre au cours du prochain siècle, qui seront provoqués par le réchauffement planétaire dû aux gaz responsables de l'effet de serre. Parmi les effets possibles du réchauffement planétaire sur les écosystèmes, mentionnons l'élargissement des prairies, la diminution de la toundra alpine et des terres humides et des changements dans la répartition des espèces végétales dans toutes les régions.

On a observé une diminution de la couche d'ozone dans le Nord canadien. Sans l'isolation assurée par la couche d'ozone, le rayonnement ultraviolet dangereux atteint en plus grande quantité la surface de la Terre. Le rayonnement ultraviolet provoque le cancer de la peau (par exemple le mélanome), la déficience immunitaire et des cataractes chez l'homme. Les cas de mélanome, forme de cancer de la peau relativement peu fréquent, augmentent chez les résidents de la Colombie-Britannique plus rapidement que tous les autres types de cancer. Quelle que soit la cause de ces augmentations récentes, ces tendances se maintiendront vraisemblablement ou s'aggraveront si la couche d'ozone continue de diminuer.

Partie V : Protéger la Colombie-Britannique — Maintenir la diversité naturelle

Les zones protégées

Au moins 652 zones protégées, d'une superficie de 6,36 millions d'hectares, appartiennent aux divers paliers de gouvernement ou à des organisations privées ou sont gérées par ceux-ci. Les zones protégées occupent environ 6.65% du territoire de la province. La plupart des secteurs protégés de la province sont petits : 81,4% ont une superficie inférieure à 10 km², et seulement 10% couvrent plus de 100 km². Le plateau intérieur, la moitié nord de la province et les écosystèmes marins ne sont pas bien représentés dans le réseau des zones protégées.

Les réserves écologiques

La Colombie-Britannique possède 131 réserves écologiques, ce qui en fait la province qui connaît la plus grande réussite à ce chapitre au Canada. Cependant, la superficie des réserves écologiques est très petite, soit, au total, seulement 2% des parcs et des zones protégées. Elles sont patrouillées, le cas échéant, par des gardes bénévoles et aucune installation matérielle ne sert à les protéger, par exemple des clôtures pour empêcher le bétail d'y pénétrer. La désignation des réserves écologiques empêche la disparition pure et simple de ces territoires au profit des activités de développement, mais celles-ci subissent encore des pressions, par exemple l'utilisation comme pâturage et pour les loisirs, l'introduction d'espèces adventices et la lutte contre les incendies de forêt.

Les aires protégées marines

Des 27,000 kilomètres de littoral en Colombie-Britannique, il n'y a pas un seul kilomètre le long duquel tous les organismes marins, commerciaux et non commerciaux, sont entièrement protégés contre les activités humaines. Au Canada, la multiplicité des compétences dans le domaine de la gestion du milieu marin (le gouvernement fédéral, les administrations provinciales et municipales et les premières nations) nuit à l'établissement et au maintien d'aires protégées marines. D'importantes portions de trois des cinq régions marines de la Colombie-Britannique sont protégées en partie grâce au secteur marin du parc national Pacific Rim, sur la côte extérieure de l'île Vancouver, et grâce à la réserve du parc national marin Moresby-Sud/Gwaii Hanass, dans le secteur sud de Haida Gwaii (îles de la Reine-Charlotte). Il existe 25 parcs provinciaux de classe A, 3 aires récréatives et un parc à l'état naturel qui se composent en partie de zones intertidales et (ou) infratidales. En septembre 1992, le gouvernement provincial a annoncé la création de neuf nouveaux parcs marins provinciaux d'une superficie totale de 18,000 hectares. L'objectif premier de ces parcs est de répondre aux besoins du plaisancier, et non pas la préservation d'écosystèmes ou de biotes marins représentatifs.

L'éthique de la biodiversité

Les systèmes vitaux fournis par l'écosphère sont tellement menacés dans certaines régions du globe que nous devons prendre des mesures positives déterminées en vue de l'établissement d'un code de comportement, ou d'un code d'éthique de la Terre, afin que notre influence sur l'écosphère ne soit plus destructrice mais plutôt constructive. Un tel code regrouperait les valeurs ou les principes moraux, lesquels s'appuient sur le fait que nous reconnaissons être des organismes dont la santé, de la naissance à la mort, dépend d'une écosphère saine. Si nous reconnaissons cela, il n'est ni difficile ni déraisonnable d'accepter que les changements que nous faisons subir aux facteurs vitaux ne devraient pas menacer la santé de l'écosphère, ni celle des divers écosystèmes qui, cumulativement, ont des répercussions sur la santé de l'ensemble. Lorsque nous aurons appris à agir comme des restaurateurs, plutôt que comme des destructeurs de la planète, nous aurons mérité notre nom scientifique, *Homo sapiens*, qui signifie «homme sage».

Introduction

This book is about biodiversity in British Columbia. Our purpose is not thoroughly to analyze biodiversity, but rather to identify large scale changes in and emerging threats to it. To this end, we discuss both the current state of biodiversity and the biophysical processes affecting its future.

Biodiversity is the diversity of living organisms. Scientists often define biodiversity at three levels of biological organization: genetic, species and ecosystem. In this book, we proceed on the assumption that genetic diversity is dependent upon the diversity of the larger ecological units, species and ecosystems. In Part I, therefore, we consider the importance of an ecosystem-centred approach to the conservation of biodiversity, explain the methods of classifying rare species, and review the vital role played by the disciplines of biosystematics and taxonomy in the study of biodiversity. In Part II, we focus on the diversity of species in British Columbia, with articles contributed by authorities in vertebrate, invertebrate and plant taxonomy. Many of these articles were originally published in *BioLine*, the official publication of the Association of Professional Biologists of British Columbia, and have been updated and reviewed by scientific peers for inclusion in this book. In Part III, we summarize information on the diversity of ecosystems in British Columbia, single out the forest, grassland, marine and urban environments for more in-depth discussion, and then look at global impacts on local ecosystem diversity. Finally, in Part IV, we discuss the various methods of protecting biodiversity and their effectiveness. We then turn back to the question of our approach to conservation and consider the need for an earth ethic. We conclude with a summary of the recommendations for the conservation of biodiversity made, on the basis of their expertise, by the authors of the various chapters.

The task of assessing biodiversity in British Columbia is daunting and a book like this only begins to scratch the surface. British Columbia is the most biodiverse province in Canada. The vertebrate fauna includes 458 fish, 20 amphibians, 19 reptiles, 448 birds and 143 mammals (Cannings and Harcombe, 1990). Of the fish, 365 are marine, about 71 (including undescribed taxa) are freshwater and 22 live in both fresh and salt water. There are at least 2,850 vascular plants, out of a Canadian total of 4,150. British Columbia's coastal waters also contain most of the 639 taxa (species and subspecies) of benthic marine algae that occur in the northeast Pacific, as well as 6,555 species of marine invertebrates (Chapters 6 and 11, this volume), making these waters one of the most biologically diverse marine environments in the world. By far the most diverse group, though, is the land insects: at least 15,000 and perhaps as many as 35,000 may inhabit British Columbia (Chapter 4, this volume). And these are just the well known groups of organisms. There are also viruses, monera (including bacteria and blue-green algae), fungi and protozoa, for which no estimates of species numbers are available.

In editing this book, we have been reminded again and again of how much there is yet to learn about the biodiversity of our province. And even as we learn, identifying new species and tracing ecological relationships, the ecosystems around us are changing. Nor is the study of biodiversity alone sufficient to assess ecosystem health or the state of the environment. Fluctuations in biodiversity are just one indicator of environmental change. For marine environments, others include: primary productivity, nutrients, species diversity, instability, disease prevalence, size spectrum, and contaminants (Rapport, 1989). For terrestrial environments, they include: primary productivity, nutrient cycling and losses, rate of decomposition, species diversity or species richness, retrogression, habitat fragmentation, minimum viable population size, minimum area requirements and population dynamics for terrestrial environments (Woodley and Theberge, 1991).

The science of assessing ecosystem health and the degree of environmental change is still young, yet decisions about the steps needed to protect the diversity of species and ecosystems need to be made now. It is this quandary that this book is meant to address: by providing an overview of the province's biodiversity and summarizing the recommendations of the experts in the various fields of biodiversity, we are trying to enable informed decision-making on the conservation of biodiversity.

References

Cannings, R.A. and A.P. Harcombe (Ed.). 1990. *The Vertebrates of British Columbia: Scientific and English Names.* Royal British Columbia Museum Heritage Record No. 20, Wildlife Branch Report No. R24.

Rapport, D.J. 1989. What constitutes ecosystem health? *Perspectives in Biology and Medicine* 33(1):120-132.

Woodley, S. and J. Theberge, 1991. Monitoring for ecosystem integrity in Canadian national parks. In: *Science and Management of Protected Areas.* Conference proceedings, Acadia University, Nova Scotia. May 14-19, 1991. Elsevier.

Part I
INTRODUCING BIODIVERSITY

Chapter 1
The Importance of Conserving Systems[1]

In a classic paper published in 1935, the British ecologist
Arthur C. Tansley discussed the ecological realities of the world:

> Though the organisms may claim our primary interest, when we are trying to
> think fundamentally we cannot separate them from their spatial environment
> with which they form one physical system. It is the systems so formed which
> from the point of view of the ecologist are the basic units of nature on the face
> of the earth (Tansley, 1935).

Tansley coined the term "ecosystem" for these fundamental
earth-surface units whose reality he suggested but made no further
attempt to define. The idea is simple yet elusive. We have not yet
learned to visualize the earth spaces in which we live as living
spaces, as vital surrounding systems that sustain us. Yet when
these living spaces are endangered so are we.

A primary challenge for everyone is to think fundamentally, to
get to the roots of our relationship with the planet, to dig below
everyday language and concepts. Tansley's invitation to question the
organic parts, to see beyond the bits and pieces that are taken to be
wholes, to understand more comprehensive surrounding realities,
has become an essential task today. The environmental ills smiting
the world, as well as those that wait threateningly in the wings, are
not acts of God sent like Job's boils to perplex humanity. Rather
they are the result of ignorance, reflecting incomplete and fragmentary concepts about the world and the place of people in it. When
conventional knowledge is wrong, wrong ideas and misdirected
activities flow from it.

Tansley's summons to identify and sympathetically understand
the "basic units of nature on the face of the earth"—the forests, the
grasslands, the lakes and streams, the mountain wildernesses, the
farmlands and the settled lands in all their three-dimensional complexity—has not yet been heeded. So far we have failed to compre-
hend and appreciate the living world as a large global ecological
system made up of smaller ecosystems.

What is important today is to change our understanding of the
world, to focus on ecosystems rather than on the individual species
and organisms that are parts of them. Such changed understanding
of surrounding realities will fundamentally affect how we live in our
planet home.

J. Stan Rowe
Professor Emeritus
Department of Plant Ecology
University of Saskatchewan
Saskatoon, Saskatchewan

Ecology Misconceived

In many ways, ecology has missed its calling. The word "ecology" is derived from the Greek "oikos" which means house or home.
Therefore, ecology literally means "the knowledge of home" or "home
wisdom." As such, it invites study of the world's living spaces and
all that is within them. Unfortunately, this insight has been

---

[1] Reprinted in part with permission from *Endangered Spaces: The Future for Canada's Wilderness* edited by Monte
Hummel, published by Key Porter Books Ltd. Copyright © 1989 by Monte Hummel.

blinkered by our focus on organisms. Species, populations and communities have drawn attention away from the larger realities of which they are a part and in which much of their meaning resides. We have not been able to see the hive for the bees nor the forest ecosystem for the trees. Hence, the conviction that the entities of prime importance on earth are plants, animals and especially people, rather than the globe's miraculous life-filled skin. Endangered species elicit torrents of public concern; endangered spaces are routinely desecrated and destroyed with scarcely a murmur of public disapproval. The priority is wrong, and from this profound error the whole world suffers.

Like a whooping crane chick hatched by sandhill cranes, ecology has not yet discovered its singularity nor declared its independence. It has not outgrown its lowly initial status as "the fourth field" of biology, the last to arrive after morphology, physiology and taxonomy. As a result, ecology has been conceived as a discipline that plays around the edges of biology rather than as a more comprehensive discipline that integrates biology with all the earth sciences.

The time is right to rethink ecology, to understand it properly. Ecology's task is not peripheral. It is central and holistic: to develop an inclusive knowledge of the ecosphere and its ecosystem parts. Biology by itself is incomplete and needs this wider framework. Organisms do not stand on their own, they evolve and exist in the context of unified ecological systems that confer those properties called life. The panda is a part of the mountain bamboo forest ecosystem and can only be preserved as such. The polar bear is a vital part of the arctic marine ecosystem and will not survive without it. Ducks are creatures born of marshes just as cacti are one with their deserts. Biology without its ecological context is dead.

The Roots of Misconception

Consider what European languages tell us about our deeply ingrained attitudes to the ecological system that envelopes us. No adequate word for it exists; only vague, nebulous terms such as "nature" and "environment." The latter is its own put-down because, based on its etymology, "environment" means "that which surrounds other things of greater importance." In other words, something else is the real centre of interest. Environment excuses itself and defers to less substantial things that make demands on it, things that falsely declare themselves more important. Such things today include the social system and the economic system: people and jobs.

What our culture needs now are words expressing the authentic, tangible reality of the surrounding ecosystems in which organisms play their appropriate roles. But words have failed us, and the reasons run deep.

Before the modern era, belief was widespread in the existence of universal orders of organization, surpassing in importance organisms in their populations and communities. The concept of all nature as an organized whole informed by reason was central to mainstream Greek natural science from Plato to the Stoics, and it carried over to such famous Romans as Cicero and, much later, to

Leonardo da Vinci (Merchant, 1980). Although gradually submerged in the Middle Ages, the idea persisted and was implicit in the counter-culture thinking of 19th-century Romantics both in Europe and North America, contributing to the philosophical framework from which concerns for conservation developed. Influential in subtle ways over the last 400 years, the idea has nonetheless been viewed as radical rather than mainstream.

The change in thinking that defamed organic nature and severed the Chosen Species' roots in nature is a legacy of the Renaissance and the Age of Enlightenment. Back then, nature was divested of mind and soul and rendered at once dead and menacing (Collingwood, 1960). The stage was set for a mechanical and materialistic view of the world consistent with the technology of the labour-saving machines and clocks that developed rapidly in the middle ages. Gradually, mechanical models became the symbols by which people understood themselves and their surroundings. The enchanted world receded as the masculine sciences took centre stage.

God was accommodated and dismissed as the Prime Mover, the clock-maker who wound up the clock of the universe, set it ticking, and then disappeared, satisfied with a job well done. Because the Judeo-Christian God is not of this world, His presumed exit was interpreted as a signal for man to take charge. In the 17th century, Descartes gave explicit form to the universe-as-mechanism idea, the key to manipulation and control. Since then, the West has practised Cartesian science, studying to discover what makes the material world tick, searching within matter for accessible levers of power. The purpose of science, said Francis Bacon, is to control Nature and force her to do humanity's bidding.

The understanding of how a clock or any other machine works is a function of its parts and their movements. Reduction is the key to comprehension. Applying this idea to the world of nature, everything is understood by dissection of smaller and smaller component parts. The physicist studying sub-atomic particles is closest to the Truth. Such a mind-set creates a fragmented world. Further, it insinuates into our minds the mischievous idea that the fragments, the parts, are somehow more important than wholes.

This mechanistic world-view delivers great power. Cartesian science has proved successful in providing knowledge-for-control in such fields as physics, molecular biology and medicine. The procedure has the appearance of being totally effective because it follows the path of least resistance, pursuing problems that yield to it while bypassing those with which it cannot cope. Disciplines where it falters or fails—psychology, anthropology, sociology, evolution, neurobiology, embryology, ecology—are written off as non-science, as "stamp-collecting," to use Lord Rutherford's phrase for all non-physics studies. "Successful scientists," said Nobel laureate Peter Medawar, "tackle only problems that successfully yield to their methodology (Medawar, 1967)."

Just because reductionism has delivered power in certain fields does not mean, however, that it opens the window on reality, that the actuality of the world is to be found in its parts, that truth resides in atomic and sub-atomic particles. The frequently asked question, "Can the whole be greater than the sum of its parts?" gives the game away, for implicit is a prior commitment to the parts. In effect, the question says we know that the parts exist, now what about their sum? Thus, the rightness of reduction is assumed by questioning whether anything other than parts can really exist.

Carried over into society, mechanistic reductionism tracks the cause of tuberculosis to a bacillus rather than a slum housing, the cause of cancer to oncogenes rather than to industrial pollution, and the cause of evolution to genetic mutations rather than to co-development with larger surrounding systems (Levins and Lewontin, 1985). Science does not entertain the awkward possibility that reality might be distorted by giving priority to parts over wholes.

Books and tracts abound explaining that the individual is more important than the social group, the person more important than the world that encapsulates her, the fetus more important than the woman that encapsulates it. Any organism, we are told, can be computed from the complete sequence of its DNA. The brain is a holograph, the body a machine. How else do we explain the success of bioengineers in replacing the grit, glue, jelly and soup of the human body with neater and more efficient metals, plastics, ceramics and semiconductors (Lenihan, 1979)?

In short, the Cartesian heritage is a fragmented perspective, focused downward rather than upward. The search for meaning at lower and lower levels of organization blunts the higher-level search for more inclusive realities. One-way vision threatens the future of the human race by blinding it to the surpassing importance of supra-organismic realities—the earth's sustaining ecosystems, the planet's skin, the ecosphere.

Ethics and the Ecosphere

Perhaps the greatest mischief of scientific materialism and explanation-by-reduction is what it does to ethical concerns. By conceiving things mechanistically and shifting meaning to their parts, modern science strips away all sense of intrinsic value (Rifkin, 1985). It destroys the intuition that things can have importance for their own sakes, independent of their parts. What real empathy can be felt for a machine-dead universe whose explanations reside in its atoms?

A meaningless universe leaves little in the conscience of people but a sense of their own diminishing importance. The last religion left to them is a slowly evaporating Humanism that isolates them from the world of nature and leaves them alone, clinging to each other in a prison of their own making, bravely repeating that only they are important, only they have souls, only they will reap rewards in the Great Hereafter. "Human rights" is the leading secular slogan.

From the precept that only humans matter, a disastrous corollary follows: The world is for exploiting. Parks are for people, animals are for shooting, forests are for logging, soils are for mining. The sole basis for ethical action is the greatest good for the greatest number of people. The values of all things lie only in their abilities to serve us.

Contemporary morality—the sense of right and wrong—is completely in-turned, completely focused on humankind. That focus makes it difficult to be sensitively concerned about the world in the face of escalating human demands. Sustainable development, we are told, must include forceful economic growth, for how else can the needs of all the world's people be met?

Lacking an ethic that attaches importance to all surrounding creation, people continue to do the wrong things for the apparent "good of humanity." People First. Five billion people going for ten, all believing in People First, increasing their wants without limit, are a sure recipe for species suicide.

If what is wrong is to see the world in fragments—atoms, species, resources—with people alone important, then what is right is to re-perceive the world as one, a whole, organically complex, beautiful beyond compare, and to reorient to it in ways that confer first importance on it. In short, replace the homocentric with the ecocentric viewpoint. Then the intrinsic values of the ecosphere and its living realm will be recognized, as will the rights of things other than humans to exist in and for themselves.

Once values are straight, everything else can fall into place. An ecocentric world-view, valuing the spinning planetary home above the organisms hitching a ride on it, elevates in importance the ecosystems that humans call land. Love of the land, love of place, love of our endangered living spaces, is the grassroots cure for the sin of species narcissism.

The world was not created for people only, but for purposes that transcend the human race with its limited foresight and imagination; therefore it behooves all conscious inhabitants of this superb planet to nurture it as a garden, maintaining it in health, beauty and diversity for whatever glorious future its denizens may together share.

Acknowledgements

Michael Dunn, Canadian Wildlife Service, reviewed the manuscript.

References

Collingwood, R.G. 1960. *The Idea of Nature.* A Galaxy Book, Oxford University Press, New York, New York.

Lenihan, J. 1979. Is man a machine? J. Lenihan and W.W. Fletcher (eds.). *The Environment and Man.* Vol. 9: The Biological Environment. Academic Press, New York, New York.

Medawar, P.B. 1967. *The Art of the Soluble.* Methuen, London.

Merchant, C. 1980. *The Death of Nature.* Harper & Row, San Francisco, California.

Rifkin, J. 1985. *Declaration of a Heretic.* Routledge & Kegan Paul, Boston and London.

Tansley, A.G. 1935. The use and abuse of vegetational concepts and terms. *Ecology* 16: 284-307.

Chapter 2
Terms of Endangerment

Classifying and categorizing taxa at risk is more difficult than it may at first appear, as the following three articles will show. Various agencies and organizations have developed a number of different systems for assigning priority to species to be conserved. We begin this chapter by looking first at one of the most widely used systems, which has been developed by the Nature Conservancy (U.S.) for their Conservation Data Centres in various jurisdictions throughout the western hemisphere, including British Columbia. Then, we consider the provincial system (the Red and Blue lists), which has recently been made compatible with the Conservation Data Centre's system. Finally, we review the national system administered by the Committee on the Status of Endangered Wildlife in Canada, which differs from the Conservation Data Centre's system in its ranking terminology and from both systems in the fact that it ranks only those species for which a detailed status report has been prepared.

Describing Rarity: the Ranking Dilemma and a Solution

In any discussion of rarity, there is usually a desire to categorize the degree of rarity, to put a label on rarity. Terms in common use include endangered, threatened, sensitive, vulnerable, unique, and rare. This need to classify stems, in part, from the need of the biological manager to determine priorities for effort (and, therefore, budget). Ranking rarity also has a geographical connotation—what might be rare in one region may be common in another. Hence, two types of rarity classification are required—relative rarity within a defined geographical or administrative region and rarity across geographical areas (global).

Andrew Harcombe
B.C. Conservation Data Centre
B.C. Ministry of Environment,
Lands and Parks
780 Blanshard Street
Victoria, B.C.
V8V 1X4

One of the problems that faces someone trying to discuss rarity in British Columbia is the plethora of systems that have been developed. The Committee on the Status of Endangered Wildlife in Canada (COSEWIC) ranks vascular plants and vertebrates based on the recommendation of individual status reports, resulting in Vulnerable, Threatened, Endangered, Extirpated and Extinct classes, as described in the last article in this section. The B.C. Endangered Plant Committee has developed a list of rare vascular and non-vascular plants and fungi, and the B.C. Rare Plant Program has developed "R" ratings for vascular plants. These different systems may be hard to evaluate objectively because the reasons for assigning a particular rank are often not documented, or have become lost or buried in subjective assessment. This subjectivity can lead to disagreements on the appropriate rank for a given species.

The British Columbia Conservation Data Centre

Initiated in February, 1991, the British Columbia Conservation Data Centre's program is designed to compile information on the status and occurrence of provincially rare vascular plants, vertebrates (excluding saltwater fish), and plant communities.

The Centre's data management system has been developed over the past 20 years by the Science Division of The Nature Conservancy (U.S.). This system can be described as a permanent and dynamic atlas and data bank that includes data on the existence, characteristics, numbers, condition, status, location, and distribution of the elements of natural diversity (Keystone Center, 1991; Jenkins, 1986; Noss, 1987). The elements of natural diversity can be a species, a plant community, or an unique feature of the natural landscape. The purpose of the Data Centre is to centralize and standardize these data, concentrating on rare animals, plants, and communities for a particular jurisdiction, such as British Columbia.

Ranking

In The Nature Conservancy's ranking system, each element of biological diversity (such as species, subspecies, plant community or special biological feature) is ranked at two levels: global (G) and provincial or subnational (S). The global rank is based on the status of the element throughout its entire range; the provincial rank is based solely on the element's status within British Columbia. A specialist from the Nature Conservancy establishes the global rank, which is primarily based on "rolling up" regional information and considering the biology of the element throughout its range (which may be great for neotropical migrants). The provincial rank cannot exceed the global rank—that is, the species cannot be more common in the province than it is globally. The provincial rank is established by the appropriate specialist in the Conservation Data Centre, usually in collaboration with other provincial experts. Recently, this ranking system has been used to establish B.C. Environment's Red and Blue lists of terrestrial vertebrates (Harper et al., this volume) and freshwater fish.

In this system, the status of an element is indicated on a scale of one to five (Table 2-1). The score is based primarily on the number of extant occurrences of the element, but other factors such as abundance, range, protection, trends, and threats are also scored and considered. For plant communities, abundance

1. **Critically imperiled:** extremely rare (5 or fewer extant occurrences or very few remaining individuals) or some factor(s) make it especially vulnerable to extirpation or extinction.

2. **Imperiled:** rare (typically 6-20 extant occurrences or few remaining individuals) or some factor(s) make it vulnerable to extirpation or extinction.

3. **Rare or uncommon:** typically 21-100 occurrences; may be susceptible to large-scale disturbances, such as loss of extensive peripheral populations.

4. **Frequent to common:** greater than 100 occurrences; apparently secure but may have a restricted distribution or future threats may be perceived.

5. **Common to very common:** demonstrably secure and essentially ineradicable under present conditions.

Table 2-1: Conservation Data Centre Ranks and Definitions

is also determined by the area of occurrence. An occurrence is defined ecologically as a location representing a habitat which sustains or otherwise contributes to the survival of a population. For example, a south-facing slope that provides winter range for ten elk would be considered a single occurrence, not ten. The greatest emphasis in the Centre's ranking system is placed on the number of extant occurrences (Jenkins, 1986), because the smaller the number of occurrences, the greater the likelihood of species endangerment. When a species' distribution is extremely limited, it is vulnerable to any number of ecological disturbances, predictable or unpredictable. Therefore, if there are fewer than 20 occurrences of a species in the province, it probably will be given a rank of one or two.

H	Historical occurrence; usually not verified in the last 40 years, but with the expectation that it may be rediscovered
X	Apparently extinct or extirpated, without the expectation that it will be rediscovered
U	Status uncertain, often because of low search effort or cryptic nature of the element; uncertainty spans a range of 4 or 5 ranks
R	Reported from the province, but without persuasive documentation for either accepting or rejecting the report
RF	Reported in error, but this error has persisted in the literature
?	No information is available or the number of extant occurrences is based on a "best guess"
A	An element (usually an animal) that is considered accidental or casual in province; a species that does not appear on an annual basis
E	An exotic or introduced species to the province
Z	Occurs in the province but as a diffuse, usually moving population; difficult or impossible to map static occurrences
T	Designates a rank associated with a subspecies
B	Breeding; the associated rank refers to breeding occurrences of mobile animals
N	Non-breeding; the associated rank refers to non-breeding occurrences of mobile animals
Q	Taxonomic validity of the element is not clear or is in question

Table 2-2: Conservation Data Centre Rank Modifiers

Absolute abundance is a secondary consideration in ranking. Some species which are extremely rare geographically may be locally common or even abundant. Examples include colonial nesting birds, such as Western Grebes (*Aechmophorus occidentalis*) or Common Murres (*Uria aalge*). In some cases, like that of the wintering Snow Goose (*Chen caerulescens*) population on the Fraser Delta, the species may be abundant enough to be legally hunted, yet still have a localized distribution that makes it extremely vulnerable to ecological disturbances, such as oil spills.

In addition to these ranks, there are several letters that either modify and provide further explanation of the rank or that further describe the type of element being ranked (see Table 2-2).

Table 2-3 provides an example of the Centre's ranking of the white sturgeon (*Acipenser transmontanus*) in B.C. Initial provincial ranking of an element is generally done by inspection; that is, a rank is assigned based on professional judgement. Ranks may be combined, such as S2S3, to illustrate the potential range based on current understanding. These preliminary ranks are used to extract from the provincial plant, animal and community lists a shorter list of priority elements. Elements on this list are further inventoried or "tracked," so this list of high priority elements is known as the

Known occurrences: B
In five major rivers and six large lakes. The occurrences, in general, and especially those in the Fraser River, cover extensive geographic areas.

Abundance: D
Lower Fraser River population abundant; sparse elsewhere. Although the Fraser population is still substantial, it is much lower than historical levels (Lane, 1991).

Range: C
Found in the main stems of the Fraser, Nechako, Stuart, lower Columbia, and upper Kootenay rivers, and in Fraser, Takla, Trembleur, Stuart, Kootenay and Williams lakes; ventures into the lower portions of some of the larger tributaries of the upper Fraser (Bowron, McGregor rivers).

Trend: B
Populations in the Fraser below Quesnel appear to be stable, or at least producing young. Other populations, such as those in the Nechako, Kootenay and Columbia rivers, appear to be in decline, with very few or no young being produced or surviving in recent years (Apperson 1992; Hildebrand 1991; personal communication with D. Ableson, 1992, B.C. Ministry of Environment Lands and Parks, Fish and Wildlife Branch, Prince George, B.C.)

Protected Occurrences: A
None protected, although the Kootenay River sport fishery is catch-and-release only.

Threats: A
With the exception of the lower Fraser populations, white sturgeon may be endangered throughout their historical range in the province, primarily because of flow alteration by dams on the Nechako, Kootenay, and Columbia rivers (Apperson, 1992; Hildebrand, 1991; personal communication with D. Ableson, 1992, B.C. Ministry of Environment Lands and Parks, Fish and Wildlife Branch, Prince George, B.C.). An apparent decline in the eulachon (*Thaleichthys pacificus*) populations in the lower Fraser River, a major food source for sturgeon, may be a cause for concern there (personal communication with M. Rosenau, 1992, B.C. Ministry of Environment Lands and Parks, Fish and Wildlife Branch, Surrey B.C.). Loss of productive slough habitat in the Fraser Valley is probably reducing the potential for young sturgeon as well (Lane, 1991). The total harvest in the lower Fraser River is "probably close to or in excess of the carrying capacity of the species" (Lane, 1991). Contamination with heavy metals and other pollutants is also of concern.

Rank: S3
Comments: Restricted to a few major water bodies in the province. Populations throughout most of the species' historical range in British Columbia are threatened by an apparent failure to spawn or by high larval mortality. Those in the lower Fraser are producing young, but face threats from habitat loss and a possible decline in food supply.

Table 2-3: B.C. Conservation Data Centre Sample Provincial Ranking Account for the White Sturgeon (*Acipenser transmontanus*). Each criterion is rated on a scale from worst to best, "A" being the worst and "D" being the best. In the case of the population trend, for example, A = declining rapidly, B = declining, C = stable, and D = increasing. Using this rating system, one can tell at a glance that the white sturgeon is unprotected and highly threatened.

tracking list. For vertebrates, the tracking list includes all taxa with provincial ranks of one to three; for plants, the tracking list includes taxa with ranks of one to two.

The Conservation Data Centre compiles all existing data on specific occurrences of these tracked elements. This data set is then used in completing computerized Provincial Ranking records, such as that shown in Table 2-3, where pertinent information for each of the ranking criteria is compiled. Thus, if 15 occurrences of an element are known, with ancillary data on potential threats to that element and recent trends in its abundance, it is possible to assign a rank, such as S2, with greater confidence.

Advantages of this Ranking System

1) The same system is used to rank vertebrates, invertebrates, vascular plants, and plant communities. In this way, it is now possible to compare the conservation value of sites containing disparate natural elements.

2) The ranking information is available to anyone who is interested in how or why a particular rank was assigned, resulting in fewer disagreements. Experts can argue about the facts rather than about mere subjective assessments.

3) Because the system uses a highly standardized methodology and terminology to process information about species and habitats, it is possible to exchange and share data across administrative boundaries, as well as to summarize range-wide information on an element. This methodology is shared with programs in all 50 American states (called Natural Heritage Programs), three other Canadian provinces, and 12 Latin American and Caribbean countries, as well as with several special regional programs (such as Great Smoky Mountains National Park, Greater Yellowstone, and the Navajo Nation).

4) The provincial and global statuses for species are given separate ranks within the system, which greatly simplifies the ranking process and clarifies the status of each ranked element.

Establishment of the Conservation Data Centre moves British Columbia into step with the majority of other jurisdictions in the western hemisphere. Using this standardized system will save time in defining terms and allow the specialists to get on with the business of conserving biodiversity.

Provincial Lists of Species at Risk

Bill Harper
Wildlife Branch
B.C. Ministry of Environment,
Lands and Parks
780 Blanshard Street
Victoria, B.C.
V8V 1X4

Sydney Cannings
B.C. Conservation Data Centre
B.C. Ministry of Environment,
Lands and Parks
780 Blanshard Street
Victoria, B.C.
V8V 1X5

David Fraser
Arenaria Research and
Interpretation
5836 Old West Saanich Road
R.R. 7
Victoria, B.C.
V8X 3X3

William. T. Munro
Wildlife Branch
B.C. Ministry of Environment,
Lands and Parks
780 Blanshard Street
Victoria, B.C.
V8V 1X4

In 1980, the first vertebrate species were designated Endangered under the Wildlife Act of British Columbia. These species were the: Vancouver Island marmot *(Marmota vancouverensis)*, sea otter *(Enhydra lutris)*, American white pelican *(Pelecanus erythrorhynchos)* and burrowing owl *(Athene cunicularia)*. They remain the only species to have ever been legally designated. The criteria used at that time to determine the species to be considered for designation were presented in Munro and Low (1980).

Since then, Red and Blue Lists of terrestrial vertebrate taxa at risk have been created and subsequently revised (Munro, 1990). These lists are designed to aid in assigning priorities for conservation of vertebrates at risk in British Columbia. Taxa not considered at risk are placed either on the Yellow List, which is for species that are actively managed at a population level, or the Green List, which is for species that are managed only by ensuring that they have adequate habitat.

The species on the Red List (Table 2-4) are Endangered or Threatened, or are being considered for such status. Any indigenous taxon (species or subspecies) threatened with imminent extinction or extirpation throughout all or a significant portion of its range in British Columbia is Endangered. Threatened taxa are those indigenous species or subspecies that are likely to become endangered in British Columbia if factors affecting their vulnerability are not reversed. For all Red-listed taxa, status reports will be written that indicate the current condition of populations within the province.

Blue-listed species (Table 2-5) are considered to be vulnerable and "at risk," but not yet endangered or threatened. Populations of these species may not be in decline, but their habitat or other requirements are such that they are vulnerable to further disturbance.

Once a species has been identified as a potential candidate for designation as Endangered or Threatened under the Wildlife Act and placed on the Red List, a detailed status report is prepared. If legal designation is merited, a brief will be prepared for Cabinet to consider in deciding whether to issue an Order in Council designating the species as either Endangered or Threatened.

Ranking

In creating the latest version of the Red and Blue Lists, the process was changed to dovetail with the ranking system used by the B.C. Conservation Data Centre and many other jurisdictions in Canada, the United States, and Central and South America, as described in the previous article.

Based on the best available information, the provincial rank is assigned to the taxon by biologists at the Conservation Data Centre. Provincial ranks are modified and updated on a continual basis. Species placement on the Red or Blue List is based on the provincial ranks only. Species with ranks of 1, 2, 1 to 2, 1 to 3, H, or X are put

Amphibians
    Ambystomatidae
        Ambystoma tigrinum   Tiger Salamander
        Dicamptodon tenebrosus   Pacific Giant Salamander
    Plethodontidae
        Plethodon idahoensis   Coeur d'Alene Salamander
        Rana pipiens   Leopard Frog

Reptiles
    Iguanidae
        Phrynosoma douglassii — Short-horned Lizard
    Colubridae
        Contia tenuis — Sharp-tailed Snake
        Pituophis melanoleucus catenifer — Gopher Snake - *catenifer* subspecies
        Hypsiglena torquata — Night Snake

Birds
    Gaviformes
        Aechmophorus occidentalis — Western Grebe
    Pelecaniformes
        Pelecanus erythrorhynchos — American White Pelican
        Phalacrocorax pelagicus pelagicus — Pelagic Cormorant - *pelagicus* subspecies
        Phalacrocorax penicillatus — Brandt's Cormorant
    Falconiformes
        Accipiter gentilis laingi — Northern Goshawk - *laingi* subspecies
        Buteo regalis — Ferruginous Hawk
        Falco peregrinus anatum — Peregrine Falcon - *anatum* subspecies
        Falco mexicanus — Prairie Falcon
    Galliformes
        Centrocercus urophasianus — Sage Grouse
    Charadriformes
        Bartramia longicauda — Upland Sandpiper
        Sterna forsteri — Forster's Tern
        Uria lomvia — Thick-billed Murre
        Uria aalge — Common Murre
        Fratercula corniculata — Horned Puffin
    Cuculiformes
        Coccyzus americanus — Yellow-billed Cuckoo
    Strigiformes
        Athene cunicularia — Burrowing Owl
        Strix occidentalis — Spotted Owl
    Piciformes
        Sphyrapicus thyroideus nataliae — Williamson's Sapsucker - *nataliae* subspecies
        Picoides albolarvatus — White-headed Woodpecker
    Passeriformes
        Eremophila alpestris strigata — Horned Lark - *strigata* subspecies
        Progne subis — Purple Martin
        Oreoscoptes montanus — Sage Thrasher
        Anthus spragueii — Sprague's Pipit
        Dendroica castanea — Bay-breasted Warbler
        Dendroica tigrina — Cape May Warbler
        Oporornis agilis — Connecticut Warbler
        Icteria virens — Yellow-breasted Chat
        Spizella breweri breweri — Brewer's Sparrow - *breweri* subspecies
        Ammodramus savannarum — Grasshopper Sparrow
        Ammodramus caudacutus — Sharp-tailed Sparrow
        Pooecetes gramineus affinis — Vesper Sparrow - *affinis* subspecies

Table 2-4: British Columbia's 1993 Red List (64 taxa in taxonomic order)

Mammals

Insectivora

Sorex tundrensis	Tundra Shrew
Sorex palustris brooksi	Water Shrew - *brooksi* subspecies
Sorex bendirii	Pacific Water Shrew
Scapanus townsendii	Townsend's Mole

Chiroptera

Antrozous pallidus	Pallid Bat
Lasiurus blossevillii	Southern Red Bat
Myotis keenii	Keen's Long-eared Myotis
Myotis septentrionalis	Northern Long-eared Myotis

Lagomorpha

Lepus americanus washingtoni	Snowshoe Hare - *washingtoni* subspecies
Lepus townsendii	White-tailed Jack Rabbit

Rodentia

Aplodontia rufa rufa	Mountain Beaver - *rufa* subspecies
Clethrionomys gapperi occidentalis	Southern red-backed vole - *occidentalis subsp.*
Microtus townsendii cowani	Townsend's Vole - *cowani* subspecies
Synaptomys borealis artemisiae	Northern Bog Lemming - *artemisiae* subspecies
Thomomys talpoides segregatus	Northern Pocket Gopher - *segregatus* subspecies
Marmota vancouverensis	Vancouver Island Marmot
Tamias ruficaudus simulans	Red-tailed Chipmunk - *simulans* subspecies
Tamias ruficaudus ruficaudus	Red-tailed Chipmunk - *ruficaudus* subspecies
Tamias minumus selkirki	Least Chipmunk - *selkirki* subspecies

Carnivora

Enhydra lutris	Sea Otter
Gulo gulo vancouverensis	Wolverine - *vancouverensis* subspecies
Mustela erminea haidarum	Ermine - *haidarum* subspecies
Mustela frenata altifrontalis	Long-tailed Weasel - *altifrontalis* subspecies

Artiodactyla

Bison bison athabascae	Wood Bison
Ovis dalli dalli	Dall's Sheep

Table 2-4: British Columbia's 1993 Red List (Continued)

on the Red List, and those with ranks of two to 3, 3, or 3 to 4 are put on the Blue List (see Tables 2-3 and 2-4). Species with ranks of 4 or 5 are not considered to be at risk in British Columbia.

All the regularly occurring vertebrate taxa in British Columbia that fall under the mandate of the Wildlife Program (except for introduced species) have been ranked and evaluated. Fish and marine mammals, except the sea otter, have not been included. In the 1993 edition of the lists, 151 taxa are listed (19% of the 850 possible species and subspecies in the province), compared with the 1991 edition (B.C. Environment, 1991), which listed 107 species (17% of the 631 possible species in the province).

Introduced Species

The Wildlife Branch does not consider for designation those species which have been introduced into British Columbia, such as the gray squirrel (*Sciurus carolinensis*) and Eurasian Skylark (*Alauda arvensis*), or which have entered B.C. following introduction

Amphibians
    Leiopelmatidae
        Ascaphus truei   Tailed Frog
    Pelobatidae
        Scaphiopus intermontanus   Great Basin Spadefoot Toad
Reptiles
    Emydidae
        Chrysemys picta   Painted Turtle
    Boidae
        Charina bottae   Rubber Boa
    Colubridae
        Coluber mormon   Western Yellow-bellied Racer
        Pituophis melanoleucus deserticola   Gopher Snake - *deserticola* subspecies
    Viperidae
        Crotalus viridis   Western Rattlesnake
Birds
    Phalacrocoracidae
        Phalacrocorax auritus   Double-crested Cormorant
    Ardeidae
        Botaurus lentiginosus   American Bittern
        Ardea herodias   Great Blue Heron
        Butorides striatus   Green-backed Heron
    Anatidae
        Cygnus buccinator   Trumpeter Swan
        Clangula hyemalis   Oldsquaw
        Melanitta perspicillata   Surf Scoter
    Cathartidae
        Cathartes aura   Turkey Vulture
    Accipitridae
        Haliaeetus leucocephalus   Bald Eagle
        Buteo swainsoni   Swainson's Hawk
    Falconidae
        Falco peregrinus pealei   Peale's Peregrine Falcon
        Falco rusticolus   Gyrfalcon
    Phasianidae
        Lagopus leucurus saxatilis   White-tailed Ptarmigan - *saxatilis* subspecies
        Tympanuchus phasianellus columbianus   Sharp-tailed Grouse - *columbianus* subspecies
    Gruidae
        Grus canadensis   Sandhill Crane
    Charadriidae
        Pluvialis dominica   Lesser Golden-plover
    Recurvirostridae
        Recurvirostra americana   American Avocet
    Scolopacidae
        Heteroscelus incanus   Wandering Tattler
        Numenius americanus   Long-billed Curlew
        Limosa haemastica   Hudsonian Godwit
        Limnodromus griseus   Short-billed Dowitcher
        Phalaropus lobatus   Red-necked Phalarope
    Laridae
        Larus californicus   California Gull
        Sterna caspia   Caspian Tern
  Alcidae
        Brachyramphus marmoratus   Marbled Murrelet
        Synthliboramphus antiquus   Ancient Murrelet

Table 2-5. **British Columbia's 1993 Blue List** (87 taxa in taxonomic order)

Alcidae con't
    Ptychoramphus aleuticus   Cassin's Auklet
    Fratercula cirrhata   Tufted Puffin
Tytonidae
    Tyto alba   Barn Owl
Strigidae
    Asio flammeus   Short-eared Owl
    Otus flammeolus   Flammulated Owl
    Otus kennicottii macfarlanei   Western Screech-owl - *macfarlanei* subspecies
    Otus kennicottii kennicottii   Western Screech-owl - *kennicottii* subspecies
    Glaucidium gnoma swarthi   Northern Pygmy-owl - *swarthi* subspecies
    Aegolius acadicus brooksi   Northern Saw-whet Owl - *brooksi* subspecies
Apodidae
    Aeronautes saxatalis   White-throated Swift
Trochilidae
    Archilochus alexandri   Black-chinned Hummingbird
Picidae
    Melanerpes lewis   Lewis' Woodpecker
    Sphyrapicus thyroideus thyroideus   Williamson's Sapsucker - *thyroideus* subspecies
    Picoides villosus picoideus   Hairy Woodpecker - *picoideus* subspecies
Tyrannidae
    Empidonax flaviventris   Yellow-bellied Flycatcher
    Empidonax wrightii   Gray Flycatcher
Corvidae
    Cyanocitta stelleri carlottae   Steller's Jay - *carlottae* subspecies
Troglodytidae
    Catherpes mexicanus   Canyon Wren
Vireonidae
    Vireo huttoni   Hutton's Vireo
    Vireo philadelphicus   Philadelphia Vireo
Emberizidae
    Dendroica virens   Black-throated Green Warbler
    Dendroica palmarum   Palm Warbler
    Wilsonia canadensis   Canada Warbler
    Chondestes grammacus   Lark Sparrow
    Calcarius pictus   Smith's Longspur
    Dolichonyx oryzivorus   Bobolink
    Pinicola enucleator carlottae   Pine Grosbeak - *carlottae* subspecies
Mammals
    Soricidae
        Sorex arcticus   Black-backed Shrew
        Sorex trowbridgii   Trowbridge's Shrew
    Vespertilionidae
        Myotis thysanodes   Fringed Myotis
        Myotis ciliolabrum   Western Small-footed Myotis
        Euderma maculatum   Spotted Bat
        Plecotus townsendii   Townsend's Big-eared Bat
    Leporidae
        Sylvilagus nuttallii   Nuttall's Cottontail
    Aplodontidae
        Aplodontia rufa rainieri   Mountain Beaver - *rainieri* subspecies
    Sciuridae
        Spermophilus saturatus   Cascade Golden-mantled Ground Squirrel
        Tamias minimus oreocetes   Least Chipmunk - *oreocetes* subspecies

Table 2-5.   British Columbia's 1993 Blue List (Continued)

Heteromyidae	
Perognathus parvus	Great Basin Pocket Mouse
Cricetidae	
Reithrodontomys megalotis	Western Harvest Mouse
Arvicolidae	
Clethrionomys gapperi galei	Southern Red-backed Vole - *galei* subspecies
Synaptomys borealis borealis	Northern Bog Llemming - *borealis* subspecies
Zapodidae	
Zapus hudsonius alascensis	Meadow Jumping Mouse - *alascensis* subspecies
Ursidae	
Ursus americanus emmonsii	Glacier Bear
Ursus arctos	Grizzly Bear
Mustelidae	
Martes pennanti	Fisher
Mustela erminea anguinae	Ermine - *anguinae* subspecies
Gulo gulo luscus	Wolverine - *luscus* subspecies
Taxidea taxus	Badger
Cervidae	
Cervus elaphus roosevelti	Roosevelt Elk
Rangifer tarandus	Woodland Caribou - southern populations
Bovidae	
Bison bison bison	Plains Bison
Ovis canadensis californiana	California Bighorn Sheep
Ovis canadensis canadensis	Rocky Mountain Bighorn Sheep
Ovis dalli stonei	Stone Sheep

Table 2-5. British Columbia's 1993 Blue List (Continued)

into neighbouring regions, such as the fox squirrel (*Sciurus niger*) and some populations of Wild Turkey (*Meleagris gallopavo*). The reasons for not including exotic species are:

1) they have been introduced by humans and are not part of British Columbia's natural wildlife heritage;

2) they may compete with native species of fauna;

3) they may not be adapted to conditions found in British Columbia; and

4) the effort to maintain viable populations of these species could be more profitably directed toward native species.

Subspecies

Previous editions of the Red and Blue Lists did not include subspecies, unless there was significant public concern for their management. However, as part of the recent change in emphasis toward managing for biological diversity, all subspecies are now considered. Most subspecies represent recognizable genetic diversity, and management at the subspecies level is a first step in preserving this diversity.

Recognizing that some genetic differences may not be expressed in gross morphological (physical) features, we must proceed with the best assessments we have, until molecular technology provides us with better measures.In some instances, taxa that are currently recognized as subspecies may well prove to be distinct species. For example, the northern pocket gopher (*Thomomys*

21

talpoides), red-tailed chipmunk (*Tamias ruficaudus*), and Brewer's Sparrow (*Spizella breweri*) all have subspecies that may prove to be separate species. Unless this variation is retained, we could lose these species even before they are formally recognized.

In general, clearly disjunct British Columbian populations are evaluated as separate taxa, and most intergrading subspecies (ones that are intermediate in form) are grouped and considered as a single species. As taxonomic studies increase our knowledge of British Columbia's fauna, these lists will be updated. In some cases, it will be wise to further investigate the taxonomy of a subspecies before initiating expensive conservation initiatives.

Peripheral Species

Peripheral species are those that are at the edge of their range. In previous editions of the Red and Blue Lists, species that were considered peripheral could be placed no higher than the Blue List, so even those that were threatened or endangered were down-listed to Blue. This policy was changed for a number of reasons.

1) There is increasing evidence that populations at the edges of their ranges possess unique genetic traits that may be important contributors to the long term survival and evolution of a species (Mayr, 1967, 1982). Scudder (1989) states, "Marginal populations have a high adaptive significance to the species as a whole, and marginal (peripheral) habitat conservation, preservation and management is one of the best ways to conserve the genetic diversity....."

2) Due to their geographic position, populations at the edges of their ranges may also be important in enabling the species as a whole to survive long term environmental changes, such as global warming.

3) Taken to its logical conclusion, the exclusion of peripheral species from legal protection (especially if applied to plants, invertebrates and plant communities) could result in a tremendous loss of British Columbian genetic resources.

4) The diversity of British Columbia's fauna is lower than that of similar geographic areas which have not been glaciated. Many species that are on our current Red and Blue Lists are taxa that have colonized (or recolonized) British Columbia relatively recently and they may still require protection to establish viable populations.

5) Responses of species to environmental changes may be more easily recognized at the periphery of their ranges than at the core and, therefore, peripheral populations should be monitored for impacts before a particular species becomes threatened or vulnerable throughout its range.

These lists will change frequently, and the status of each species is not permanent. The ecological systems in which these species live are dynamic; populations will grow and shrink. Species may be added to these lists as a result of habitat loss and fragmentation. The status of a species may change as we gather more information on its abundance, distribution, and taxonomy, and some species will respond to management efforts and be down-listed or delisted.

The Red and Blue Lists play an important role in facilitating initial identification of endangered, threatened, and vulnerable species, but serve only as a first step in proper conservation of British Columbia's biota which are at risk.

National Criteria for the Designation of Endangered and Threatened Species

The Committee on the Status of Endangered Wildlife in Canada (COSEWIC) was formed in 1977 as a result of a recommendation of the Federal-Provincial Wildlife Conference. It arose from the need for a single official, scientifically sound national listing of wild species at risk in Canada. Its mandate is to determine the national status of wild species, sub-species and geographically separate populations in Canada. All native plants and animals, except invertebrates, fall under COSEWIC's purview.

COSEWIC consists of one representative from the wildlife agency of each province or territory, one from each of four federal agencies (Canadian Wildlife Service, National Museums, Department of Fisheries and Oceans, and Parks Canada), and one from each of three nationally-based private conservation agencies (Canadian Nature Federation, Canadian Wildlife Federation, and World Wildlife Fund Canada). COSEWIC elects its own chairperson for a two year term. Sub-committees carry out the work of bringing status reports on species from the various biological groups (birds, terrestrial mammals, fish, marine mammals, amphibians, reptiles, and plants) to the attention of COSEWIC. For each sub-committee, COSEWIC appoints a chairperson who is an expert in that field and who, in turn, recruits knowledgeable sub-committee members.

These sub-committees obtain, review and present to COSEWIC the reports upon which all determinations of species status are based. Each report provides an up to date description of the distribution, abundance and population trends of a species. These status reports may be submitted to the appropriate sub-committee by individuals, COSEWIC members, or a contractor hired by COSEWIC to prepare the report. It is the sub-committees' job to ensure scientific quality, propose an appropriate national status for the species, subspecies or population in question, and present the status report to COSEWIC membership for formal assignment of status.

COSEWIC meets annually in April to declare official status for all the species for which status reports have been prepared and circulated. Following each annual meeting, a news release is issued listing new designations. In addition, up to date lists of all species evaluated and their new designations are available annually from the COSEWIC Secretariat. The status reports for designated species can be obtained from the Canadian Nature Federation in Ottawa. COSEWIC recognizes five categories of species at risk: vulnerable, threatened, endangered, extirpated and extinct. Definitions of these categories are provided in Table 2-6. Sometimes, when a status

W.T. Munro
Wildlife Branch
B.C. Ministry of Environment,
Lands and Parks
780 Blanshard Street
Victoria, British Columbia
V8V 1X4

Species
"Species" means any species, subspecies or geographically separate population.

Rare
Term has not been assigned by COSEWIC since 1989; replaced by "vulnerable", which permits inclusion of rare species which are vulnerable because of their rarity, and those species where declining numbers may indicate a potential threat. Species assigned as "rare" prior to 1989 will remain so until their status is reviewed.

Vulnerable species
Any indigenous species of fauna or flora that is particularly at risk because of low or declining numbers, occurrence at the fringe of its range or in restricted areas, or for some other reason, but is not a threatened species.

Threatened species
Any indigenous species of fauna or flora that is likely to become endangered in Canada if the factors affecting its vulnerability do not become reversed.

Endangered species
Any indigenous species of fauna or flora that is threatened with imminent extinction or extirpation throughout all or a significant portion of its Canada range, owing to human action.

Extirpated species
Any indigenous species of fauna or flora no longer existing in the wild in Canada but occurring elsewhere.

Extinct species
Any species of fauna or flora formerly indigenous to Canada but no longer existing anywhere.

Table 2-6: Definitions of Conservation Status

report is reviewed, it is decided that the species is not at risk. In such cases, COSEWIC notes that the report on the species has been accepted and no designation is required. Since its inception, COSEWIC has evaluated 303 species. For 68 species, it found that no designation was required. The remaining 230 species were classified as shown in Table 2-7 and the designations of those that occur in British Columbia are shown in Table 2-8.

In 1988, the Council of Wildlife Ministers established a national strategy for the recovery of endangered species. Working from the COSEWIC lists, the Committee for the Recovery of Nationally Endangered Wildlife (RENEW) establishes recovery teams, approves recovery plans, encourages the cooperation and support of various government and non-government organizations, and publishes an annual report on national efforts to recover endangered and threatened terrestrial vertebrates. While RENEW can aid recovery efforts

Category (Status)	Birds	Mammals (Terrestrial)	Mammals (Marine)	Fish	Plants	Amphibians & Reptiles	Total
Vulnerable	19	15	5	35	27	7	108
Threatened	8	5	3	10	25	2	53
Endangered	10	5	6	3	21	4	49
Extirpated	1	3	2	2	2	1	11
Extinct	3	1	1	4	-	-	9
Totals	41	29	18	54	76	14	230

Table 2-7: Number of Species in Each Category Designated by COSEWIC

by establishing national priorities and standards, the participating government agencies remain responsible for implementing recovery plans for populations and species that fall within their jurisdictions.

The status of each species listing is not permanent, as it is based on the best information available at the time of study. It may be reviewed whenever new information is presented. It is possible to change a species designation to a more critical one, should its prospects for continued survival deteriorate, or to a less critical one, should risks diminish or recovery through the RENEW program be successful. Examples of criteria used for changing status are shown in Table 2-9.

It should be emphasized that COSEWIC is concerned only with drawing official attention to the possible loss of wild species that have historically maintained populations in Canada. It is not mandated to take any action which would alter the fortunes of species beyond establishing their status and publishing the information upon which that status is based. There are no legal consequences or requirements following upon declaration of status. The purpose of COSEWIC and of declaring status is to provide a national scientific consensus which may be used by jurisdictions in the exercise of their mandates.

Acknowledgements

Michael Dunn, Canadian Wildlife Service, reviewed an earlier version of the manuscript.

References

Describing Rarity: the Ranking Dilemma and a Solution

Apperson, K. 1992. *Kootenai River White Sturgeon Status Report.* Unpublished report. Idaho Department of Fish and Game, Coeur d'Alene, Idaho.

Hildebrand, L. 1991. *Lower Columbia River Fisheries Inventory. 1990 Studies.* Vol I: Main report. B.C. Hydro, Environmental Resources, Vancouver, B.C.

Jenkins, R.E. 1986. Information management for the conservation of biodiversity. In: *Biodiversity.* E.O. Wilson (ed.). National Academic Press, Washington, D.C. pp. 231-239.

Keystone Center. 1991. *Final Consensus Report of the Keystone Policy Dialogue on Biological Diversity on Federal Lands.* The Keystone Center, Keystone, Colorado.

Lane, E.D. 1991. Status of the White Sturgeon, *Acipenser transmontanus,* in Canada. *Canadian Field-Naturalist.* 105:161-168.

Noss, R.F. 1987. From plant communities to landscapes in conservation inventories: a look at the nature conservancy (USA). *Biological Conservation* 41:11-37.

Provincial Designation of Threatened and Endangered Species

B.C. Environment. 1991. *Managing Wildlife to 2001: A Discussion Paper.* B.C. Ministry of Environment, Lands and Parks, Victoria, B.C.

Mayr, E. 1967. Population size and evolutionary parameters. In: *Mathematical Challenges to the Neo-Darwinian Interpretation of Evolution.* P.S. Moorehead and M.M. Kaplan. (eds.). Wistar Institute Symposium Monographs 5. pp. 47-58.

Mayr, E. 1982. Adaptation and selection. *Biologische Zentralblatt* 101:161-174.

Status	Scientific Name	Common Name
Extirpated	*Phrynosoma douglasii*	Pygmy Short-horned Lizard
Endangered	*Adiantum capillus-veneris*	Southern Maidenhair Fern
	Dermochelys coriacea	Leatherback Turtle
	Falco peregrinus	Anatum Peregrine Falcon
	Oreoscoptes montanus	Sage Thrasher
	Strix occidentalis	Spotted Owl
	Catostomus sp.	Salish Sucker
	Enhydra lutris	Sea Otter
	Balaena glacialis	Right Whale (Pacific population)
	Marmota vancouverensis	Vancouver Island Marmot
Threatened	*Epipactus gigantea*	Giant Helleborine
	Azolla mexicana	Mosquito Fern
	Iris missourensis	Western Blue Flag
	Speotyto cunicularia	Burrowing Owl
	Buteo regalis	Ferruginous Hawk
	Lanius ludovicianus	Loggerhead Shrike
	Brachyramphus marmoratus	Marbled Murrelet
	Picoides albolarvatus	White-headed Woodpecker
	Gasterosteus spp.	Enos Lake Stickleback
	Cottus confusus	Shorthead Sculpin
	Megaptera novaeangliae	North Pacific Humpback Whale
	Bison bison athabascae	Wood Bison
Vulnerable	*Limnanthes macounii*	Macoun's Meadowfoam
	Cephalanthera ausitinae	Phantom Orchid
	Dicamptodon ensatus	Pacific Giant Salamander
	Sterna caspia	Caspian Tern
	Tyto alba	Common Barn Owl
	Accipiter cooperii	Cooper's Hawk
	Otus flammeolus	Flammulated Owl
	Strix nebulosa	Great Gray Owl
	Numenius americanus	Long-billed Curlew
	Falco peregrinus pealei	Peale's Peregrine Falcon
	Cygnus buccinator	Trumpeter Swan
	Falco peregrinus tundrius	Tundra Peregrine Falcon
	Gasterosteus aculeatus	Charlotte Unarmored Stickleback
	Gasterosteus sp.	Giant Stickleback
	Acipenser medirostris	Green Sturgeon
	Acipenser transmontanus	White Sturgeon
	Lampetra macrostoma	Lake Lamprey
	Sardinops sagax	Pacific Sardine
	Rhinichthys osculus	Speckled Dace
	Rhinichthys umatilla	Umatilla Dace
	Myotis thysanodes	Fringed Myotis
	Myotis keenii	Keen's Long-eared Bat
	Antrozous pallidus	Pallid Bat
	Mustela erminea haidarum	Queen Charlotte Islands Ermine
	Euderma maculatum	Spotted Bat
	Gulo gulo	Wolverine-Western Population
	Rangifer tarandus caribou	Western Woodland Caribou
	Ursus arctos	Grizzly Bear

Table 2-8: Species Designated by the Committee on the Status of Endangered Wildlife in Canada Which Occur in British Columbia

Extirpated to Endangered

Downlisting

a) Species, subspecies or population is successfully reproducing at the F2 (second) generation in free-ranging or fenced conditions.

b) Survival of offspring to reproductive age is greater than mortality over that time period.

c) Habitat essential to species' survival is adequately protected in the foreseeable future (30 years) through management or preservation.

d) Population is stable or increasing but numbers still very small.

e) A backup source of breeding age individuals is available in Canada (captive, cultivated or wild individuals from another country).

Uplisting

a) No record for 3-5 years of individuals or populations in suitable habitat in spite of investigation.

Endangered to Threatened

Downlisting

a) At least 3 populations separated spatially by some considerable distance.

b) All sites are adequately protected and regulated areas.

c) Most or all populations are showing significant increase with natural expansion of range occurring, or all existing habitat at site is occupied.

d) Habitat, in addition to that currently occupied, is available for continued range expansion.

e) Threat is still evident in part or most of species' former range in Canada.

f) Captive or cultivated stock for wild release maintained if needed.

Uplisting

a) Very few native populations exist (perhaps 3 or less remaining, protected or not) and populations are declining drastically due to habitat loss, excessive harvest, natural catastrophes, environmental stresses or other factors caused by man.

Threatened To Vulnerable

Downlisting

a) Populations have recovered or increased to a point where extinction or extirpation is unlikely as long as currently available habitat is preserved or managed.

b) All or most wild, free-ranging or fenced populations are increasing or stable with no evidence of a decrease for the last 3-5 years.

c) Populations are sufficiently low so the species is uncommon within its range or confined to a small geographic area—otherwise, the species is delisted.

Uplisting

a) Several populations have decreased drastically in numbers, perhaps only 1/2 of those existing 5-10 years ago remain.

b) Rapid decline in numbers caused by environmental factors, pollutants, or other detrimental factors caused by man.

c) Factors causing decline are still evident.

Vulnerable to No Designation Required

a) Significant numerical increases in known population(s) found or established that substantially increases the number previously known on the area of its documented range.

b) Population currently occupies most suitable habitat remaining in its former geographical range.

c) A harvestable portion could be removed for subsistence, commercial or recreational use, if necessary, without seriously reducing the population size or affecting its potential for sustained growth.

Table 2-9: Examples of Criteria for Changing Status of Species listed by the Committee on the Status of Endangered Wildlife in Canada

Munro, W.T. and D. Low. 1980. Preliminary plan for the designation of threatened and endangered species. In: *Threatened and Endangered Species and Habitats in British Columbia and the Yukon*. R. Stace-Smith, L. Johns and P. Joslin. (eds.). B.C. Ministry of Environment, Victoria, B.C. pp. 65-74.

Munro, W.T. 1990. Criteria for the designation of endangered and threatened species under the B.C. Wildlife Act. *BioLine* 9:12-14.

Scudder, G.G.E. 1989. The adaptive significance of marginal populations: a general perspective. In: *Proceedings of the National Workshop on Effects of Habitat Alteration on Salmonid Stocks*. C.D. Levings, L.B. Holtby and M.A. Henderson (eds.). Canadian Special Publication of Fisheries and Aquatic Sciences 105. pp. 180-185.

Chapter 3
A Rose by any Other Name...

The biodiversity crisis is upon us, as remarkable and unique kinds of plants and animals become extinct daily around the world. We need to act quickly to preserve key portions of the world's biological diversity. Regrettably, the information available for making decisions critical to the conservation and management of living resources often is poor. Even in relatively well known areas, like British Columbia, our knowledge of the basic units of biological diversity generally is not very good.

Biological systematics (or biosystematics) and taxonomy[1] are fundamental to discovering, documenting, describing, and interpreting the elements of biological diversity. They provide the structure we need to understand evolutionary relationships. Taxonomists, for example, assign standardized scientific names to organisms, thereby promoting unambiguous communication about species, subspecies and other formally recognized levels of biological diversity. A rose by any other name might smell as sweet, but only by referring to its scientific name can we be sure that we're all talking about the same rose and the same smell. In addition, biosystematists and taxonomists offer to the public and to other scientists various species identification services, including species identification itself, reference specimens, identification keys, field guides, and distributional atlases.

In this paper, we outline how scientific names are structured and assigned, then discuss the biological significance of subspecies. Next, we look at voucher specimens, what they are and why they are needed. Finally, we discuss the changing roles of museums in supporting and participating in biodiversity initiatives.

Edward H. Miller
Biology Department
University of Victoria
P.O. Box 1700
Victoria, B.C.
V8W 2Y2

Geoffrey G.E. Scudder
Zoology Department
University of British Columbia
6270 University Boulevard
Room 2534
Vancouver, B.C.
V6T 1Z4

Names and Naming
Essentials of Nomenclature

"Nomenclature" refers to the scientific naming of plants and animals, as well as to the system of names that results (Jeffrey, 1989). Nomenclature aims to provide uniformity, universality, accuracy and stability in names for different kinds of organisms. Naming of organisms follows formal rules that are laid out in three "codes of nomenclature" governing bacteria, animals and plants (including fungi) (American Society for Microbiology, 1975; International Trust for Zoological Nomenclature, 1985; International Association for Plant Taxonomy, 1988). The codes differ from one another in detail, but essentially are structured along similar lines.

---

[1] "Systematics" refers broadly to the science concerned with biological diversity, such as species richness or geographic distribution. "Taxonomy" refers more specifically to the theory and practice of naming and classifying organisms (Mayr and Ashlock, 1991).

A "taxon" (plural, "taxa") refers to any particular taxonomic group, such as the rainbow trout (a species) or roses (a family). In classifications, taxa are arranged in a hierarchy, the levels of which are called "categories." The species is the basic unit for accurate scientific identification. Thus, the Steller's Jay, British Columbia's Provincial Bird, is classified as in Table 3-1.

This species' two-part name, or "binomen," is *Cyanocitta stelleri*. It is unique and is, for most purposes, a sufficient designation.[2] When subspecies are recognized (see below), the name consists of three words, such as *Cyanocitta stelleri carlottae*, and is called a "trinomen." Other widespread conventions are given below.

(a) Genus, species and subspecies names are italicized or underlined (as above).

(b) The name of the person(s) who first formally published the original description and name of a species or subspecies (or other taxon) may be given, especially when the scientific name is first used in a publication. The date of first publication of the scientific name may also be indicated, separated by a comma. Hence: "*Cyanocitta stelleri carlottae* Osgood, 1901" or "*Cyanocitta stelleri carlottae* Osgood." The names of some well known scientists may be abbreviated (such as "L." for "Linnaeus," as in "*Homo sapiens* L.").

(c) If a species has been moved to a genus different from the one in which it was placed originally, then the name of the person(s) who described it first is placed in parentheses. Thus, the Steller's Jay is called *Cyanocitta stelleri* (Gmelin, 1788), or *Cyanocitta stelleri* (Gmelin), because Gmelin described the species originally as *Corvus stelleri*. In botanical nomenclature, the author responsible for the new combination is included also. The Douglas-fir tree is, therefore, known as *Pseudotsuga menziesii* (Mirbel) Franco.

(d) To refer to several species within a genus, the genus name is given, followed by the abbreviation spp., as in *Cyanocitta* spp. To refer to a single species whose name is unknown or uncertain, the genus name may be given, along with the abbreviation sp., as in *Cyanocitta* sp.

For plants, several categories below the species level are recognized under the Botanical Code: *subspecies*, *variety* and *form*. To avoid ambiguity in the use of these categories, the insertion of abbreviations that indicate rank is mandatory under the Botanical Code, for example lodgepole pine, *Pinus contorta* var. *latifolia*; balsam poplar, *Populus balsamifera* ssp. *balsamifera* (or, equivalently, subsp. *balsamifera*); and common camass, *Camassia quamash* var.

Category	Taxon	Common Name
Class	Aves	Birds
Order	Passeriformes	Songbirds
Family	Corvidae	Crows, jays, etc.
Genus	*Cyanocitta*	Jays (some)
Species	*stelleri*	Steller's Jay

Table 3-1. Example of the Taxonomic Classification of the Steller's Jay

[2] Unfortunately, different terminology is used under the different codes. In the Botanical Code, the complete binary name is referred to as the "specific name," and the second term as the "specific epithet." In the Zoological Code, the complete binary name is referred to as the "binomen," and the second term as the "specific name" (Jeffrey, 1989:57).

quamash f. *albiflora* (the white-flowered form of the variety; Scoggan, 1978). Names of plant taxa below the species level generally are reduced to the trinomial form unless it would cause ambiguity (Jeffrey, 1989). In the last example, all four names are necessary (in botany, authors of names invariably are given along with the scientific name, as for the Douglas-fir above; in the preceding examples we have omitted them for clarity).

It is extremely important for authors of technical or scientific reports and publications to cite the proper scientific name for species they mention. Organisms that are well known to the general public and have stable scientific names often have English names as well: birds, butterflies, and flowers are examples. English names are easier to remember, after all, and are very effective in communication because they make more sense to most people (Mayr, 1990). In ornithology, the first letter of such names usually is capitalized by convention, as in Blue Jay. In other disciplines, however, common names are not capitalized routinely unless a proper name such as that of the species' discoverer forms part of the name, so that we have, for example, Douglas-fir, silver fir, Engelmann spruce, spotted frog and moose. Formal systems of English names must be linked to scientific nomenclature, and reference to a standard work is essential. Otherwise, confusion and ambiguity inevitably arise (at least 10 English names have been applied to the Steller's Jay, for example, including Black-headed Jay and Long-crested Jay [Banks, 1988]).

Taxonomic Changes

Advances in biosystematics and taxonomy often necessitate taxonomic changes. The simplest sorts of change involve a new status or new synonymy (when forms with different names are recognized to be the same). For example, in 1925, Swarth and Brooks described a new species of British Columbian sparrow, and placed it in the already-recognized genus *Spizella*. They named it the Timberline Sparrow *(Spizella taverni)* in honour of the Canadian ornithologist Percy Taverner. This newly recognised species thus became an addition to the list of species then accepted within *Spizella*, which included Tree Sparrow (*S. arborea*) and Chipping Sparrow (*S. passerina*). Subsequently, most ornithologists concluded that *S. taverni* really belonged to the same species as Brewer's Sparrow *(S. breweri)*. *Spizella breweri* had been formally described and named much earlier (1856) by John Cassin, so its name took precedence and, for most of this century, *S. taverni* was relegated to a subspecies of Brewer's Sparrow, with the default name *S. breweri taverni* (indeed, even the taxon's namesake reckoned that eventually it would be relegated to subspecies status [Taverner, 1928:293]). Current opinions are split, with one recent authority again recognizing *S. taverni* as a full species, based on its ecological, genetic and behavioural features (Sibley and Monroe, 1990).

This example illustrates that there may be differing opinions about the scientific name for a genus, species or subspecies, and about the level at which a form of plant or animal should be recognized. Such uncertainty, combined with changes to classification, makes nomenclature unstable and makes it more difficult to communicate information about biological diversity. Refinements to classification and nomenclature below and at the level of species occur particularly often as our understanding of relationships at these levels grows. We discuss this subject next.

Subspecies

The subspecies has been called "the most critical and disorderly area of modern systematic theory," with a "demonstrably flimsy conceptual basis" (Wilson and Brown, 1953:100, 107; Baker, 1985; Panchen, 1992). Scientists are particularly interested in investigating the evolution of subspecies because subspecies may be in the early stages of speciation (the formation of new species), and they represent genetic diversity within species.

To understand the subspecies issue, we need first to back up and look more closely at species. Species can be defined as "groups of interbreeding natural populations that are reproductively isolated from other such groups" (Mayr and Ashlock, 1991:429). Defined in this way, the species category reflects biological reality for most

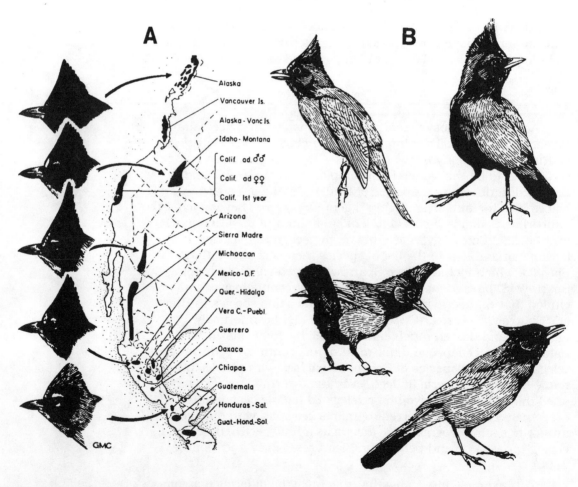

Figure 3-1. Geographic Variation in British Columbia's Provincial Bird, the Steller's Jay (*Cyanocitta stelleri*), Which Has Six Described Subspecies (American Ornithologists' Union 1957). Plumage of the face and crest exhibits striking variation across the species' range, examples of which are shown in part A (shaded areas represent sample localities in the study). This species uses the face and crest extensively in communication, as shown in part B. Sources: Brown, 1963 and 1964.

groups of sexually reproducing organisms, so it is useful and valuable for describing the natural world. All species exhibit many kinds of variation in relation to, for example, season (like the white fur coats grown by some species in winter) or geography (like the regional differences in the plumage of the face and crest in the Steller's Jay [Figure 3-1]) (Mayr, 1963). Geographic variation alone may be considerable and is used in the formal recognition of subspecies by scientists. Some species, for example, occur as small isolated populations or in restricted habitats (or both), while others occupy large and continuous geographic areas that often contain many different habitats. Given the enormous variety in size, shape, connectedness and ecological attributes of species' ranges, it is not surprising that species differ greatly in whether they exhibit geographic variation at all and, if they do, how that variation is expressed.

The category of subspecies was first employed by ornithologists, mainly in North America, to designate geographically differentiated "races" within species (Sibley, 1954; Stresemann, 1975). Though the definition of subspecies is still a matter of debate, it is reasonable to think of a subspecies as having some unique trait or characteristic that is permanent and passed on from generation to generation. The debate about whether subspecies should be considered a separate taxonomic category usually can be reduced to two questions (Simpson, 1961), which we will discuss in turn: 1) Does the subspecies have an objective basis; and 2) Is it useful to formally recognize subspecies?

The Hermit Thrush (*Catharus guttatus*) illustrates how difficult it is to answer the first of these questions. This thrush is widely distributed in North America, and varies in plumage and size across its range. In his detailed analysis of geographic variation, Aldrich (1968) identified ranges for 10 subspecies of this thrush in continental North America, plus several islands (Figure 3-2). One subspecies, *nanus* ("2" in Figure 3-2), breeds only on the Queen Charlotte Islands and outer islands of the Alexander Archipelago in southeastern Alaska. Another, *vaccinius* ("3" in Figure 3-2), occurs on Vancouver Island "and in a very limited area on the mainland near Vancouver" (Aldrich, 1968:20). These two island forms have fairly natural boundaries to their breeding ranges, so they fit well with the notion that subspecies have an objective basis. However, like the eight other subspecies of Hermit Thrush that breed from Alaska to Newfoundland, most geographic variation is not structured so simply (Figure 3-3). Where geographic variation is extensive, it can often usefully be summarized, however crudely, in terms of subdivisions of a species (Stuessy, 1990). Where it is not extensive, or where other complications occur (such as species that migrate and associate with individuals from different breeding or wintering ranges) other characteristics, such as seasonal or genetic variation, need to be examined to identify subspecies.

However they are defined, though, the number and distinctiveness of subspecies provide a useful index of genetic diversity within a species (Figure 3-3). Biosystematists are interested in tracking genetic diversity to determine evolutionary relationships among categories of organisms. Consequently, much work has been done to identify "evolutionarily significant units," and subspecies have been

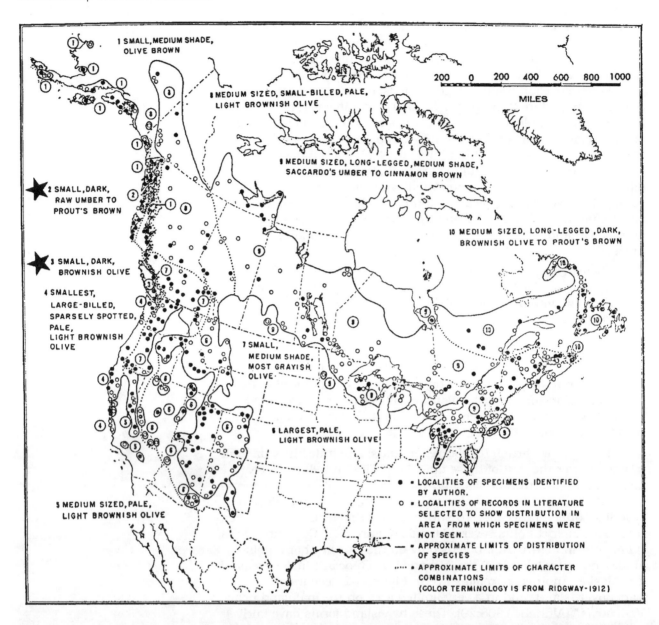

Figure 3-2. Breeding Distribution of the Hermit Thrush (*Catharus guttatus*) and its Subspecies. Each subspecies is represented by a number. The suggested range boundaries between them are shown as dotted lines. The two subspecies referred to in the text are *nanus* ("2") and *vaccinius* ("3"), and are marked by stars. Source: Aldrich, 1968.

found to be useful in approximating such units (Ryder, 1986), though they do not substitute for more rigorous or direct measures of geographic variation or genetic diversity. For example, some stocks of salmon that look the same may differ genetically and, therefore, be evolutionarily significant units, though they do not differ enough to be defined as separate subspecies. Dizon et al. (1992) have proposed a more extensive approach to identifying "stocks" that are evolutionarily significant units. To employ their concept, data are required on a species' distribution, population characteristics (such as demography and social structure), physical characteristics or "phenotype" (such as size and colour patterns), and genetic attributes. Species that are highly fragmented geographically, and whose local populations differ from one another in

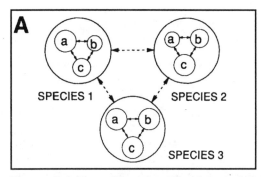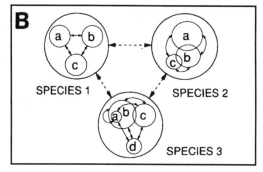

Figure 3-3. Two Conceptualizations of the Structure of Species. Species can be considered to be distinct from one another, each composed of subspecies (a, b and c) that do not overlap geographically, as in model A. Extremely low genetic interchange, or none at all, is symbolized by broken arrows; solid arrows represent genetic interchange between subspecies. A more realistic model (B) shows species composed of entities with different range sizes and locations, and with more complex patterns of geographic overlap and genetic interchange. The hypothetical species 3 in part B resembles the Hermit Thrush in some ways (compare with Figure 3-2). Source: Dizon et al., 1992.

population characteristics, phenotype and genotype, represent one extreme, "Phylogeographic Type I" (Figure 3-4), while those that are not geographically fragmented and differ little between populations represent the other extreme, "Phylogeographic Type IV." The population subdivisions, or stocks, of a Phylogeographic Type I species represent high genetic diversity, and demand separate conservation and management strategies.

So, despite the complications of the subspecies category, it is useful because it offers an approximation of genetic diversity within species and it can often (though not always) be said to have an objective basis. The subspecies is, therefore, important in conservation and a useful tool in management (Bunnell and Williams, 1980; Bunnell, 1990).

THEORETICAL CLASSIFICATION		Differential Selection		OPERATIONAL CLASSIFICATION	
		Little	Great		
GENE FLOW	little or none	III	I	great	GEOGRAPHIC LOCALIZATION
	high	IV	II	little or none	
		Little	Great		
		Proxies for Differential Selection			

Figure 3-4. Characterization of Subdivisions, or Stocks, of a Species by Distinctiveness and Genetic Isolation (or Phylogeographic Type, "PT"). In an ideal situation (the "theoretical classification"), we would know how much gene flow occurs among stocks, and why and how they have evolved differently. In practice (the "operational classification"), genetic interchange usually must be estimated by the degree to which ranges have become geographically localized, and evolutionary distinctiveness must be approximated by features like external characteristics and habitat, which are called "proxies for differential selection." For example, a stock that is highly geographically localized and has very distinctive characteristics ("proxies for differential selection") would be classified as phylogeographic type I ("PTI"). Source: Dizon et al., 1992.

Voucher Specimens: a Scientific and Ethical Imperative

Specimens must usually be collected and preserved before the species to which they belong can be identified. Accurate identification relies on the availability of expert taxonomists, and of published identification keys or field guides, and usually involves the use of reference specimens like pressed plants or pinned insects. After specimens have been identified, representative samples of them should be set aside as "voucher specimens." These voucher specimens can then later be used to independently verify an investigator's claims, to re-evaluate the species or life stages present in a sample in light of improved knowledge, to make historical comparisons, or to trace ecological changes (Figure 3-5).

Voucher specimens can be defined as "representative specimens or samples that are obtained in field surveys and research, and that are preserved to permit independent verification of results and to allow further study" (Miller and Nagorsen, 1992). The term "specimen" includes conventional preparations like skins, skulls, pressed plants, or dead animals in preserving fluids, as well as documentation, such as photographs and audio recordings. Tape

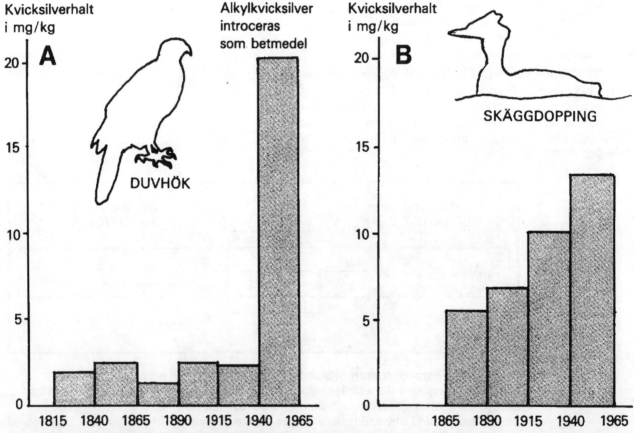

Figure 3-5. Example of the Usefulness of Voucher Specimens. The histograms summarize information on long-term changes in mercury levels in the feathers of (A) Northern Goshawk *(Accipiter gentilis)*, and (B) Great Crested Grebe *(Podiceps cristatus)*. In neither case were early collections made for the purposes to which they were eventually put; indeed, the uses and analytic techniques employed could not have been predicted so long ago. Source: Odsjö, 1991.

recordings of songs and calls provide acceptable evidence for the occurrence of many species of birds, frogs and toads, for example; it is not necessary to kill any if the objective is solely to document occurrence.

The nature of the evidence to be obtained must be worked out in advance of a study or survey, on the premise that independent examination of the material may be undertaken to corroborate the results. A photograph, audio recording or sight record (for easily distinguished species, like Bald Eagles *[Haliaeetus leucocephalus]*) is adequate for simply documenting a bird species' presence in an area. If one were interested in determining whether Hermit Thrushes migrating through an area belonged to the subspecies *nanus*, however, probably it would be necessary to collect and preserve some individuals as study skins. Thus, the kind and number of voucher specimens and the nature of accompanying documentation (such as date and locality), must be tailored to each study, and must be planned to meet the basic scientific criterion of independent verification.

Investigators also have an ethical obligation to use fully the creatures that they collect. Extremely rare and long-lived plants or animals, like the White Pelican (*Pelecanus erythrorhynchos*), normally should never be killed for scientific specimens, regardless of how scientifically valuable they may be, because to do so would have a severe impact on their population. However, every effort should be made to salvage specimens of such species that have died of other causes.

The scientific importance of voucher specimens can be illustrated by reference to small forest mammals. They include species that are difficult or impossible to identify except as scientific specimens, including many bats, shrews and mice. Indeed, some forms have only been recognized as distinct species recently. For example, the deer mouse (*Peromyscus maniculatus*) and Columbian mouse (*P. oreas*) were confirmed as different species only in 1988; and dusky and vagrant shrews (*Sorex monticolus* and *vagrans*) were not confirmed as distinct from one another until 1977 (Nagorsen, 1990). Regrettably, some important past surveys of small mammals did not preserve voucher specimens, so it is impossible to know which of these species was trapped, or in what proportions or habitats. Our knowledge of the geographical distributions and ecological requirements of these important species has been limited as a result (Figure 3-6; Miller, 1993). The forest bat, *Myotis keenii*, is another example. The range of this species is unclear. It is so uncommon, and possibly endangered, that it would be unethical to go out and collect live specimens; consequently, it is expected that few specimens will become available in the future to document its occurrence. As a result, the current understanding of its British Columbia range must be based on re-examination of all known museum specimens. Almost all previously published records have turned out to be for the commoner western long-eared myotis (*M. evotis*); only three *M. keenii* specimens are known (Nagorsen and Brigham, 1993)!

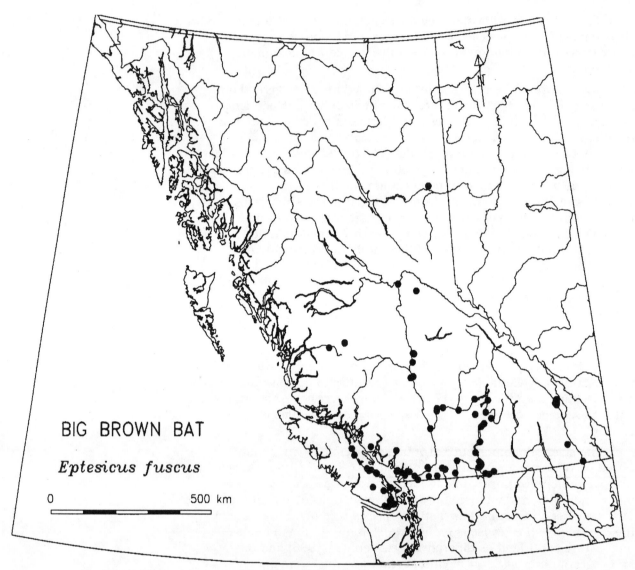

Figure 3-6. Geographic Distribution of the Big Brown Bat (*Eptesicus fuscus*) in British Columbia, Based on Museum Specimens. Many bat species can be identified only by experts, and certain ones can be identified only on the basis of specimens like skins and skulls. Source: Nagorsen and Brigham, 1993.

Biological Systematics and the Importance of Museums

The disciplines of biosystematics and taxonomy are in trouble. They are receiving decreasing research funding and fewer staff positions in museums, universities and other research institutions (Culotta, 1992). The declining resources affect the integrity of biosystematics and taxonomy as disciplines, and will have severe effects on their conceptual development and on the development and application of new techniques (O'Hara et al., 1988). As well, the disciplines will not be able to continue producing and updating materials such as lists of flora and fauna, identification manuals and keys, and identification services, which are essential to a broad range of biodiversity initiatives. We need, therefore, to expand research and training in biological systematics and taxonomy, and to survey biological diversity throughout the province. As well, we must continue to refine our conceptual and theoretical understanding of the elements of biodiversity, as even common words like "species" are used in different ways (Dowling et al., 1992; Rojas, 1992).

Practical needs for end products, on the one hand, and the strength and maturation of biosystematics and taxonomy as disciplines, on the other, are interrelated. Biodiversity initiatives demand a sound taxonomic and biosystematic underpinning. Yet, we still don't know very much about the most diverse group of organisms on the planet, the invertebrates (Kosztarab and Schaefer, 1990), and we lack many fundamental sorts of information about vertebrate biodiversity, as evidenced by the recent discovery of the spawning grounds of the culturally important and long-exploited Japanese eel (*Anguilla japonica*) (Tsukamoto, 1992). Conservation and biodiversity initiatives demand information that requires focussed empirical or conceptual research of a very sophisticated kind. Genetic and evolutionary analyses of endangered species are well known examples (Barrowclough, 1984; Ledig, 1993). One such investigation suggests that the endangered Spotted Owl (*Strix occidentalis*) actually comprises two species, not just one (Barrowclough and Gutiérrez, 1990; Figure 3-7). Soltis et al. (1992) have shown that the herbaceous plant, *Bensoniella oregona*, a species with an extremely narrow distribution in Oregon and California, is genetically depauperate and, thus, may be endangered because it may be unable to adapt to environmental change. Finally, the Atlantic codworm (*Pseudoterranova decipiens)*, a roundworm parasite of commercially important fish that causes millions of dollars in damage to the fishing industry each year, has been found to comprise several cryptic species with entirely different distributions and life histories (Bristow and Berland, 1992; Miller, 1993; and included references). Such investigations will be increasingly difficult to carry out if resources and scholarly training in biosystematics and taxonomy continue to decline. This taxonomic impediment is being addressed in many countries, but not in Canada thus far (Krebs, 1992; U.K. House of Lords, 1992).

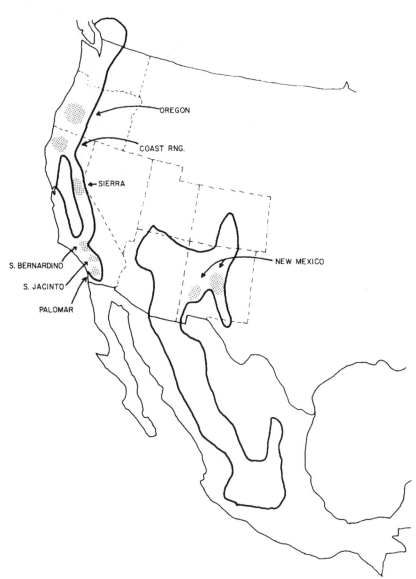

Figure 3-7. Geographic Distribution of the Spotted Owl (*Strix occidentalis*). The populations in the two parts of the range are so different genetically that they could be considered separate species. Localities from which genetic samples were taken are shaded. Source: Barrowclough and Gutiérrez, 1990.

39

Indeed, in 1993, the Canadian Museum of Nature set back biosystematics in Canada by discontinuing research in ornithology, mammalogy and herpetology as part of a plan to move toward inter-disciplinary studies in "solution-oriented research."

Support for the financial aspects of collections maintenance is declining (Danks, 1991; Wiggins et al., 1991). The diverse costs of collections include those of chemicals for preserving collections, storage cabinets, information management systems, staff to care for and document collections, packaging and shipping of loans, and storage space. These basic costs must be met if collections are to be maintained for posterity and made accessible to the growing number of people who want to use, examine or undertake research on the collections.

Natural history museums originated from private collections that were later amalgamated into institutions. Since the nineteenth century, such institutions have been centres of research, though the most successful of them also provided public education (Alexander, 1979; Farber, 1982). Today's large natural history museums continue to fulfill the three basic functions of collections storage, collections study and public education. It is in the areas of collections conservation and biosystematic and taxonomic research that museums contribute most to the understanding of biodiversity. Natural history museums should further broaden their usefulness to potential users in fields outside biosystematics and taxonomy, particularly in environmental fields like pollution biology, applied entomology and wildlife biology (Miller, 1985), by:

○ diversifying preservation techniques;

○ making collections more accessible to users; and

○ increasing training in specimen preparation and collections use.

The recent discovery that DNA can be extracted from dried and preserved museum specimens (Kocher et al., 1989; Kocher, 1992), and even from 30-million-year-old fossil termites (DeSalle et al., 1992), underscores the contribution of natural history museums to the preservation and understanding of biodiversity. Such discoveries are possible only where collections have grown vigorously and been maintained properly. It is through such discoveries that we gain the understanding of the natural world that is vital to our continued existence on this planet.

Acknowledgements

We thank Richard Hebda for his help and suggestions, Emily McCullum for her patience and editorial skills, and Stein Glommen for translating Odsjö's paper.

References

Alexander, E.P. 1979. *Museums in Motion. An Introduction to the History and Function of Museums.* American Association for State and Local History, Nashville, Tennessee.

Aldrich, J.W. 1968. Population characteristics and nomenclature of the Hermit Thrush. *Proceedings of the United States National Museum,* No. 3637.

American Ornithologists' Union. 1957. *Check-list of North American Birds*. Fifth edition. American Ornithologists' Union, Baltimore, Maryland.

American Society for Microbiology. 1975. *International Code of Nomenclature of Bacteria*. American Society of Microbiology, Washington, D.C.

Baker, A.J. 1985. Museum collections and the study of geographic variation. In: *Miller* (1985). pp. 55-77.

Banks, R.C. 1988. *Obsolete English Names of North American Birds and Their Modern Equivalents*. U.S. Fish and Wildlife Service, Resource Publication No. 174.

Barrowclough, G.F. 1985. Museum collections and molecular systematics. In: *Miller* (1985). pp. 43-54.

Barrowclough, G.F. and R.J. Gutiérrez. 1990. Genetic variation and differentiation in the Spotted Owl (*Strix occidentalis*). *Auk* 107:737-744.

Bristow, G.A. and B. Berland. 1992. On the ecology and distribution of *Pseudoterranova decipiens* C (Nematoda: Anisakidae) in an intermediate host, *Hippoglossoides platessoides*, in northern Norwegian waters. *International Journal of Parasitology* 22:203-208.

Brown, J.L. 1963. Ecogeographic variation and introgression in an avian visual signal: the crest of the Steller's Jay, *Cyanocitta stelleri*. *Evolution* 17:23-39.

Brown, J.L. 1964. The integration of agonistic behavior in the Steller's Jay *Cyanocitta stelleri* (Gmelin). *University of California Publications. Zoology* 60:223-328.

Bunnell, F.L. 1990. Biodiversity: what, where, why, and how. In: *Wildlife Forestry Symposium Proceedings: A Workshop on Resource Integration for Wildlife and Forest Managers*. A. Chambers (ed.). Canada-British Columbia Forest Resource Development Agreement Report No. 60. pp. 29-45.

Bunnell, F.B. and G.R. Williams. 1980. Subspecies and diversity—the spice of life or prophet of doom. In: *Threatened and Endangered Species and Habitats in British Columbia and the Yukon*. R. Stace-Smith, L. Johns and P. Joslin (eds.). Ministry of Environment, Victoria, B.C. pp. 246-259.

Culotta, E. 1992. Museums cut research in hard times. *Science* 256:1268-1271.

Danks, H.V. 1991. Museum collections: fundamental values and modern problems. *Collection Forum* 7:95-111.

DeSalle, R., J. Gatesy, W. Wheeler and D. Grimaldi. 1992. DNA sequences from a fossil termite in Oligo-Miocene amber and their phylogenetic implications. *Science* 257:1933-1936.

Dizon, A.E., C. Lockyer, W.F. Perrin, D.P. Demaster and J. Sisson. 1992. Rethinking the stock concept: a phylogeographic approach. *Conservation Biology* 6:24-36.

Dowling, T.E., B.D. DeMarais, W.L. Minckley, M.E. Douglas and P.C. Marsh. 1992. Use of genetic characters in conservation biology. *Conservation Biology* 6:7-8.

Farber, P.L. 1982. *The Emergence of Ornithology as a Scientific Discipline: 1760-1850*. D. Reidel, Dordrecht, Holland.

International Association for Plant Taxonomy. 1988. *International Code of Botanical Nomenclature*. Koelz Scientific Books, Königstein, Germany.

International Trust for Zoological Nomenclature. 1985. *International Code of Zoological Nomenclature*. University of California Press, Berkeley, California.

Jeffrey, C. 1989. *Biological Nomenclature*. Third edition. Edward Arnold, London, England.

Kocher, T.D. 1992. PCR, direct sequencing, and the comparative approach. *PCR Methods and Applications* 1:217-221.

Kocher, T.D., W.K. Thomas, A. Meyer, S.V. Edwards, S. Pääbo, F.X. Villablanca and A.C. Wilson. 1989. Dynamics of mitochondrial DNA evolution in animals: amplification and sequencing with conserved primers. *Proceedings of the National Academy of Science USA* 86:6196-6200.

Kosztarab, M. and C.W. Schaefer (eds.). 1990. *Systematics of the North American Insects and Arachnids: Status and Needs*. Virginia Agricultural Experiment Station, Virginia Polytechnic Institute and State University, Blacksburg, Virginia.

Krebs, J.R. (Chairperson). 1992. *Evolution and Biodiversity--The New Taxonomy*. Report of the Committee set up by the Natural Environment Research Council. Natural Environment Research Council, London, England.

Ledig, F.T. 1993. Secret extinctions: the loss of genetic diversity in forest ecosystems. In: *Our Living Legacy: Proceedings of a Symposium on the Biological Diversity of British Columbia*. M.A. Fenger, E.H. Miller, J.F. Johnson and E.J.R. Williams (eds.). Ministry of Environment, Lands and Parks, Ministry of Forests, and Royal B.C. Museum, Victoria, B.C. pp. 127-140.

Main, B.Y. 1990. Restoration of biological scenarios: the role of museum collections. *Proceedings of the Ecological Society of Australia* 16:397-409.

Mayr, E. 1963. *Animal Species and Evolution*. Belknap Press, Cambridge, Massachusetts.

Mayr, E. 1990. A natural system of organisms. *Science* 348:491.

Mayr, E. and P.D. Ashlock. 1991. *Principles of Systematic Zoology*. Second edition. McGraw-Hill, New York, New York.

Miller, E.H. (ed.). 1985. *Museum Collections: Their Roles and Future in Biological Research*. B.C. Provincial Museum Occasional Paper No. 25.

Miller, E.H. and D.W. Nagorsen. 1992. Voucher specimens: an essential component of biological surveys. In: *Methodology for Monitoring Wildlife Diversity in B.C. Forests*. L. Ramsay (ed.). B.C. Ministry of Environment, Lands and Parks, Victoria, B.C. pp. 11-15.

Miller, E.H. 1993. Biodiversity research in museums: a return to basics. In: *Our Living Legacy: Proceedings of a Symposium on the Biological Diversity of British Columbia*. M.A. Fenger, E.H. Miller, J.F. Johnson and E.J.R. Williams (eds.). Ministry of Environment, Lands and Parks, Ministry of Forests, and Royal B.C. Museum, Victoria, B.C. pp. 141-173

Nagorsen, D.W. 1990. *The Mammals of British Columbia*. A taxonomic catalogue. Royal B.C. Museum Memoir No. 4.

Nagorsen, D.W. and R.M. Brigham. 1993. *Bats of British Columbia*. University of British Columbia Press, Vancouver, B.C.

Odsjö, T. 1991. Miljöprovbanken. *Fauna och Flora* 3:112-125.

O'Hara, R.J., D.R. Maddison and P.F. Stevens. 1988. Crisis in systematics. *Science* 241:275-276.

Osgood, W.H. 1901. Birds of the Queen Charlotte Islands. *North American Fauna* 21:38-50.

Panchen, A.L. 1992. *Classification, Evolution, and the Nature of Biology*. Cambridge University Press, Cambridge, England.

Rojas, M. 1992. The species problem and conservation: what are we protecting? *Conservation Biology* 6:170-178.

Ryder, O.A. 1986. Species conservation and systematics: the dilemma of subspecies. *Trends in Ecology and Evolution* 1:9-10.

Scoggan, H.J. 1978. *The Flora of Canada. Part 2—Pteridophyta, Gymnospermae, Monocotyledoneae*. National Museum of Science, Ottawa, Ontario, Publications. in Botany, No. 7(2), Ottawa, Ontario.

Sibley, G.C. 1954. The contribution of avian taxonomy. *Systematic Zoology* 3:105-110.

Sibley, C.G. and B.L. Monroe, Jr. 1990. *Distribution and Taxonomy of Birds of the World*. Yale University Press, New Haven, Connecticut.

Simpson, G.G. 1961. *Principles of Animal Taxonomy*. Columbia University Press, New York, New York.

Soltis, P.S., D.E. Soltis, T.L. Tucker and F.A. Lang. 1992. Allozyme variability is absent in the narrow endemic *Bensoniella oregona* (Saxifragaceae). *Conservation Biology* 6:131-134.

Stresemann, E. 1975. *Ornithology from Aristotle to the Present*. Harvard University Press, Cambridge, Massachusetts.

Stuessy, T.F. 1990. Plant Taxonomy. *The Systematic Evaluation of Comparative Data.* Columbia University Press, New York, New York.

Swarth, H.S. and A. Brooks. 1925. The Timberline Sparrow, a new species from northwestern Canada. *Condor* 27:67-69.

Taverner, P.A. 1928. *Birds of Western Canada.* National Museum of Canada, Bulletin No. 41.

Tsukamoto, K. 1992. Discovery of the spawning area for Japanese eel. *Nature* 356:789-791.

U.K. House of Lords (Select Committee on Science and Technology) 1992. *Systematic Biology Research, Vol 1--Research.* HL Paper 22-I. Her Majesty's Stationery Office, London, England.

Wiggins, G.B., S.A. Marshall and J.A. Downes. 1991. The importance of research collections of terrestrial arthropods. *Bulletin of the Entomological Society of Canada* 23(2), Suppl.

E.O. Wilson and W.L. Brown, Jr. 1953. The subspecies concept and its taxonomic application. *Systematic Zoology* 2:97-111.

Part II
SPECIES DIVERSITY

Chapter 4

Endangered Terrestrial and Freshwater Invertebrates in British Columbia[1]

Sydney Cannings
B.C. Conservation Data Centre
B.C. Ministry of Environment,
Lands and Parks
780 Blanshard Street
Victoria, B.C.
V8V 1X5

In 1981, in the middle of a hot and sunny summer on the British Columbia coast, my brother Rob and I journeyed to the Brooks Peninsula, a finger of land jutting into the Pacific Ocean on the northwest coast of Vancouver Island. We went there looking for endemic insects, since it was thought that all or part of the peninsula had escaped glaciation, and plants formerly known only from the Queen Charlotte Islands had recently been discovered there. In two weeks of regular collecting, using nothing fancier than butterfly nets and yogurt bins, we collected 3600 specimens of terrestrial invertebrates. Two years later, after we had received them back from specialists around the world, we compiled a list of 519 species in 190 families. Of the 519, 31 were undescribed species and an additional 34 were species previously unknown to Canada!

A few years later, Rob put a Malaise trap (a tent-like affair designed to catch all manner of flying insects) in a second-growth alder forest near Sooke, as part of a survey of gall midges. After four months, the trap had caught (in addition to a wealth of other insects) 100 species of gall midges, 90 of which were new to science!

These two stories illustrate the present state of our knowledge of terrestrial invertebrates in British Columbia. Although by far the greatest part of British Columbia's biodiversity is made up of invertebrates, they are the least known of any major biological group. Our knowledge of the distribution and abundance of insects and other invertebrates in British Columbia is so incomplete and fragmentary that, except for some abundant species, it is usually impossible to say from the data at hand what the true status of a species is. This is the case even for small, supposedly well-known groups such as butterflies or dragonflies.

The diversity of invertebrates makes them good indicators of small, unique habitats, but it also makes it very difficult for biologists to identify and study them. We often hear that many of the species in tropical forests will disappear before they are discovered, and this could be equally true in our province. A good guess at the number of insect species in British Columbia is 35,000; only about 15,000 have been found so far. The mind-boggling diversity makes it very difficult for entomologists to do a literature or collection survey to determine which species are endangered or threatened. Specialized, detailed surveys will be required for almost every species that is suspected of being endangered.

Because we lack baseline knowledge, even good local surveys are difficult, if not impossible, to interpret. For example, we don't know whether the new species we found on the Brooks Peninsula

[1] Originally published in the Spring 1990 (Volume 9, Number 2) issue of *BioLine*, official publication of the Association of Professional Biologists of British Columbia. Revised and updated for this publication.

are endemic, endangered, or widespread, simply because no similar surveys have been done up and down the mainland coast. There is hope, however, that this situation may change in the coming years. Despite budgetary constraints, there has been increasing interest in invertebrate surveys, as well as increasing cooperation between government ministries and universities in planning them.

Many large birds and mammals have been hunted or in some way exploited to the point of extinction or extirpation, but this is not usually the case for insects and other terrestrial invertebrates. In British Columbia, the small number of collectors and the generally high reproductive capacity of invertebrates means that invertebrates (at least terrestrial ones) are essentially immune to human predation. Certain accessible populations of rare butterflies (the Eversmann's parnassians on Pink Mountain, for instance) may be in danger of local extermination. But for most threatened and endangered invertebrates the problem is habitat destruction, and despite our general ignorance about invertebrate distribution, we do know a number of species that are confined to threatened habitats of very limited extent in the province.

A quick perusal of well-known groups of invertebrates produces a list of about 50 species—all insects—that could be considered threatened or endangered because of their rarity and restricted distribution (Table 4-1). Unfortunately, freshwater molluscs are not included in this list because we know little of their distribution and status in British Columbia, though elsewhere in North America they are among the most vulnerable of invertebrate groups to environmental degradation. The information on this list is gleaned from collections at the Royal British Columbia Museum in Victoria, the University of British Columbia in Vancouver, and the Canadian National Collection in Ottawa.

This list is dominated by species represented in the province by peripheral populations. But we shouldn't ignore their plight here just because they may be more common, or even widespread, elsewhere. For one thing, habitat destruction is occurring everywhere and, if we think that others are going to take care of our endangered species, we are fooling ourselves. Also, we should take the advice of Scudder (1989): "Marginal [peripheral] populations have a high adaptive significance to the species as a whole, and marginal [peripheral] habitat conservation, preservation and management is one of the 'best' ways to conserve the genetic diversity and resources of ... species. Marginal habitats are an essential prerequisite for the maintenance of this diversity and versatility."

Here are eight examples of insect species that are endangered or threatened in British Columbia.

Argia vivida (vivid dancer): Threatened. A beautiful, intensely blue damselfly, the vivid dancer is found only in a few scattered springs in British Columbia. All but one of the localities are creeks and ponds associated with hotsprings, which are of course under tremendous development pressure. Most accessible hotsprings in British Columbia have already lost their unique flora and fauna by water diversion into swimming pools. The only population of vivid dancers not associated with a hotspring is in a small cool spring near Penticton, which is being severely impacted by cattle trampling around its margins.

Dragonflies
 Argia vivida   Vivid Dancer
 Ischnura damula   Plains Forktail
 Erythemis collocata   Western Pondhawk
Mantises
 Litaneutria minor   Ground Mantis
Crickets
 Neduba steindachneri   a cricket
Grasshoppers
 Dissosteira spurcata   a grasshopper
True bugs
 Harmostes dorsalis   a rhopalid bug
 Coriomeris insularis   a coreid bug
 Heterogaster behrensii   a seed bug
 Neosuris castanea   a seed bug
 Nysius paludicolus   a seed bug
 Nabis lovetti   a damsel bug
 Phytocoris occidentalis   a plant bug
 Ioscytus politus   a shore bug
 Notonecta spinosa   a backswimmer
 Carsonus aridus   a leafhopper
Beetles
 Cicindela parowana   Parowana Tiger Beetle
 Eleodes extricatus   a darkling beetle
 Euphoria inda   a scarab beetle
Wasps
 Anoplius depressipes   a spider wasp
 Eucerceris vittatifrons   a sand wasp
Flies
 Nemomydas pantherinus   a mydas fly
 Apiocera barri   an apiocerid fly
 Efferia okanagana   a robber fly
 Comantella pacifica   a robber fly
 Cyrtopogon anomalus   a robber fly
 Dicolonus simplex   a robber fly
 Eucyrtopogon spiniger   a robber fly
 Lasiopogon fumipennis   a robber fly
 Lasiopogon pacificus   a robber fly
 Lasiopogon willametti   a robber fly
 Myelaphus lobicornis   a robber fly
 Megaphorus willistoni   a robber fly
 Nicocles rufus   a robber fly
 Ospriocerus aecus   a robber fly
 Scleropogon bradleyi   a robber fly
 Laphria ventralis   a robber fly
Butterflies
 Apodemia mormo   Mormon Metalmark
 Gaeides xanthoides dione   Large Copper
 Mitoura barryi   Barry's Hairstreak
 Mitoura johnsoni   Johnson's Hairstreak
 Icaricia icarioides blackmorei   Icarioides Blue
 Plebejus saepiolus insulanus   Greenish Blue
 Occidryas chalcedona perdiccas   Perdiccas Checkerspot*
 Occidryas editha taylori   Taylor's Checkerspot
 Basilarchia archippus   Viceroy*
 Euchloe ausonides   Island Large Marble[x]
 Papilio machaon pikei   Pike's Swallowtail
Wind scorpions
 Eremobates gladiolus   a wind scorpion
Spiders
 Hexura picea   a folding door spider

Table 4-1: Rare and Threatened Terrestrial and Freshwater Invertebrates of British Columbia. An incomplete, preliminary list. Key: * = extirpated; [x] = extinct.

Ischnura damula (Plains forktail): Endangered. A smaller, daintier damselfly than the previous species, the Plains forktail is represented in British Columbia by a single, relict population in the Liard River hotsprings. These springs, although they are within a provincial park, will be flooded if a planned hydro-electric project of British Columbia Hydro on the Liard River goes ahead.

Litaneutria minor (ground mantis): Threatened. This, the only native praying mantis in Canada, is restricted to the dry shrub-steppes of the southern Okanagan Valley. Despite recent entomological attention to this habitat in the last decade, the ground mantis has been found only a handful of times.

Notonecta spinosa (backswimmer): Threatened. In Canada, this aquatic bug is restricted to valley bottom wetlands in the Okanagan Valley, which are disappearing rapidly.

Cicindela parowana (tiger beetle): Endangered. This beautiful tiger beetle is known only from alkaline flats in Penticton, Okanagan Falls, and Oliver.

Eucyrtopogon spiniger (robber fly): Endangered. This species is only known from the holotype specimen collected in Victoria. Its habitat of dry, garry oak woodland is under heavy development pressure.

Mitoura johnsoni (Johnson's hairstreak): Threatened. Rare throughout its Pacific Northwest range, this attractive little butterfly had not been recorded in Canada for many years when it was rediscovered in the University Endowment Lands (now Pacific Spirit Park) last year. Since then, a number of lepidopterists have been searching for it and it has turned up in Stanley Park and Lynn Creek Headwaters Park. Its larval food plant is the western hemlock mistletoe.

Papilio machaon pikei (sage swallowtail): Threatened. This recently-discovered subspecies is known only from populations along south-facing riverbanks on a 500 kilometre stretch of the Peace River in British Columbia and Alberta, and from one population in the Kleskum Hills badlands of Alberta. Another population near Findlay Forks on the Peace River was probably flooded out by the Bennett Dam; the present British Columbian populations are threatened by dams planned for this river.

The Entomological Society of Canada has established a standing committee on endangered species, and one of its aims is to press for the recognition of endangered invertebrates in all provinces. In British Columbia, invertebrates cannot even be nominated as threatened or endangered species because they are not considered wildlife and are therefore excluded from the provincial Wildlife Act's endangered species provisions. I believe that it is essential that we recognize in legislation their importance to the natural diversity of our province.

Acknowledgements

Crispin Guppy, Royal British Columbia Museum, reviewed the manuscript.

References

Scudder, G.G.E. 1989. The adaptive significance of marginal populations: a general perspective. In: *Proceedings of the National Workshop on Effects of Habitat Alteration on Salmonid Stocks.* C.D. Levings, L.B. Holtby, and M.A. Henderson. (eds.). Canadian Special Publication of Fisheries and Aquatic Sciences 105. pp. 180-185.

Editor's Note: Dr. G.G.E. Scudder of the Zoology Department at U.B.C. has recently produced a report for the Wildlife Branch of B.C. Environment entitled: **Priorities for Inventory and Descriptive Research on British Columbia Terrestrial and Freshwater Invertebrates**. *This report details the rarest invertebrates in the province and recommends inventory priorities to increase our knowledge about the status and distribution of these species.*

Chapter 5

British Columbia's Butterflies and Moths

Butterflies and moths are the best documented group of insects in British Columbia. Even so, our knowledge of butterflies is barely adequate to provide a tentative assessment of their conservation status, and most moths are still so poorly understood that not even a preliminary assessment of their abundance can be made. The historical and modern distributions of butterflies and moths, as with all insects, are documented through collections of specimens deposited in private and public museums. It is ironic that public concern about the preservation of biodiversity is rapidly increasing at the same time as public support for the collections and research which document biodiversity is rapidly decreasing.

Our present knowledge of British Columbian butterflies and moths which are not forest or agricultural pests is primarily due to the efforts of amateur collectors. There are only ten significant collections of B.C. butterflies and moths in the world, eight in B.C., one in Ottawa and one in the USA. Only five are public collections, the other five are private collections which may eventually be deposited in public museums. The five private collections contain about 50% of the existing specimens of butterflies and 25% of the moths from B.C. The private collections are therefore an extremely important resource for documenting the diversity of butterflies and moths of British Columbia.

The collections of butterflies and moths contain a wealth of data, including specimens and their data labels. These labels identify the specimens, who collected them and where and when they were collected. This information can generally only be obtained by examining individual specimens and labels, since few of the collections are catalogued, either on paper or by computer, and very little of the data has been published. There is a computerized partial catalogue of the insect collection of the Royal British Columbia Museum, and Jon Shepard has privately compiled a personal card catalogue of butterfly distribution records and "dot" range maps for moths.

The existing collections document the presence in B.C. of about 174 species and an additional 90 subspecies of butterflies, and over 2000 species of moths. Additional species and subspecies are discovered in B.C. every year, many of them new to science. A preliminary assessment of the conservation status of B.C.'s butterflies indicates that there are 28 species and 21 subspecies of provincial conservation concern (Table 5-1). One subspecies is extinct, one species and one subspecies are extirpated from B.C., four species and four subspecies are endangered, eight species and three subspecies are threatened, and ten species and nine subspecies are vulnerable. An additional eight species and subspecies are known from only one or two populations in B.C., but the areas they occur in are too poorly sampled to make an assessment of their status.

Crispin Guppy
Royal British Columbia Museum
675 Belleville Street
Victoria, B.C.
V8V 1X4

Jon Shepard
R.R. 2, Sproule Creek Road
Nelson, B.C.
V1L 5P5

Status		Regions of British Columbia					All Regions	
		South Coast	Okanagan Similkameen	Kootena & Rockies	Peace River Lowland	Montane N.B.C.	Total	Grand Total
Extinct	Species	-	-	-	-	-	0	1
	Subspecies	1	-	-	-	-	1	
Extirpated	Species	-	1	-	-	-	1	2
	Subspecies	1	-	-	-	-	1	
Endangered	Species	2	1	1	-	-	4	8
	Subspecies	4	-	-	-	-	4	
Threatened	Species	-	4	-	4	-	8	11
	Subspecies	-	-	-	3	-	3	
Vulnerable	Species	2	4	2	-	2	10	19
	Subspecies	2	-	1	5	1	9	
Status Unclear	Species	-	-	3	-	2	5	8
	Subspecies	-	-	1	-	2	3	
Total	Species	4	10	6	4	4	28	49
	Subspecies	8	0	2	8	3	21	
Grand Total	Species & Subspecies	12	10	8	12	7	49	

Table 5-1: The Number of Butterflies of Conservation Concern in British Columbia.
Regions not listed lack butterflies of provincial level conservation concern, although they may be of local interest. The status of each species was assessed according to the criteria of the Committee on the Status of Endangered Wildlife in Canada.

The extinct subspecies is an undescribed subspecies of the large marble *(Euchloe ausonides)*. Only thirteen specimens exist in museums around the world, and the last one was collected in 1908. The only three populations ever known anywhere in the world formerly occurred on Vancouver Island and the adjacent Gabriola Island. The reasons for the extinction are unknown, but may have resulted from a combination of habitat alteration and introduction of parasites and avian predators.

The extirpated species is the viceroy *(Basilarchia archippus)*, which formerly occurred throughout the southern interior. The viceroy was apparently extirpated due to pesticide applications on apple orchards. A subspecies of the Chalcedon checkerspot *(Euphydryas chalcedona peridiccas)*, formerly known in Canada from only two localities on Vancouver Island, has been extirpated through unknown causes. Five of the endangered species and subspecies occur on southern Vancouver Island, one occurs near Vancouver, one in the Similkameen Valley (Keremeos) and one in the Kootenays (Cranbrook).

The ranges of all of British Columbia's endangered and threatened species and subspecies of butterflies are shown as the shaded areas on the south coast, the south Okanagan and Similkameen, and the Peace River in Figure 5-1 along with the immediate vicinities of Merritt and Cranbrook. To create the figure, we compiled all the butterfly collection sites onto a single map. Because many of the dots represent the collection of only a single specimen of one species, this map exaggerates the distribution and thoroughness of collection activities in the province. It is not currently possible to compile a similar map for moths but, since 90% of the specimens come from perhaps 40 localities, extensive inventory work is needed to determine their abundance and distributions.

Collecting of butterflies has been concentrated in southern B.C., with very large areas of B.C. completely untouched in the north, in the Coast Range, and on northern Vancouver Island. The lack of collections on the Queen Charlotte Islands reflects a very real lack of butterflies. Collection sites are clustered along river systems, which correspond to the highway network. The cost of travel by float plane or helicopter to wilderness areas is prohibitive to amateurs; hence, those areas remain unsurveyed.

There is a notable lack of publications about the butterflies and moths of B.C.. For butterflies, the best that are available are a handful of popular books on the butterflies of North America, such

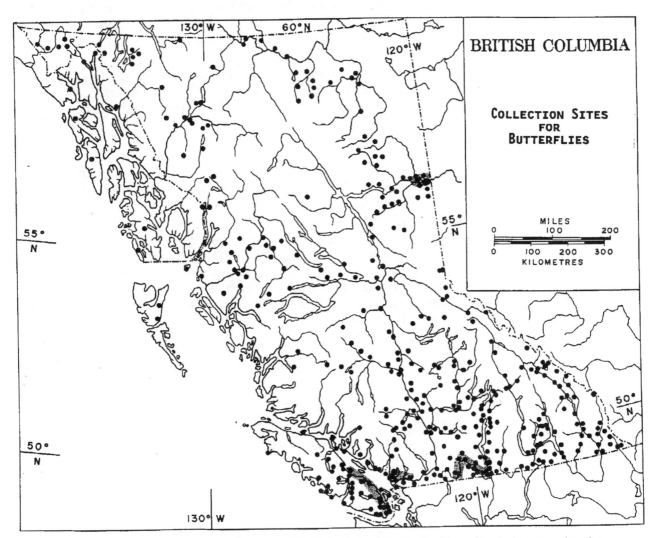

Figure 5-1. Collection Sites for Specimens of British Columbia Butterflies. Shaded areas, plus the immediate vicinity of Merritt and Cranbrook, indicate the ranges of all the endangered and threatened species in B.C. Source: original figure provided by authors.

as Scott (1986), Pyle (1981), Tilden and Smith (1986), and Howe (1975). These general books are difficult to use in B.C. because the emphasis is usually on other parts of North America. For moths the only general book, which is still widely used despite being outdated and incomplete, was published in 1903 (Holland, 1903)! Otherwise, an extensive search through the literature must be made for the various treatments of families, subfamilies or genera. Even then, identification of most moths requires a specialist and a very extensive reference collection, and data on distribution can be obtained only through a search of collections and literature.

A number of areas in the province are suffering from extensive habitat degradation and destruction due to settlement and industry. Southern Vancouver Island and the Gulf Islands, the lower Fraser Valley, the Okanagan Valley, the Flathead region of the Kootenays, the Peace River lowlands, and many other smaller areas are being impacted. There is an immediate need in these areas, especially on southern Vancouver Island, to map the distribution and abundance of the butterflies and moths. With this knowledge, we can minimize the adverse effects of our activities, especially on the species which are rare or endangered locally or provincially.

The unsampled areas of the province need to be inventoried, so that the status of many of the species presently thought to require conservation can be clarified. We removed the Arctic fritillary (*Clossiana chariclea*), from our list of butterflies requiring conservation as the result of a partial inventory of the Tatshenshini River drainage in 1992 by C. Guppy (as part of a Ministry of Forests project funded by the B.C. Provincial Government Corporate Resource Inventory Initiative). Similar inventories of other parts of the province are needed to clarify the status of many species of butterflies, as well as most species of moths.

There is much to be concerned with regarding the status of butterflies and moths in B.C., but there is also hope. Public awareness of the economic and intrinsic value of biodiversity is rapidly increasing, and that awareness is beginning to result in increased support by government and industry for research and protection of non-game species of wildlife, such as butterflies and moths. Further research and growing public awareness of the problems should help to keep most of the endangered and threatened butterflies and moths of British Columbia off the extirpated list.

References

Holland, W.J. 1903. *The Moth Book*. Doubleday, Page & Co., New York, New York.

Howe, W.H. (ed.) 1975. *The Butterflies of North America*. Doubleday & Co., Garden City, New York.

Pyle, R.M. 1981. *The Audubon Society Field Guide to North American Butterflies*. Alfred A. Knopf, New York, New York.

Scott, J.A. 1986. *The Butterflies of North America*. Stanford University Press, Stanford, California.

Tilden, J.W. and A.C. Smith. 1986. *A Field Guide to Western Butterflies*. Houghton Mifflin Co., Boston, Mass.

Chapter 6
Biodiversity of Marine Invertebrates in British Columbia

In the darkened hall, a slide projector hums in the background and the glow from the screen reflects off the faces of a rapt audience. As each new image appears, there are murmurs of recognition interspersed with gasps of amazement. After years of giving slide shows about the colourful underwater life of British Columbia, it still thrills me to hear those reactions. But it also concerns me that much of the life that inhabits the fertile waters of B.C. is, for most people, "out of sight out of mind." A recent book by Wilson (1988) contains 57 chapters on biodiversity, only two of which deal directly with the marine environment. Does this mean that the loss of diversity is not a problem in the sea? Is it a case of lack of data or are we simply dealing with the apparently more urgent terrestrial and fresh-water problems first? I suspect it is a little of each.

In the following pages I will describe the rich diversity of marine invertebrates in British Columbia and outline some of the threats to this diversity.

Philip Lambert
Royal British Columbia Museum
675 Belleville St
Victoria, B.C.
V8V 1X4

State of Knowledge

Investigations of biodiversity in the ocean are hampered by our sampling methods. For the most part, we still drop a scoop down on the end of a long wire and blindly drag it along the bottom, pulling it up to see what we accidentally captured. In shallow areas we now have the use of SCUBA for studying the flora and fauna directly, but even then our stay is brief because of the limitations of our air-breathing physiology. Generally, our knowledge of an area is directly related to its proximity to population centres. Regions that are easy to get to are studied first, with outlying regions left for the occasional expensive collecting trip.

Taxonomic experts usually limit themselves to one group of invertebrates. Thus, our knowledge can be detailed for one taxon and poor in others, depending on the interests of the scientists that have worked in a geographic area. For example, Dr. McLean Fraser, former Director of the Pacific Biological Station, was world renowned for his taxonomic work on a group of colonial animals called hydroids. Similarly, crabs and their distributions are well known through the work of Dr. J.F.L. Hart. However, other groups, like sea cucumbers, still need taxonomic revision and documentation. Compounding these gaps in our knowledge is a dwindling supply of competent taxonomists. As they switch to high tech fields of research, universities are producing fewer biologists trained in the field of taxonomy.

Causes of Marine Invertebrate Biodiversity in British Columbia

The richness of the marine invertebrate fauna in British Columbia is probably due to one or all of three causes: the age of the Pacific basin, water temperatures on the Pacific coast, and the geographic diversity of that coast.

The Pacific basin is older than the Atlantic. As a result, species have had time to evolve and diversify. Sea stars, for example, number 68 species on this coast, versus 42 on the east coast of the U.S.

Water temperatures are generally milder here than on the east coast of Canada. Our average water temperature in winter is 6-8°C and the intertidal zone seldom freezes. In the Maritimes, the water temperature hovers around 0°C and the intertidal zone often freezes. Fewer species can survive those harsh conditions.

Our coast is also physically very diverse. Hundreds of inlets, many deeper than the adjacent continental shelf, penetrate the coast range. Thousands of islands impede the tidal floods, creating swift currents through narrow channels. Strong tidal jets stir up the nutrients brought down by rivers and streams. On the western edge of the continental shelf, upwelling during the summer months brings nutrient-rich waters to the surface. Blooms of phytoplankton provide a kick start for the rest of the food chain, upon which our commercial fisheries depend.

Marine Invertebrate Diversity

I can usually impress an audience with a list of record-sized marine invertebrates from the west coast that Guinness would envy—the largest chiton in the world, the largest octopus, the largest sea slug, the heaviest sea star, the biggest barnacle. I can also boast 68 species of sea stars, over 600 amphipod crustaceans, 75 sea anemones and their relatives, 478 species of polychaete sea worms, 111 species of nudibranchs, and the list goes on. The west coast of Canada is generally recognized as being exceedingly rich in marine species when compared to other temperate regions in the world. In Table 6-1, I have summarized the numbers of species in some invertebrate groups in B.C. and elsewhere. It is difficult to find comparative figures for all invertebrate groups in each geographic area, and the data vary greatly, as some are taken from old literature, some from scientific papers, and others from popular books. I have facilitated comparison of the figures in Table 6-1 by calculating the number of species per degree of latitude (shown in brackets). For most groups, B.C. is richer in species than the Atlantic regions but, for the three groups for which I have information, B.C. is less rich than New Zealand.

About 6,555 species of marine invertebrates (animals without backbones) are documented for B.C. (Austin 1985). When we compare this number with that of the much better known vertebrate species—400 fishes, 161 marine birds and 29 marine mammals (Cannings & Harcombe, 1990)—it becomes obvious that marine invertebrates make up a major portion (92%) of our marine animals.

Group	BC 48 to 55 degrees latitude	Atlantic Canada 42 to 60 degrees latitude	Atlantic US 35 to 45 degrees latitude	British Isles 50 to 60 degrees latitude	New Zealand 42 to 50 degrees latitude
Hydroids (Hydrozoa)	149 (21)	50 (3)	250 (25)	NA	NA
Sea anemones (Anthozoa)	75 (11)	NA	56 (6)	73 (7)	NA
Moss animals (Bryozoa)	81 (12)	NA	127 (13)	81 (8)	132 (17)
Sea worms (Polychaeta)	478 (68)	NA	327 (33)	NA	NA
Sea slugs (Opisthobranchia)	111 (16)	NA	98 (10)	134 (13)	NA
Beach hoppers (Amphipoda)	600(86)	313 (5)	NA	NA	NA
Sea lice (Isopoda)	100 (14)	75 (4)	51 (5)	80 (8)	NA
Shrimps, Crabs & Relatives (Decapoda)	133 (26)	89 (5)	NA	NA	NA
Shrimps (Natants)	81 (12)	60 (3)	38 (4)	41 (4)	NA
True crabs (Brachyura)	35 (5)	14 (1)	53 (5)	67 (6)	59 (7)
Sea stars (Asteroidea)	68 (10)	NA	42 (4)	50 (5)	NA
Sea squirts (Tunicata)	55 (8)	NA	48 (5)	60 (6)	76 (10)

Table 6-1: Numbers of Species from Selected Temperate Regions of the World.
Data in brackets = numbers of species per degree of latitude. NA = data not available. Sources: Austin, 1985; Bousfield, 1981; Butler, 1980; Gordon, 1986; Gosner, 1971; Hart, 1982; Ingle, 1980; Lambert, 1981; Manuel, 1981; Millar, 1970 & 1982; Mortensen, 1977; Naylor, 1972; Rafi, 1985; Ryland & Hayward, 1977; Shih, 1977; Smaldron, 1979; Squires, 1990; Thompson & Brown, 1976.

If we were to add the undocumented species, and there are probably many, to this estimate of marine invertebrate species numbers, it would rise to well over 95%. Yet, it is the vertebrate 5% that traditionally gets most of the press. Further, these numbers refer to multicellular invertebrates and do not include a huge group of single-celled protozoa, such as bacteria and ciliates. Marine animals with a backbone (vertebrates) constitute a single subgroup of the Phylum Chordata. Those without a backbone (invertebrates) make up the remainder of the phylum Chordata and over 30 additional phyla.

The marine vertebrates may be more visible, but they are no more fascinating or varied than the invertebrates and, certainly, no more ecologically or evolutionarily important. Species diversity in the sea is actually lower than on land; however, at higher taxonomic levels (class and phylum) marine ecosystems are significantly more diverse. Almost every major taxonomic category is represented in the sea, and only about half as many on land (Earl, 1991). For example, there are no sea stars or sponges on land. Ray (1991) estimated that the total number of phyla and classes in the sea is approximately 20 times the total on land. The majority of marine phyla occur in the coastal zone, so that area could be considered the most biologically diverse realm on the planet.

The sea's greater biological diversity at higher taxonomic levels reflects its longer evolutionary history—organisms have had time to develop very different life forms and functions. Loss of one of these groups would, therefore, be of major significance. Peanut worms (sipunculids), for example, are among the many small phyla that have no terrestrial counterparts. Though there are only 320 species of peanut worms in the world, they have a body plan that differs from all other phyla. Similarly, inn-keeper worms (Echiuroidea), phoronid worms, moss animals (Bryozoa), lamp shells (Brachiopoda), and comb jellies (Ctenophora) are all relatively small groups, but each is unique in its body plan and, therefore, is classified in a separate phylum. Earl (1991) feels that the genetic information in the groups that represent major evolutionary divisions and contain only a few species is especially precious. She uses the example of the horseshoe crab (*Limulus*), an ancient relative of the spider. There are now only four species left in the world (none in B.C.) and most are dependent on shallow polluted bays on the east coast of the U.S. If we have to choose, it is these small relict groups that should be the target of conservation efforts.

Floating at the mercy of the currents is a diverse group of invertebrates and larval fish, which we call zooplankton (as opposed to phytoplankton, which are plants), from the Greek word planktos meaning "wandering." Rather than being a taxonomic category, plankton is simply a term for small organisms that float or drift in water, especially at or near the surface. Every phylum has some representative in the plankton; most are minute in size but some, like large jellyfish, are somewhat bigger. The data on this group are scattered among references to various taxa, making it difficult to determine the total number of species. However, LeBrasseur's (1964) preliminary checklist of zooplankton from his own samples contained a total of 196 invertebrates and 50 larval fish. This list probably covers only the most common species of zooplankton in B.C. Larval forms of all the invertebrates make up a large percentage of the zooplankton at certain times of the year, creating a veritable living soup. Like so many other invertebrates, zooplankton are a critical component of the marine food web.

Marine Invertebrate Habitats in B.C.

Invertebrates have invaded just about every possible ecological niche. But this evolutionary divergence has taken place over millions of years of relatively slow change. The next 100 years, in which humans are likely to cause relatively rapid changes in the environment, will be a real test of the adaptability of the invertebrates.

Four different examples will help to give a picture of invertebrate diversity on the coast of B.C.: Barkley Sound, fiords like Jervis Inlet, the Fraser River Estuary and deep-sea hot vents.

Barkley Sound, on the west coast of Vancouver Island, is a microcosm of the rich marine habitats of the British Columbian coast. Austin et al. (1982) list 1,500 species of marine invertebrates found there, excluding bacteria and other microorganisms. Some of the main groups are shown in Table 6-2.

The richness of the fauna is closely correlated with the variety of habitats in Barkley Sound. These habitats range from the placid waters of protected inlets to the ocean swells of exposed coastline.

Their diverse substrates include: mud, sand, gravel, rocks, boulders, reefs, eel grass, or kelp. Each of these substrates is preferred by a specific community of animals that is often limited to that habitat. As we shall see later, loss of habitat is the primary cause of species loss.

Group	Number of Species
Sponges (Porifera)	88
Sea anemones & relatives (Cnidaria)	150
Sea stars & relatives (Echinodermata)	85
Sea squirts (Tunicata)	50
Moss animals (Bryozoa)	105
Molluscs (Mollusca)	380
Segmented worms (Annelida)	183
Crustaceans (Crustacea)	355
Other	85
Total	**148**

Table 6-2: Some of the Main Groups of Marine Invertebrates in Barkley Sound . Source: Austin et al., 1982

Hundreds of fiords cut deep into most of the Coast Range. They have a characteristic shape—steep sided, long, narrow, and deep, with a shallow entrance called a sill. Tidal currents, which are usually light in a fiord, often increase at the sill where the large volume of water in the inlet must squeeze over a shallow lip. Sea whips and sea fans favour this location. Often, a river drains into the head of the fiord, bringing quantities of silt from the surrounding land. Animals living in this environment must be able to attach to solid rock and withstand a rain of sediment. Those that can filter large quantities of water to extract their food do quite well here. I have seen a cloud sponge (*Aphrocallistes sp.*) as big as a small car, boot sponges (*Rhabdocallyptus*) five feet long, and slender white organ-pipe sponges (*Chonelasma*) suspended from a vertical cliff in the still, dark waters of Jervis Inlet. Brilliant red sea fans (*Paragorgia*) reaching three feet in height project from a vertical precipice. They could have been growing there for hundreds of years.

Some mid-water animals, like eel-pouts (*Lycodapus*), are reminiscent of deep sea species. In fact, many mainland fiords are much deeper (Jervis Inlet is 725 metres deep) than the adjoining continental shelf, which is less than 200 metres deep. Gorgonian sea fans, common on the walls of some inlets, are also collected in deep water off the continental shelf. Glass sponges, normally associated with polar or deep seas, are common in fiords. Tube worms, sponges, brachiopods, cup corals, gorgonian sea fans, anemones, and squat lobsters (*Munida*) are some of the dominant organisms on the vertical rock walls (Levings et al, 1983).

The semi-protected waters of the Strait of Georgia support communities which are intermediate between those of the exposed coast and protected fiords. In waters which are between 50 and 300 metres deep, 71% of the bottom consists of sediment which is dominated by burrowing invertebrates (infauna), such as sea cucumbers, clams, heart urchins (*Brisaster*), polychaetes and brittle stars.

Within the Fraser River Estuary alone, there are 500 kilometres of shoreline, most of which is highly productive mud. Estimates of biomass for the 14,000 hectares of Sturgeon and Roberts Banks, the main mud flats on the outer shores of Lulu Island, are shown in Table 6-3 (Levings, unpublished data).

These data show that plants make up the greatest bulk of the living material on the mud flats. But it is the invertebrate activity that keeps the system going. The many different layers of invertebrates decompose the organic residue of the plants and convert it

Group	Biomass (kcal/m²)
Vascular plants (eel grass)	1320
Microbenthic algae (diatoms on bottom)	65
Benthos (invertebrates in mud)	40
Phytoplankton (microscopic floating plants)	15
Zooplankton & surface invertebrates	0.6

Table 6-3: Estimates of the Plant and Invertebrate Biomass in Sturgeon and Roberts Banks. Source: Levings, unpublished data.

back into living tissue. The numbers of different types of bacteria at the surface of the mud (they decrease with depth) have been estimated at up to 17,000,000,000 per cubic centimetre (Rublee, 1982).

Ciliates and nematodes, two other groups of invertebrates, prey on these bacteria. Ciliates are single celled animals which use tiny hairlike projections, or cilia, to move. Their numbers have been estimated at up to 20,000,000 per square metre. Nematodes are cylindrical unsegmented worms and "are probably the most abundant metazoans in the biosphere..." (Heip et al., 1985). Near the high tide mark in salt marshes, as many as 5,000,000 nematodes have been recorded per square metre, but this estimate is probably quite low as even a sieve of 50 micrometre mesh size does not catch all the nematodes. The number of nematode species is, therefore, poorly known. It is not uncommon to obtain 50 or more species of nematodes in a single ten square centimetre core. Ciliates and nematodes comprise the bulk of the meiofauna (the microscopic organisms that live between grains of sediment) and come from many taxa, including nemertean worms, gastrotrichs, kinorhynchs, bryozoans, gastropods, brachiopods, polychaete and oligochaete worms, ostracods, and tardigrades.

The meiofauna of these mud flats support small bottom-dwelling (benthic) crustaceans, such as amphipods, harpacticoid copepods, isopods, mysids, cumaceans, shrimp and juvenile crabs. Although we do not have reliable data on the numbers of all the benthic crustaceans in the mud flats, we do know that, province-wide, this diverse group of crustaceans includes the numerous species of amphipods, isopods, shrimp and crabs noted in Table 6-1 and, in addition, 111 species of harpacticoid copepods, 40 species of shrimp-like mysids, and 24 species of mud-dwelling cumaceans (Austin, 1985).

Deep-sea hydrothermal vents provide a fourth example of a place with a rich and unique diversity of marine invertebrates. Geysers of sea water, superheated by volcanic sources beneath the earth's crust, have been discovered at several deep sites off the coast of British Columbia. Biologists have identified 53 species of invertebrates living around vents between northern California and British Columbia, 49 of which are found only at such vents (Tunnicliffe, 1988). A newly discovered vent off B.C. has yielded 8 new species known from nowhere else (endemic) (Tunnicliffe, 1991a).

These hot vent ecosystems are highly significant for two reasons. First, they are totally independent of the sun's energy. The bacteria which form the base of the food chain at these sites use energy from hydrogen sulfide to convert carbon dioxide into organic compounds. Secondly, many of the animals show primitive features, which suggest that they have been genetically isolated from other animals since the age of the dinosaurs (Tunnicliffe, 1991b).

Compared to Barkley Sound, Jervis Inlet, and the Fraser Delta, hot vent discoveries have added not only to the list of species but also to the diversity of families and genera found in B.C.

Threats to Marine Invertebrate Species Diversity

Habitat loss and marine pollution are the two main threats to marine invertebrate species diversity.

More than 50% of the world's population inhabits a narrow band beside the ocean, endangering the most productive parts of the marine environment—the coastal wetlands and the fish habitat of the continental shelf. The populations of coastal areas have been growing at three times the rate of national populations (World Resources Institute, 1990). Population trends in B.C. are similar to those of the rest of the world, though so far marine habitat loss in B.C. has been largely confined to areas close to shore, especially estuaries.

Few people know that estuaries are twice as productive (in terms of the amount of living material produced every square metre) as the richest farmland and that the insect larvae and plants produced by estuarine salt marshes feed salmon and other fish. So there has been little objection when areas are dyked, marshes filled in, shallow bays dredged or marinas built, blocking the sunlight that drives the productivity of the invertebrates and plants in the mud. It is encouraging to see that attitudes are changing and people are beginning to protest against developments which threaten such habitats.

Habitats can be modified or rendered unusable through pollution. For example, in Victoria Harbour, some sediment is so heavily polluted with lead, mercury, cadmium and polychlorinated biphenyls (PCBs) from previous industrial sites that it cannot be moved for fear it will poison other parts of the ocean floor (Macdonald, 1991). The extension of farmland through the use of dykes to reclaim marshlands can indirectly cause over-fertilization of surrounding waters. Regular flooding of land by the Fraser River provided a natural source of nutrients, which has since been replaced by artificial fertilizers made from non-renewable resources. Excessive use of these fertilizers can cause an imbalance in nutrient ratios resulting in over-enrichment (eutrophication) of the waters receiving the run-off. Such eutrophication has already been documented for the Baltic Sea, North Sea, Adriatic Sea, parts of coastal India, and the Black Sea (Patin, 1985; Mee, 1992). It is true that most of these examples are in poorly flushed basins. But if the volume of effluent is great enough, even our relatively well-flushed water bodies could be taxed to their limit in time.

The population emptying wastes into British Columbian coastal waters is increasing at an alarming rate, but the volume of the water body remains the same. In 1973, there were 85 marine discharge permits issued for B.C. In 1986, the province issued an additional 245 permits for waste discharge into the Fraser River Basin, which eventually empties into the Strait of Georgia Basin. By 1988, the number of permits had risen to 393. Eighty-one percent of the 41 municipalities discharging sewage into the Fraser River system have only primary treatment (Kay, 1989). The sea is not an infinite sink and sooner or later the input will be more than the system can deal with or carry away. And where is "away"? The open ocean? Already chemicals like DDT and PCBs have been detected in remote parts of the ocean.

Marine pollution, including eutrophication and sedimentation from coastal run-off, may outweigh harvesting, habitat destruction, species introductions, atmospheric effects, and climate change as a threat to biological diversity in the oceans (Eiswerth, 1990). Pollution tends to reduce the diversity of an ecosystem. Species which cannot tolerate the toxins, chemicals or extreme conditions either die, leave the area or fail to reproduce successfully. With few surviving predators, a handful of tolerant species can quickly become over-abundant, making the ecosystem as a whole less resistant to normal stresses.

This reduction of diversity has been documented for a number of sites along the coast. In Alice Arm, the lowest species diversity occurred near the discharge of a molybdenum mine, but improved with distance from the source (Brinkhurst et al., 1987). This effect was observed after only 18 months of discharge. Four years after the discharge was stopped, diversity had recovered to levels similar to those of adjacent inlets, except at the stations nearest the mine, where different species were observed. In an inlet receiving wastes from a groundwood plant, worms were virtually non-existent near the plant, but increased to 91 species at the mouth of the fiord. In a nearby pristine inlet, 126 species were recorded (Fournier & Levings, 1982). The diversity of bottom fauna was lowest near the Britannia Beach copper mine discharge and gradually increased from shallow to deep water stations and from tailings to non-tailings areas. Even 12 years after the discharge ended, diversity had not returned to normal (Ellis & Hoover, 1990). The mine had discharged tailings for 75 years. Areas directly under fish farms have depauperate benthic fauna, while the perimeter is characterized by an abnormal benthic assemblage dominated by one species of worm (Cross, 1990).

Some direct effects of toxins have also been documented. PCBs are detrimental to the reproductive systems of marine mammals. Levels in many marine mammals exceed 50 parts per million, the amount above which most goods are considered toxic waste! These high values are the result of a process called bioaccumulation. PCBs in the sediment are first ingested by the benthic invertebrate fauna. As the toxin moves through each successive level in the food pyramid, it becomes concentrated tenfold, with top carnivores receiving the highest dose. Cummins (1988) estimates that, of the 1.2 million tons of PCBs in the world, 31% has already been released into the environment. Releasing the remainder would eliminate a wide range of marine mammals. Tributyltin (TBT), one active ingredient in antifouling paint, has been shown to cause the development of male genitals on female snails (imposex) (Alvarez & Ellis, 1990). The incidence of imposex is directly correlated with proximity to marinas. Schiewe et al. (1991) have shown a direct cause and effect between aromatic hydrocarbons (components of petroleum containing a benzene ring, such as benzopyrene) and liver tumours in bottomfish. A correlation between the two has been known for years. In Puget Sound, a once thriving Native oyster (*Ostrea luridea*) industry came to an end due to sulphite waste liquor from pulp mills and over-exploitation (McKernan et al., 1949). Fiddler crabs near an oil-spill site showed long term reductions in recruitment and population density, behavioural changes, and higher winter mortality (Capuzzo, 1990). In these examples, the way in which the toxins entered the species studied was not documented. It is likely that the toxins were first incorporated into the tissues of plants, or herbivorous and detritus-eating invertebrates which are in the diet of these species. Adverse effects and corrective measures are usually only noted when higher animals that serve as our food become contaminated. The sublethal effects to all the intermediaries are poorly known and constitute a major gap in our knowledge.

Since our knowledge of marine invertebrates is still at a pioneering stage in B.C., we really don't know whether we have endangered or made extinct any species of marine invertebrates. Glynn and DeWeerdt (1991), however, have documented the first extinction of a coral species, those colonial invertebrates whose carefully constructed homes make up the bulk of all tropical reefs. In the sea, coral reefs correspond to rain forests in species richness, so the loss of a coral species could be highly significant. Glynn and DeWeerdt (1991) expressed concern that this event may be the harbinger of future extinctions, should habitat destruction and coral reef bleaching continue. It remains to be seen whether the relatively few documented cases of marine species extinction, mostly vertebrate, are merely the result of poor monitoring rather than the resistance of marine life to extinction. Only recently have scientists begun ringing the alarm bells about the need for more study of marine biodiversity (Vermeij, 1989; Eiswerth, 1990; Beatley, 1991; Earl, 1991; Ray & Grassle, 1991; Steele, 1991).

Much needs to be done to identify and document distributions of species in all our coastal waters. To illustrate the state of our knowledge, in 1974 a group of us from the Provincial Museum spent two months SCUBA diving and collecting in the area around Prince Rupert. During that time we collected at 34 sites. I paid particular attention to the nudibranchs. Seventeen (55%) of the 31 species collected had not been recorded in that area before (Lambert, 1976)—and nudibranchs are a popular and presumably better known group!

Looking on the Bright Side

On a more positive note, many useful chemicals have been derived from marine organisms and, since few, if any, of these organisms have yet been made extinct, we still have an opportunity to discover a great many more. Almost a quarter of all medical prescriptions originate from plants or microorganisms (Ehrlich & Wilson, 1991). Much publicity has been given to the potential of higher plants in tropical rain forests to provide many more chemicals for use in medicine. Because plants are sessile and cannot escape predators, they have evolved a multitude of toxic chemicals that repel plant eaters. These toxins are the sources of the beneficial chemicals. Marine organisms provide this same potential, with one difference—a large proportion of sessile organisms in the sea are animals. Sponges, sea squirts, hydroids, corals, and sea anemones—to name a few—contain chemicals that repel predators.

Scientists are only just beginning to assess the potentially useful molecules that may be present in marine species. In *A Wealth of Wild Species*, Myers (1983) presents an impressive list of chemicals already isolated from these animals. An extract from octopus relieves hypertension. A Caribbean sponge produces a compound that acts against diseases caused by viruses. An enzyme, chitosanase, from shells of shrimps, crabs and lobsters, helps to prevent fungal infections. Didemnin, derived from sea squirts, appears to attack two classes of viruses. It is also reported to double the life expectancy of animals suffering from leukaemia. Cold causing viruses can be inhibited by an agent derived from seaweed. One type of flu virus is resisted by extracts from three species of sea stars. A Caribbean sponge produces a compound (cytarabine) which is effective against herpes and encephalitis, and treats leukaemia. One hundred and twenty species of sponges have been screened for bio-active chemicals—almost half contain antibiotic substances. An extract from clams, oysters, and abalone, called paolin I, arrests many harmful bacteria including streptococci. A related agent, paolin II, inhibits herpes viruses and reduces some tumours. Of the more than 2,000 species of marine life tested, about 40 % of the species with bio-active chemicals came from corals and their relatives. In the northeast Pacific, antimicrobial activity was found in 28 of 40 species of marine sponge from the San Diego region (Thompson et al., 1985). Scheuer (1990) reviewed some recent progress in assessing the biomedical potential of marine organisms and found that the potential chemical storehouse in the sea has barely been tapped. A number of screening programs are now underway. For example, a company in Washington state has begun testing marine species from the west coast (personal communication on July 17, 1991 with J. Majnarich, Bioconsultants Inc.), as has a federal lab in Halifax. Ideally, we will find a way to artificially synthesize useful compounds, once they have been identified. In this way, we can benefit from the millions of "clinical trials" that have occurred in nature through evolution.

Summing it Up

The ecosystems in which we live are the product of millions of years of evolution. We are still totally dependent on the biosphere to nourish ourselves. Every piece of food we put in our mouths comes from the land or the sea. It is at our peril that we poison our environment. The spread of DDT and PCB throughout the globe should be proof enough that what we do locally can have global consequences. After any mass extinction, it is the dominant life form that usually succumbs. The loss of biodiversity caused by our activities is now proceeding at a rate greater than during any previous so-called mass extinction. Being the dominant predator on the planet, we can control the extinction of our own species by changing our behaviour before it is too late. How do we identify incorrect behaviour? Aldo Leopold (1968) summed it up this way. "A thing is right when it tends to preserve the integrity, stability, and beauty of the biotic community. It is wrong when it tends otherwise."

Fortunately, in British Columbia much of our coastline is still relatively pristine and unaffected by population pressure. Since most threatened areas are near coastal cities and industrial sites, protection of these areas should be our first priority.

As I have indicated in this paper, knowledge of marine invertebrates is variable. Nor is it likely to improve if funding for taxonomic research remains at its current low levels. Budget cutbacks have limited the amount of research at natural history museums and shifted funding priorities at universities toward non-taxonomic research, so that few students are being trained in taxonomy. Consequently, at a time when questions about biodiversity are high on the political agenda, the people and facilities needed to provide answers about marine biodiversity are in short supply.

Acknowledgements

Colin Levings, Fisheries and Oceans Canada, reviewed the manuscript.

References

Alvarez, M.M.S. and D.V. Ellis. 1990. Widespread neogastropod imposex in the northeast Pacific: Implications for TBT contamination surveys. *Marine Pollution Bulletin* 21:244-247.

Austin, W.C., M.V. Preker, A. Bergey and S. Leader. 1982. *Marine Invertebrates Recorded From the Barkley Sound Region.* Report for Bamfield Marine Station contract 81-82.

Austin, W.C. 1985. *An Annotated Checklist of Marine Invertebrates in the Cold Temperate Northeast Pacific, Vol. 1-3.* Khoyatan Marine Laboratory, Cowichan Bay, B.C.

Beatley, T. 1991. Protecting biodiversity in coastal environments: Introduction and overview. *Coastal Management* 19:1-19.

Bennett, E. W. 1964. The marine fauna of New Zealand: Crustacea brachyura. *New Zealand Department of Science and Industry Research Bulletin* 153:1-120.

Bousfield, E.L. 1981. Evolution in North Pacific Coastal Marine Amphipod Crustaceans. In: *Evolution Today.* G.G.E. Scudder and J.L. Reveal (eds.). Hunt Institute for Botanical Documentation, Pittsburgh, Pennsylvania.

Brinkhurst, R.O., B.J. Burd and R.D. Kathman. 1987. Benthic studies in Alice Arm, B.C. during and following cessation of mine tailings disposal, 1982 to 1986. *Canadian Technical Report of Hydrography and Ocean Sciences* 89:1-45.

Butler, T.H. 1980. Shrimp of the Pacific Coast of Canada. *Canadian Bulletin of Fisheries and Aquatic Sciences* 202:280.

Cannings, R.A. and A.P. Harcombe. 1990. *The Vertebrates of British Columbia, Scientific and English Names.* Royal British Columbia Museum, Victoria, B.C.

Capuzzo, J.M. 1990. Biological effects of petroleum hydrocarbons: Predictions of long-term effects and recovery. *Northwest Science* 64:247-249.

Cross, S.F. 1990. *Benthic Impacts of Salmon Farming in British Columbia, Vol. 1— Summary Report for B.C. Ministry of the Environment.* Aquametrix Research Ltd., Sidney, B.C.

Cummins, J.E. 1988. Extinction: the PCB threat to marine mammals. *Ecologist* 18:193-195.

Earl, S.A. 1991. Sharks, squids, and horseshoe crabs—the significance of marine biodiversity. *Bioscience* 41:506-509.

Ehrlich, P.R. and E.O. Wilson. 1991. Biodiversity studies: Science and policy. *Science* 253:758-762.

Eiswerth, M.E. 1990. (ed.). *Marine Biological Diversity: Report of a meeting of the Marine Biological Diversity Working Group*. Technical Report. Woods Hole Oceanographic Institution, Woods Hole, Mass.

Ellis, D.V. and P.M. Hoover. 1990. Benthos on tailing beds from an abandoned coastal mine. *Marine Pollution Bulletin* 21:477-480.

Fournier, J.A. and C.D. Levings. 1982. Polychaetes recorded near two pulp mills on the coast of northern British Columbia: A preliminary taxonomic and ecological account. *Syllogeus* 40:1.

Glynn, P.W. and W.H. de Weerdt. 1991. Elimination of two reef-building hydrocorals following the 1982-83 El Nino warming event. *Science* 253:69-71.

Gordon, D.P. 1986. The marine fauna of New Zealand: Bryozoa:Gymnolaemata (*Ctenostomata and Cheilostomata Anasca*) from the western South Island continental shelf and slope. *New Zealand Oceanographic Institute Memoir* 95:1-121.

Gosner, K.L. 1971. *Guide to Identification of Marine and Estuarine Invertebrates: Cape Hatteras to the Bay of Fundy*. John Wiley and Sons. Toronto, Ontario.

Hart, J.F.L. 1982. *Crabs and Their Relatives of British Columbia*. Handbook No. 40. British Columbia Provincial Museum, Victoria, B.C.

Heip, C., M. Vincx and G. Vranken. 1985. The ecology of marine nematodes. *Oceanography and Marine Biology: An Annual Review* 23:399-489.

Ingle, R.W. 1980. *British Crabs*. Oxford University Press, Toronto, Ontario.

Kay, B.H. 1989. *Pollutants in British Columbia's Marine Environment*. A State of the Environment Fact Sheet No. 89-2. Environment Canada.

Lambert, P. 1976. Records and Range Extensions of Some Northeastern Pacific Opisthobranchs (Mollusca: Gastropoda). *Canadian Journal of Zoology* 54 (2):293-300.

Lambert, P. 1981. The Seastars of British Columbia. *B.C. Provincial Museum Handbook* No. 39:1-153.

LeBrasseur, R.J. 1964. *Data Record: A Preliminary Checklist of Some Marine Plankton from the North Eastern Pacific Ocean*. Fisheries Research Board of Canada, Manuscript Report 174. p. 1-14.

Leopold, A. 1968. *A Sand County Almanac and Sketches Here and There*. Paperback edition. Oxford University Press, New York, New York.

Levings, C.D., R.E. Foreman and V.J. Tunnicliffe. 1983. Review of the benthos of the Strait of Georgia and contiguous fjords. *Canadian Journal of Fisheries and Aquatic Sciences* 40B:1120-1141.

MacDonald, G. 1991. Victoria's toxic harbours. Monday Magazine:12 September.

Manuel, R.L. 1981. *British Anthozoa, Vol. 18*. Academic Press, Toronto. 241 pp. (Synopses of the British fauna [New Series])

McKernan, D.L., V. Tartar and R. Tollefson. 1949. An investigation of the decline of the native oyster industry of the State of Washington, with special reference to the effects of sulphite pulp mill waste on the Olympia Oyster (Ostrea lurida). *Washington State Department of Fisheries, Biological Report* 49A.

Mee, L.D. 1992. The Black Sea in crisis: A need for concerted international action. *Ambio* 21:278-286.

Millar, R.H. 1970. *British Ascidians—Tunicata: Ascidiacea*. Synopses of the British fauna No. 1. Academic Press, New York, New York.

Millar, R.H. 1982. The marine fauna of New Zealand: Ascidiacea. *New Zealand Oceanographic Institute Memoir* 85:1-117.

Mortensen, T. 1977. *Handbook of the Echinoderms of the British Isles*. Dr. W. Backhuys, Uitgever, Rotterdam. (Reprint of original 1927 book)

Myers, N. 1983. *A Wealth of Wild Species: Storehouse for Human Welfare*. Westview Press, Boulder, Colorado.

Naylor, E. 1972. *British Marine Isopods*. Synopses of the British Fauna. No. 3. Academic Press, London.

Patin, S.A. 1985. Biological consequences of global pollution of the marine environment. In: *Global Ecology*. C.H. Southwick (ed.). Sinauer Associates. Sunderland, Mass.

Rafi, F. 1985. *Synopsis Speciorum. Crustacea: Isopoda et Tanaidacea. Bibliographia Invertabratorum Aquaticorum Canadensium* National Museums of Canada. Ottawa, Ontario 4:1-51.

Ray, G.C. 1991. Coastal-zone biodiversity patterns. *Bioscience* 41:490-498.

Ray, G.C. and J.F. Grassle. 1991. Marine biological diversity. *Bioscience* 41:453-457.

Rublee, P.A. 1982. Bacteria and microbial distribution in estuarine sediments. In: *Estuarine Comparisons*. V.S. Kennedy (ed.). Academic Press. Toronto, Ontario.

Ryland, J.S. and P.J. Hayward. 1977. *British Anascan Bryozoans. Cheilostomata: Anasca.* Synopses of the British fauna. No. 10. Academic Press. London.

Scheuer, P.J. 1990. Some marine ecological phenomena: Chemical basis and biomedical potential. *Science* 248:173-177.

Schiewe, M.H., D.D. Weber, M.S. Myers, F.J. Jacques, W.L. Reichert, C.A. Krone, D.C. Malins, B.B. McCain, S. Chan and U. Varanasi. 1991. Induction of foci of cellular alteration and other hepatic lesions in English sole *(Parophrys vetulus)* exposed to an extract of an urban marine sediment. *Canadian Journal of Fisheries and Aquatic Sciences* 48:1750-1760.

Shih, C.T. 1977. *A Guide to Jellyfish of Canadian Atlantic Waters, Volume Natural History Series No.5.* National Museums of Canada. Ottawa, Ontario.

Smaldron, G. 1979. *British Coastal Shrimps and Prawns—Keys and Notes for the Identification of the Species.* Academic Press, New York, New York. (Synopses of the British fauna [New Series] No. 15)

Squires, H.J. 1990. Decapod crustacea of the Atlantic coast of Canada. *Canadian Bulletin of Fisheries and Aquatic Sciences* 221:1-532.

Steele, J.H. 1991. Marine functional diversity. *Bioscience* 41:470-474.

Thompson, J.E., R.P. Walker and D.J. Faulkner. 1985. Screening and bioassays for biologically-active substances from forty marine sponge species from San Diego, California, USA. *Marine Biology* 88:11-21.

Thompson, T.E. and G.H. Brown. 1976. *British Opisthobranch Molluscs—Mollusca: Gastropoda.* Synopses of the British fauna. No. 8. Academic Press. London.

Tunnicliffe, V. 1988. *Biogeography and Evolution of Hydrothermal-Vent Fauna in the Eastern Pacific Ocean.* Proceedings of the Royal Society of London, B233, 347-366.

Tunnicliffe, V. 1991a. *Biodiversity: The Marine Biota of British Columbia. Our Living legacy, a Conference on the Biodiversity of British Columbia.* In press.

Tunnicliffe, V. 1991b. *The Nature and Origin of the Modern Hydrothermal Vent Fauna.* Palaios. In review.

Vermeij, G.J. 1989. Saving the sea: what we know and what we need to know. *Conservation Biology* 3:240-241.

Wilson, E.O. 1988. *Biodiversity.* National Academy Press. Washington, D.C.

World Resources Institute. 1990. *World Resources 1990-91.* Oxford University Press. Oxford, England.

Chapter 7
Rare and Endangered Bryophytes in British Columbia[1]

The bryoflora (mosses, liverworts and hornworts) of British Columbia are the most diverse in Canada. Indeed, more than three-fourths of the total bryoflora of Canada are represented in British Columbia: 30 genera and 160 species are found only in this province. The documented moss flora of B.C. numbers approximately 190 genera, comprising approximately 700 species, while liverworts and hornworts number approximately 76 genera, comprising approximately 250 species.

Not all of these species are endemic to British Columbia. British Columbia is at the southern or northern limit of the ranges of a number of bryophytes which are endemic to arctic regions, temperate near-coastal regions of the United States, or semi-arid regions of the United States. The few bryophytes confined to British Columbia are found mainly in near coastal areas and are best represented in the Queen Charlotte Island archipelago and adjacent island groups.

The combination of bryophytes at the limits of their ranges and our use of their habitats has resulted in several bryophytes becoming rare and/or endangered. A rare bryophyte is defined here as one that shows a restricted geographic range, usually of less than five square kilometres, or one that is known from fewer than five localities, even when those localities are widely distant from each other. An endangered bryophyte is one that is found in a site vulnerable to disturbance through development, resource exploitation, or natural causes. The latter would involve mainly landslides or avalanches or the natural expansion of seed plant vegetation to occupy a specialized site of limited extent.

A few bryophytes have been inadvertently introduced to the province. Some introduced bryophytes that would have been considered rare a decade ago are now relatively common. Among these are the lawn-weed moss (*Pseudoscleropodium purum*), and a moss of poor disturbed soil of urban areas, *Pseudocrossidium hornschuchianum*. Other introduced species that were relatively frequently encountered a decade ago are now extremely rare. A good example is *Pottia truncata*, a moss which is confined mainly to clayey agricultural soil that has escaped disturbance for more than three years and which can be rapidly crowded out by invasive seed plants.

W.B. Schofield
Department of Botany,
University of British Columbia
6270 University Boulevard,
Suite 3529
Vancouver, B.C.
V6T 1Z4

[1] Originally published in the Summer 1990 (Volume 9, Number 3) issue of *BioLine*, official publication of the Association of Professional Biologists of British Columbia. Revised and updated for this publication.

Sensitive Bryophyte Habitats

Most bryophytes found in forest habitats are widespread species. Consequently, forest destruction is unlikely to result in rarity of many bryophyte species within the forest as a whole. On cliffs and along watercourses within forests, however, a number of species are likely to be drastically reduced or completely extinguished through forest destruction. Particularly vulnerable are species confined to a very specific substratum or microclimate, such as the extremely humid climatic area of the coast. It is in this area that many bryophytes of unusual phytogeographic significance are concentrated.

Another climatic (and geographic) region in which uncommon bryophytes are vulnerable to extinction or extreme reduction is the semi-arid interior, where agriculture and pastureland have drastically reduced their habitat. Among the rare bryophytes of this region are several mosses of unusual phytogeographic significance, including species extremely rare throughout their world range, whose nearest populations outside the province are sometimes on another continent. Examples are *Pterygoneurum kozlovii*, otherwise known from Czechoslovakia and the Ukraine, *Crossidium rosei*, otherwise found in Peru, and *Phascum vlassovii*, otherwise encountered in Ukraine, Armenia and central Asia.

Rare and Endangered Bryophytes

The bryophytes within British Columbia that can be considered rare or endangered are mainly those that are: at the extremes of their northern or southern ranges; restricted endemics; or rare throughout their range. Their rarity tends to reflect very restricted habitat specificity and limited reproductive potential. Those that are endangered show these same features, but are especially vulnerable because their habitat is ephemeral or in areas likely to be developed or otherwise disturbed.

Among the rarest bryophytes are those that are endemic to the province or are restricted to western North America. Fortunately, a number of these species are found in ecological reserves, and in some cases, the only populations known are confined to reserves. Examples are *Sphagnum schofieldii* and *S. wilfii*, both found on poor fens. Those found in unprotected localities include: *Cephaloziella brinkmanii*, which grows on rotten wood; *Seligeria careyana*, which is confined to humid, shaded limestone; and *Solenostoma schusterana* and *Scapania hians salishensis*, which are both found on the damp soil of open sites.

A number of bryophytes at the northern limits of their ranges are endangered by their reproductive inefficiency and their vulnerability to human disturbance. These species are from warmer climates, and are therefore confined mainly to milder coastal areas. For example, the moss, *Alsia californica*, grows either on rocks or on other plants in coastal areas. The populations growing on rocks are unable to reproduce sexually, while the epiphytic populations (those that grow perched on other plants) can reproduce by bearing

72

sporangia. As another example, the moss, *Bartramia stricta*, is confined to exposed outcrops associated with Garry oak *(Quercus garryana)* thickets and its few known populations are all vulnerable to human disturbance. The moss, *Fissidens pauperculus*, is represented by a single tiny population, fortunately in a city park. *Phaeoceros hallii*, a hornwort, is also known from a single population—in a roadside ditch—and is, thus, vulnerable to destruction, if not already destroyed.

Species of more arctic and subarctic distribution also extend southward into mountains in the north of the province. In most instances, these bryophytes are probably less rare than the available collections would suggest. Further exploration in the Cassiar and northern Rocky Mountains is likely to show them to be neither rare nor endangered. Among the hepatics (liverworts) which are presently thought to be rare are: *Peltolepis quadrata, Sauteria alpina, Arnellia fennica, Marsupella revoluta, Radula prolifera* and *Scapania simmonsii*. Among the mosses are: *Andreaeobryum macrosporum, Cinclidium arcticum, Hygrohypnum polare, Oreas martiana, Psilopilum cavifolium,* and *Tetraplodon pallidus*.

Several bryophytes show extraordinarily discontinuous world distributions and are considered rare throughout their range. In the province, three climatic regions harbour these bryophytes. They are most richly represented in the extreme oceanic climates near the coast. Those whose main range is in southeast Asia and whose North American range is confined to near the north Pacific coast include, among the hepatics: *Apotreubia nana, Chandonanthus hirtellus, Cololejeunea macounii, Dendrobazzania griffithiana, Gymnomitrion pacificum, and Radula auriculata;* and among the mosses: *Brachydontium olympicum, Bryhnia hultenii, Claopodium pellucinerve, Gollania turgens, Iwatsukiella leucotricha, Philonotis yezoana, Pleuroziopsis ruthenica, Sphagnum junghuhnianum* and *S. subobesum*. In the same coastal areas, another group of species otherwise confined to Europe, where it is rare, includes the following: *Kurzia trichoclados, Daltonia splachnoides, Hymenostylium insigne, Leptodontium recurvifolium, Rhodobryum roseum* and *Zygodon gracilis*.

Other bryophytes showing unusually interrupted world distributions are represented by isolated British Columbian populations, which are sometimes the only ones known in North America. In all those cited here, the populations are limited and their areas are not protected. *Andreaea mutabilis* is a moss widely distributed in the Southern Hemisphere, but represented by very few populations in the Northern Hemisphere, one of which is in the Queen Charlotte Islands. Although not protected by legislation, this population is sufficiently remote that it is unlikely to be destroyed. The moss, *Bartramia halleriana*, although widely distributed in tropical and subtropical latitudes, shows only three populations in North America, two of which are in British Columbia and the third of which is near the boundary of Alberta and British Columbia. Fortunately, the largest population is protected in a national park.

Three bryophytes appear to have been extinguished from the flora of British Columbia within the past decade. Each of these, however, occupies a very specialized and ephemeral habitat; thus, the species may be found in new localities in the future. The moss,

73

Discelium nudum, appears to be confined to open earth banks that are subject to invasion by seed plants or to natural slumping. All known populations of this species in the province have disappeared in nature. The same is true of the extremely tiny mosses, *Micromitrium tenerum* and *Pseudephemerum nitidum*, both of which occupy clayey soil. Their sites have been destroyed through human activity, which is particularly unfortunate because theirs were the only known populations in North America. Two other bryophytes, the hornwort, *Phaeoceros hallii*, and the moss, *Fissidens pauperculus*, may become extinct in the province through natural invasion of their sites by other plants or through erosion of the site.

Of those bryophytes confined to western North America, several appear to be rare, and if unprotected, are vulnerable to extinction in British Columbia. Those endemic to British Columbia and restricted to fewer than five localities include the hepatics: *Cephaloziella brinkmanii, Solenostoma schusterana, Scapania hians ssp. salishensis* and *Frullania hattoriana*; and the mosses: *Dicranella stickinensis, Seligeria careyana, Sphagnum schofieldii*, and *S. wilfii*. Among these, only the type specimens (the specimen on which the original description of each species is based) are known for the *Cephaloziella, Dicranella* and the two *Sphagna*. The status of the former two taxa is unknown, while the sites of the two *Sphagna* are protected in the Vladimir J. Krajina Ecological Reserve.

Other species confined to western North America are the mosses, *Trematodon boasii* and *Tortula bolanderi*. Each of these species appears to be very restricted in British Columbia, and the habitats can be considered vulnerable to destruction through natural events, including the slumping of cliffs (*Tortula*) and the erosion of banks in subalpine areas (*Trematodon*).

The Future

Fortunately, many of the rare and endangered species of bryophytes are protected to some degree within parkland. Others, however, are in considerable danger of extinction. Those found in semi-arid regions and along watercourses within old growth forest are in greatest need of protection.

Most rare bryophytes are likely to be discovered only by knowledgeable amateurs or experts; thus, their documentation relies on information accumulated by a very restricted number of researchers. A number of the conclusions reported here may ultimately prove, therefore, to be in error. It is hoped that many of the taxa will prove to be more common than the current data indicate.

But we do not need accurate data to realize that, when sensitivity toward the natural environment is reflected in wise planning, populations of all biotic diversity, including rare and endangered bryophytes, will be protected. This goal can most successfully be attained through effective education and intelligent legislation that do not threaten the survival of the organisms upon which we depend.

Dr. Robert R. Ireland, Canadian Museum of Nature, Ottawa, reviewed the manuscript.

Chapter 8
Rare and Endangered Lichens in British Columbia

The latest inventory of the lichens and allied fungi of British Columbia appeared in 1987 (Noble et al., 1987), and listed 1,013 species in 205 genera, as well as 18 subspecies, 33 varieties, and 2 forms. Impressive though these figures appear, they conceal the disturbing fact that many records are based on one-time-only collections. What is more, scores of other lichens, some of them undescribed, remain to be added to future editions of the British Columbia checklist.

Until more is known about the lichen flora of British Columbia, it will be impossible to decide, with few exceptions (Goward, 1993), which species deserve special status as rare or endangered. Experience in boreal and temperate countries elsewhere, however, allows us to at least point to certain salient features of rare and endangered lichens, and to identify those land use practices most likely to threaten their continued existence.

In Sweden, for example, seventeen lichen species are considered to have disappeared since 1850, while another two hundred are now endangered (Floravårdskommittén för Lavar, 1987). A breakdown of Sweden's rare lichens by habitat reveals that by far the greatest number (60% to 70%) are restricted to forest habitats. Of the rest, about 20% occur in agricultural and grassland settings, 10% to 15% occur on rocky outcrops below treeline, and another 2% to 3% are restricted to alpine localities. These percentages, which probably hold also for British Columbia's lichens, differ significantly from comparable figures for vascular plants. In particular, a much lower percentage of endangered vascular plants is dependent on undisturbed forest ecosystems.

Also noteworthy is the fact that roughly half of Sweden's extirpated lichens are crustose species. In Britain, too, crustose species account for about 20 of the 40 species that have disappeared during the past century (Hawksworth et al., 1974). It can be inferred, therefore, that crustose lichens, which represent roughly 60% of British Columbia's total lichen flora, are probably no less sensitive to disturbance than are the macrolichens; any attempt to maintain lichen diversity in British Columbia at its present level must be based on a careful examination of the entire lichen flora—not just the more conspicuous fruticose and foliose species.

In both Sweden and Britain, the greatest single threat to lichens is widely held to be air pollution (Floravårdskommittén för Lavar, 1987; Hawksworth et al., 1974). Because most lichens are adapted to receive their mineral requirements directly from the air, they tend to be highly efficient accumulators of atmospheric impurities, including sulphur dioxide and its byproducts—concentrating

Trevor Goward
University of British Columbia
C/o Edgewood Blue
Box 131
Clearwater, B.C.
V0E 1N0

77

them to levels beyond the tolerance of many lichen species. This sensitivity to air pollution accounts, in part, for the impoverished species diversity characteristic of most cities.

Though air pollution currently poses little threat to lichens in most regions of British Columbia, it apparently does affect the lichens of the southwest corner of the province where, in some areas, an estimated twenty kilograms of sulphates per hectare are deposited annually. Observations in Burns Bog, for example, suggest that many lichen species have already gone into decline (Goward and Schofield, 1983). Particularly disturbing is the absence in this vicinity of *Lobaria pulmonaria*—a widespread species known to be highly sensitive to air pollution.

Southwest British Columbia doubtless contains a disproportionately large percentage of the province's rare lichens, owing to the presence of several Mediterranean species which are here at or near the northern edge of their range (Noble, 1982). Particularly rich in rarities are the Gulf Islands and the east coast of Vancouver Island. For this reason, it is disturbing to see these areas being subjected to increasing demands for housing and recreational development. Any successful attempt to preserve British Columbia's lichen diversity at its current level must include the establishment of sizable nature preserves in this corner of the province. What is more, these preserves must somehow be safeguarded against increasing levels of air pollution.

Far more significant, however, than the currently localized effects of air pollution and urban sprawl, are the already province-wide ravages of logging. Logging is unquestionably the largest single threat to British Columbia's lichens. As already mentioned, 60% to 70% of British Columbia's rare lichens probably occur in forested ecosystems, where they are highly vulnerable to current forestry practices. In particular, the younger, managed forests that are replacing British Columbia's original forests at the rate of 200,000 to 250,000 hectares per year (McKinnon, 1994) are unlikely to provide the ecological stability required by many lichen species.

In Britain, lichenologists have long recognized that many lichens occur only in forests that have been undisturbed for hundreds or possibly thousands of years (Rose, 1976). Such lichens can reliably be used as indicators of environmental continuity in forests. The limited distribution, slow growth, and frequent inability of many lichen species to colonize disturbed habitats virtually ensure that clear cutting and associated practises (such as scarification and burning) on the scale practised in British Columbia will result in the extirpation of many epiphytic lichen species. Loss of these lichens means also loss of their ecosystem services, such as organic matter decay, nutrient recycling, and wildlife food. Here, then, is a new perspective on the Ministry of Forests' current practice of "liquidating" "overmature" forests to make way for new forest plantations (Ministry of Forests, 1984).

While politicians are finally awakening to the need to preserve habitat for wildlife, their commitment to the preservation of Canada's flora is tentative at best. Even the full name of our national forum for rare and endangered species is Committee on the Status of Endangered *Wildlife* in Canada (emphasis added), commonly known by the acronym COSEWIC (Cook and Muir, 1984). Its Plants

Subcommittee, formed in 1980, was something of an afterthought. By plants, furthermore, COSEWIC means only vascular plants: lichens lie entirely outside its purview.

To date, there have been few attempts (and none at all in Canada) to accord conservation status to North American lichens. The most ambitious is the Rare Lichens Project of the Smithsonian Institution, in Washington, D.C. (Pittam, 1991). In this program, each candidate lichen is placed in one of five categories, depending on: 1) the estimated number of sites in which it occurs; 2) its abundance within those sites; 3) its global range; and 4) its vulnerability to existing or potential threats.

Only six lichens occurring in British Columbia have been assigned conservation status under this program. These designations, however, are meant to indicate global status, and therefore do not always reflect the status of these species in British Columbia. In short, the problem remains that too little is known about the abundance of most of British Columbia's rare and endangered lichens.

To help correct this situation and ensure that future land-use decisions do not inadvertently bring about the extirpation of lichens currently present in British Columbia, I here propose four courses of action. Several other, more detailed recommendations are included in Goward (1993).

First, inventory lichen populations in critical portions of the province—especially the Lower Mainland, the West Coast, the Bunchgrass Zone, and the old-growth Interior Cedar-Hemlock forests of the Southern Interior.

Second, incorporate lichen protection into resource planning and management. Steps to this end include: assembling a database of probable rare and endangered lichens; encouraging naturalist groups to prepare and maintain species lists of lichens in carefully delimited areas; ensuring that the selection of future protected areas is consistent with the requirements of lichen conservation; and coordinating lichen conservation efforts in British Columbia with similar initiatives in other parts of the world. These steps might be best accomplished through the appointment of a provincial coordinator of lichens.

Third, set aside sizeable ecoreserves in critical areas in the southwest corner of the province, as well as in lowland old-growth forests elsewhere.

Fourth, identify and implement some means of maintaining air quality in forests intended to preserve lichen diversity. For example, British Columbian parks and ecoreserves might be granted some degree of jurisdiction over air quality in their preserves.

Whether British Columbia's original lichen flora remain intact, or whether some species have already been lost through logging, agriculture, pollution or urban sprawl is not certain. What is certain is that British Columbia's lichen flora still remain *essentially* intact—a claim that applies to very few regions on earth. Will British Columbia's lichen flora remain essentially intact thirty years from now? The answer to that question depends on whether British Columbians take up the challenge and responsibility of actively preserving the lichen diversity of this province. For if we do not do so soon, we will have forfeited the opportunity to do so forever.

Acknowledgements

Dr. Bruce McCune, Oregon State University, reviewed the manuscript.

References

Cook, F.R. and D. Muir. 1984. The Committee on the Status of Endangered Wildlife in Canada (COSEWIC): History and Progress. *Canadian Field Naturalist* 98:63-70.

Floravårdskommittén för Lavar. 1987. A preliminary list of threatened lichens in Sweden. *Svensk Botanisk Tidskrift* 81:237-256.

Goward, T. 1993. *Lichen Inventory Requirements for British Columbia.* Unpublished report. B.C. Ministry of Environment and Parks. Victoria, B.C.

Goward, T. and W.B. Schofield. 1983. The lichens and bryophytes of Burns Bog, Fraser Delta, southwestern British Columbia. *Syesis* 16: 53-69.

Hawksworth, D.L., B.J. Coppins and F. Rose. 1974. Changes in the British lichen flora. In: *The Changing Flora and Fauna of Britain.* D.L. Hawksworth (ed.). Academic Press. London, England. pp. 47-48.

McKinnon, A. 1994. British Columbia's old growth forests. *Bioline* 11(2):1-9.

Ministry of Forests. 1984. *Kamloops Timber Supply Area Strategic Plan.*

Noble, W.J. 1982. *The Lichens of the Coastal Douglas-fir Dry Subzone of British Columbia.* Ph.D. Thesis. University of British Columbia. Vancouver, B.C.

Noble, W.J., T. Ahti, G.F. Otto and I.M. Brodo. 1987. A second checklist and bibliography of the lichens and allied fungi of British Columbia. *Syllogeus* 61:1-95.

Pittam, S.K. 1991. The rare lichens project: a progress report. *Evansia* 8:45-47.

Rose, F. 1976. Lichenological indicators of age and environmental continuity in woodlands. In *Lichenology: Progress and Problems.* D.H. Brown, D.L. Hawksworth and R.H. Bailey (eds.). Academic Press. London, England.

Chapter 9

Macrofungi of British Columbia

Fungi are an all too easily overlooked, yet vital, component of British Columbia's ecosystems. Macrofungi are those fungi which form large fructifications visible without the aid of a microscope. This artificial but convenient grouping is here defined to include fungal families or genera where the majority of included species produce fruit bodies greater than one centimetre in diameter. Unlike microfungi, which are made conspicuous by the diseases, decay and moulding they cause, macrofungi are the ones most likely to be seen directly by the public. They are also the most likely to be either indicator or threatened beneficial species. This chapter, therefore, focuses on explaining what we know about the macrofungi.

The number of macrofungi in British Columbia may well exceed the number of species of vascular plants but, with a few spectacular exceptions, the macrofungi are largely inconspicuous or are lumped together in the public's mind as mushrooms, toadstools, conks or puffballs, if differentiated even this far. The loss of any one, again with a few exceptions, such as chanterelles (*Cantharellus spp.*) or pine mushrooms (*Tricholoma magnivelare*), would not be viewed with alarm by the general populace. In fact, a few macrofungi are either aggressive plant pathogens (disease agents) or agents of destruction of wood structures, and their eradication or control, like that of weeds, may well be a legitimate goal.

However, macrofungi are extremely important and beneficial organisms for many reasons. Excepting cedars (*Chaemacyparis, Calocedrus, Thuja*) and maples (*Acer*), all major timber trees and many ornamentals are dependent on ectomycorrhizal fungi (fungi which live in close association with roots), most of which are macrofungi. These mycorrhizal fungi help provide nutrients, water and protection from root pathogens to their host plants. Therefore, their reduction or elimination will lead to loss or deterioration of host trees and, consequently, the timber industry and wilderness habitats.

In addition, macrofungi directly benefit higher trophic levels. Mushrooms form a part of the diet of native animals such as squirrels, voles and deer. Truffle-like fungi, for example, are obligately dependent on animal ingestion and dispersal; even as some of these animals are largely dependent on the fungi as food. Others, through causing wood decay, create essential habitats for a variety of animals by either making cavities in trees or logs or preparing the wood for colonization by insects and, indirectly, by larger animals. The fungal role in decomposing plant matter is vital to the recycling of both natural and industrial forest waste and dead wood. As a by-product of this ability to degrade complex polymers (lignin and cellulose), some fungi can be used to decontaminate soil or ground water containing some types of pollutants.

Humans also benefit financially and otherwise from macrofungi. Macrofungi such as morels (*Morchella* spp.), false

Scott. A. Redhead
Centre for Land and Biological
Resources Research
Research Branch,
Agriculture Canada
Ottawa, Ontario
K1A 0C6

81

morels (*Gyromitra esculenta*), pine mushrooms, chanterelles, and king boletes (*Boletus edulis*) are now commercially harvested directly from natural habitats, supporting a multimillion dollar industry. Macrofungi generate masses of pharmaceutically active chemicals such as antibiotics, anticarcinogens, hormones, pheromones, toxins, carcinogens, enzymes, and pigments. Each species presents a unique combination of these features and, therefore, represents potential benefits.

Finally, aesthetically, some macrofungi are among the most picturesque, colourful, and delicate formations in nature. A profusion of large mushrooms, coral fungi, and bracket fungi along a woodland trail can make an area a wilderness wonderland, worth preserving for the sake of beauty alone.

Status of Knowledge about Macrofungi in British Columbia

Unfortunately, the present state of knowledge of the macrofungal flora, or macromycota, does not easily lend itself to detailed analysis of the type envisaged for biodiversity preservation legislation. Current data are insufficient for comparison of the different ecoregions with themselves and other regions in Canada or North America. This is not to say there is a paucity of information; rather, the existing information is too fragmented and incomplete. For more than 90% of the province, less than 1% of the macrofungal flora has been systematically documented. There are no published monographs, keys, or lists for identification of the bulk of British Columbian macrofungi. Most await documentation of their distribution in the province, a costly and time-consuming task even without comparing regions.

For this report, an effort has been made to provide a guide to the published information on several of the major taxa comprising the B.C. macromycota. Macrofungi are classified, based on the way they form sexual spores, into basidiomycetes, such as mushrooms, and ascomycetes, such as morels. Basidiomycetes are the largest and best known of these groups.

Polypores are the basidiomycetes which primarily cause tree decay and, as such, represent the most documented macromycota under consideration. They form multiporoid fruitbodies which do not decay easily like the fleshy poroid boletes. Two examples of common polypores are the artist's fungus (*Ganoderma applanatum*) and turkey tails (*Trametes versicolor*). Although there are additional sources for some species, the recent monograph of the polypores by Gilbertson and Ryvarden (1986-1987) offers a standard for the group. They report 162 species from British Columbia but, with few exceptions, do not give precise distributional information. Their maps merely indicate the presence of polypores within broad political boundaries. More detailed information is available in the form of decay studies which took place in the 40's and 50's and were mainly published in the Canadian Research Journal series (Botany).

Other major taxonomic groups which decay wood are found among the nonpolypore Aphyllophorales, which is a heterogeneous assemblage of species. Data on these fungi are scattered in hundreds of publications. A compilation of all recognized North American taxa associated with wood, along with pertinent data on distribution and hosts has been prepared by Ginns and Lefebvre (Centre for Land and Biological Resources Research, unpublished) for Agriculture Canada. There are 364 species. They surveyed 662 references continent-wide to come up with this information. Again detailed information on occurrence within the province itself is largely lacking.

There is no accurate listing of Canadian mushrooms (basidiomycetes in the orders Agaricales, Boletales, Cantharellales). The most comprehensive treatment of mushrooms in Canada is the one for Quebec by Pomerleau (1980). In Pomerleau's publication, 70 species of Cortinarius are documented. Cortinarius is the largest genus of agarics (of the order Agaricales), and well over 1000 species are thought to exist in North America. Pomerleau's treatment is very incomplete, however. In the 1970's, Dr. J. Ammirati of the University of Washington, the North American expert on Cortinarius, was a member of a field expedition to boreal Quebec and collected over 100 species in a three week period in one forest zone. A similar number could be expected to be found in the northeastern boreal corner of B.C. alone. Yet, for the entire province of B.C., only 21 species have been documented!

Agriculture Canada has been developing a comprehensive database on all published reports of agarics, boletes and chanterelles in Canada. In this database, 464 species (identified from 221 references) were listed for B.C. As noted for Cortinarius above, this inventory is very incomplete. A reasonable estimate of the number of species in this group would be between 1500 and 2000.

Additionally, there are less numerous groups of macrofungi such as gasteromycetes (puffballs, birds nest fungi, false truffles, and stinkhorns), hydnoid fungi (such as *Hydnum*, *Sarcodon*, and *Hydnellum*), and terrestrial coral fungi. Fleshy and stromatic ascomycetes comprise yet another group of macrofungi. References to these groups have been compiled by Ms. L. L. Norvell of the University of Washington and the Oregon Mycological Society in a master index to the macrofungi of the Pacific Northwest, including B.C. Over 180 ascomycetes are cited, along with over 100 miscellaneous basidiomycetes.

Two other sources of British Columbian macrofungi documentation are: Canada's two fungal/host indices (Conners, 1967; Ginns, 1986a) and listings of holdings in herbaria (such as Lowe, 1969). Many of these data refer to parasitic microfungi, however. Nonetheless, some "microfungi," such as the rusts and gall producers, form rather large, conspicuous fructifications. Publications, such as Funk's (1985), *Foliar Fungi of Western Trees*, and Ziller's (1974), *The Tree Rusts of Western Canada*, contain descriptions of such fungi from B.C.

These information sources represent the bulk of the published records of macrofungi for British Columbia. In total, over 1,250 individual species are more or less documented for the province. This figure covers only a fraction of the species actually present, since even some common species have not been documented in the literature.

Status of Individual Species in B.C.

In North America, documentation of the status of individual species is usually not possible on a large scale. Exceptions have been made for economically important species which are particularly distinctive, such as Indian paint fungus (*Echinodontium tinctorium*), a wood decaying member of the Aphyllophorales (Thomas, 1958). For the most part, however, we must rely on field observations by trained biologists to determine the status of individual macrofungi. It is significant that only a single fungus, namely the fuzzy sandozi (*Oxyporus nobilissimus*), appears on any endangered species list (the Oregon Natural Heritage Program List) in North America (Christy, 1991). This fungus is a very large, conspicuous polypore known to be associated with old growth noble fir (*Abies procera*) in Washington and Oregon. It has not been found in B.C. Both the U.S.D.A. Forest Products Laboratory, Wisconsin, and the Department of Botany, University of Washington, are involved in documenting this species in the Pacific Northwest (Coombs, 1991). There are no other North American fungi on endangered lists!

Though not officially endangered, *Polyporoletus sublividus*, another polypore, may be designated rare and, perhaps, endangered in B.C. It appears to have been collected only twice in Canada, both times on Vancouver Island—once in the vicinity of Lake Cowichan in 1929, and once near Courtenay, B.C. in 1963 (records in the National Mycological Herbarium [DAOM] in Ottawa). Unfortunately, one cannot point to an existing Canadian population with any certainty (30 years have elapsed since the last sighting); once located again, this species should be well plotted by the next researcher. This species was recently rediscovered in the Cascade Range in Washington State (specimens at the University of Washington and DAOM). A second British Columbian polypore, *Albatrellus caeruleoporus*, is rare in the Pacific Northwest and was only recently found in B.C. (Spahats Creek Provincial Park) by T. Goward (Ginns, unpubl.). It had not been reported to be in western North America by Gilbertson and Ryvarden (1986-87). The fungus is a distinctive species, having an overall blue coloration.

Among the non-poroid Aphyllophorales, *Stereopsis humphreyi*, a white spatula-shaped fungus, is conspicuous, unusual, and rare, so that it, too, might be considered endangered. Only two localities are known with certainty, one on the Olympic Peninsula in Washington and one in Naikoon Provincial Park near Tow Hill on the Queen Charlotte Islands, B.C. (Redhead and Reid, 1983). *Typhula mycophaga*, a coral fungus which is parasitic on puffballs, is known only from the Beaver River Valley of Glacier National Park (Berthier and Redhead, 1982). It may be rare, as nothing like it has been seen on other expeditions in western North America. Another easily characterized species, *Mycena tubarioides*, an agaric found in the same valley and also in the nearby, extensive Moberly Marsh in the Columbian River valley (Rocky Mountain Trench), is unknown elsewhere in North America. Its restricted habitat, the bases of decaying cattails *(Typha sp.)*, no doubt obscures it from ready detection (Redhead, 1984). The agaric, *Xeromphalina campanelloides*, occurs in B.C. in two locations on Vancouver Island: at Lake Cowichan in Honeymoon Bay Provincial

Park, and in Gold Stream Provincial Park. It has been collected outside B.C. on the Olympic Peninsula and in restricted sites in eastern North America in disjunct populations (Redhead, 1988). Unfortunately, *X. campanelloides* resembles *X. campanella*, an exceedingly common species, hence it is only recognized with difficulty in the field. Among the agaric species more conspicuous to the trained eye are those in the genus *Phaeocollybia* (see example in Figure 9-1). This genus reaches its northernmost limit in western North America on Vancouver Island in the Carmanah Valley. One new species, *P. carmanahensis*, was discovered in old growth in the upper Carmanah Valley (Redhead and Norvell, 1993). Six species in total are known from the valley, the only site in western Canada where this genus is found (unpublished data by Norvell and Redhead). The other five species occur in greater abundance in the Pacific Northwest. It should be noted, however, that adjacent valleys have not been searched (with the exception of a one day trip to the Walbran).

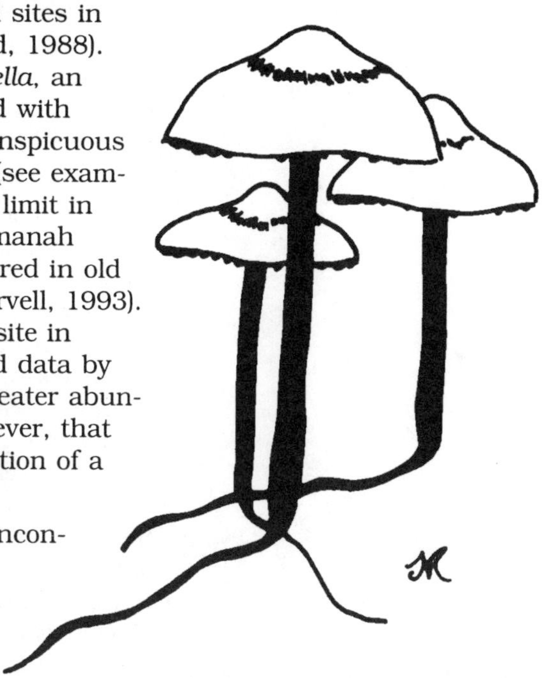

Rare, or rarely seen species in genera which form inconspicuous or nondescript fructifications abound in the province, even within populated areas. Some examples are: *Tetragoniomyces uliginosus*, known worldwide from three sites, one in Finland, one in Germany, and one on the University of British Columbia endowment lands (Oberwinkler and Bandoni, 1981); *Dacrymyces aquaticus*, known only from the University of British Columbia endowment lands (Bandoni and Hughes, 1984); *Hypochnopsis mustaliensis*, known only from one collection from the Lake Cowichan area in B.C. (Ginns, 1989); *Syzygospora subsolida*, known to science only from one collection in the Beaver River Valley, Glacier National Park (Ginns, 1986b). Many other Aphyllophorales fall into this category.

Figure 9-1. *Phaeocollybia attenuata.* This mushroom genus reaches its northernmost limit in western North America in the Carmanah Valley on Vancouver Island, where it was recently discovered. Source: drawing by Scott Redhead.

Future Studies

It is obvious that an immense amount of work would be required to properly inventory the macrofungal population of British Columbia. Even if funds were available for this research, there would be a shortage of trained researchers in Canada. Therefore, several different types of approaches would have to be considered.

One tactic is to develop a grid of comparable permanent plots for long term study in selected habitats. This grid approach would allow comparison between different vegetation zones. In particular, such plots should be established in undisturbed areas, including old growth of each forest type, as well as bunchgrass and alpine and tundra regions. Data from such plots should then be compared to that from second growth or disturbed areas. Such studies need to be funded on a long term basis. They could be combined with studies of the effects on biodiversity of harvesting commercial species. Studies similar to the one recommended here have been initiated in Washington, Oregon and California by Dr. J. Ammirati (University of Washington), Ms. L. L. Norvell (along with the Oregon Mycological Society in Portland), and Dr. David Largent (Humboldt State University). It should be noted, however, that there are definite limitations to the sampling methods used to gather data from plots, and that rare fungi are, therefore, likely to be overlooked. De Vries (1990) demonstrated that the number of species of wood-inhabiting macrofungi increased continually with increasing plot sizes in European temperate forests, a result of the fact that there is high fungal biodiversity.

A second approach is to promote general "floristic" studies within the province. The large number of species which require rudimentary documentation and the existence of many undescribed species are major hurdles to a more detailed inventory. Such floristic studies can be promoted in several ways, which must take into account the shortage of trained individuals. A general position, such as a museum biologist with a specialty in fungi could be created at the provincial level; a joint university/provincial position could be created; linkages could be made with established national research centres to ensure fungal expertise is directed towards B.C.; funds could be established to support student research at a university; funds or support in kind could be used to tap expertise among advanced amateurs, such as those belonging to the Vancouver Mycological Society; and/or British Columbian field work by North American experts on different groups of fungi could be funded.

A third option would be to concentrate on fungi that are thought to be rare and try to document their occurrence in greater detail, as is being done for fuzzy sandozi (Coombs, 1991). This approach may prove to be futile unless the fungus is large, easily characterized, and produces long-lived fructifications (increasing the likelihood of discovery). *Oxyporus nobilissimus* is one fungus that fulfils these requirements.

A compilation of the reported macrofungi from B.C. was prepared in 1993 for the province of British Columbia (Redhead, 1993). This list establishes a basis for further studies. Many more species need to be annotated. To be effective, reference material should be maintained in the province. Currently, there are major mycological collections at the University of British Columbia (Vancouver) and at the Pacific Forestry Centre (Victoria). Measures should be taken to ensure that both collections remain in B.C. and that each is actively curated and funded. Strong links to the National Mycological Herbarium at the Centre for Land and Biological Resources Research in Ottawa exist and should be encouraged.

Potential Threats to Fungi

The greatest threat to the native mycota, in particular macrofungi, is habitat destruction. Destruction of any one type of habitat, usually climax vegetation, will inevitably lead to the creation of secondary habitats, which will actually encourage an increase in some fungal species, especially opportunistic ones, decayers of slash, parasites of weedy species, many saprophytic (decomposer) molds, and pioneering mycorrhizal species. Competition with these fungi may then cause the species which survived the initial habitat destruction to decline. Interestingly enough, some habitats maintained by humans now harbour rare species, and these habitats, such as mowed meadows in Poland, must be continually mowed to "preserve" the habitat (Guminska, 1992). Such habitats were probably historically rejuvenated by natural disturbances (such as fire or grazing by wild animals), but now require our intervention.

We impact negatively on the mycota in several other ways. Air pollution, which has frequently been linked to damage to trees, has been shown to affect mycorrhizal fungi and, as a consequence, fungal diversity in Europe. Usually, this effect is the result of acidification of sensitive soils. Some genera and families of ectomycorrhizal macromycetes are more sensitive to pollution than others and disappear first (Arnolds, 1988, 1989, 1991, 1992). In severely affected areas, the mycorrhizal fungi are so "sick" that they are making their host trees unhealthy. Ground or groundwater pollution may also come in the form of enrichment, particularly from nitrogenous sources, such as farm runoff or fertilization of forests with sewage. Symbiotic relationships are often finely balanced, so if a group of plants can grow prolifically in the absence of its mycorrhizal partners, it may well shuck them off. Ultimately, the enrichment may eliminate some mycorrhizal species which are unable to regenerate when the ecosystem reverts back to its unenriched status (Arnolds, 1988). Runoff water contaminated by fungicides may have similar effects in eliminating mycorrhizal species.

Theoretically, overharvesting of commercial species could lead to decimation of the population. However, "overharvesting" is a comparative term—nobody knows how much harvesting a mycelium (the vegetative part of a fungus), a forest or a region can sustain. Commercial harvesting of fungi in B.C. was the topic of discussion in March, 1992, in Victoria (de Geus et al., 1992). At this meeting it was recognized that legitimate concerns about the biological effects of physically nondestructive harvesting could not be answered because of the lack of data. Long term studies have not been in place long enough to show significant trends. Studies of the effects of harvesting edible mushrooms have been initiated in Oregon (*Cantharellus cibarius*) (Norvell, 1992) and in California (*Boletus edulis*, *Cantharellus cibarius*, and *Tricholoma magnivelare*) (personal communication in 1992 with D. Largent, Humbolt State University, 1993). Commercial harvesting is concentrated in western North America, so that, in Canada, British Columbia has the most to lose or gain from the wild mushroom harvesting industry.

Given the facts that British Columbia probably has the highest or second highest diversity of macrofungi in Canada, that macrofungi are being commercially exploited most in B.C., and that scientific documentation of the macrofungal species is relatively poor, additional ecological and floristic studies are recommended. The limited resources available in both federal and provincial institutes would be most effectively utilized if linked to a resident mycologist charged with exploring British Columbia's mycota.

Acknowledgements

Dr. James Ginns and Ms. Louise Lefebvre (Agriculture Canada) allowed use of their unpublished manuscript on Aphyllophorales of North America. Ms. Lorelei L. Norvell (University of Washington) sent data from her master index for Pacific Northwest fungi.Data on agarics was extracted from an unpublished compilation by S.A. Redhead and Elizabeth Fox (Agriculture Canada). Dr. Shannon Berch, University of British Columbia, reviewed the manuscript.

References

Arnolds, E. 1988. The changing macromycete flora in the Netherlands. *Transactions of the British Mycological Society* 90:391-406.

Arnolds, E. 1989. A preliminary Red Data list of macrofungi in the Netherlands. *Persoonia* 14:77-125.

Arnolds, E. 1991. Mycologists and nature conservation. In: *Frontiers in Mycology.* D.L. Hawksworth (ed.). C.A.B. International, London. p. 243-264.

Arnolds, E. 1992. Mapping and monitoring of macromycetes in relation to nature conservation. *McIlvainea* 10:4-27.

Bandoni, R.J. and G.C. Hughes. 1984. A new Dacrymyces from British Columbia. *Mycologia* 76:63-66.

Berthier, J. and S.A. Redhead. 1982. Presence de Typhula mycophaga sp. nov. sur Lycoperdon en Colombie-Britannique. *Canadian Journal of Botany* 60:1428-1430.

Christy, J. 1991. The most noble polypore—endangered, or already gone to passenger pigeon-land? *Mushroom, the journal* 9(3):11.

Conners, I.L. 1967. *An Annotated Index of Plant Diseases in Canada and Fungi Recorded on Plants in Alaska, Canada and Greenland.* Canadian Department of Agriculture Publication 1251. Ottawa, Ontario.

Coombs, D.H. 1991. Was it endangered? Perhaps even extinct? Looking for a Big Fuzzy one. *Mushroom, the journal* 9(4):5-8.

Funk, A. 1985. *Foliar Fungi of Western Trees.* Canadian Forestry Service Publication BC-X-265. Victoria, B.C.

de Geus, N., S.A. Redhead and B. Callan. 1992. *Wild Mushroom Harvesting Discussion Session Minutes.* B.C. Ministry of Forests, Victoria, B.C.

Gilbertson, R.L. and L. Ryvarden. 1986-1987. North American polypores. *Fungiflora* Vols. 1-2. Oslo.

Ginns, J. 1986a. Compendium of plant disease and decay fungi in Canada 1960-1980. *Agriciculture Canada Publication 1813.* Ottawa, Ontario.

Ginns, J. 1986b. The genus Syzygospora (Heterobasidiomycetes: Syzygosporaceae). *Mycologia* 78:619-636.

Ginns, J. 1989. Descriptions and notes for some unusual North American corticioid fungi (Aphyllophorales, Corticiaceae). *Memoirs of the New York Botanical Gardens* 49:129-137.

Guminska, B. 1992. Macromycetes of the Pieniny National Park (S. Poland). *Veröffentlichungen des Geobotanischen Institutes der ETH, Stifftung Rübel in Zürich* 107:238-252.

Lowe, D.P. 1969. *Check List and Host Index of Bacteria, Fungi, and Mistletoes of British Columbia.* Forest Research Laboratory, Victoria, British Columbia. Information report BC-X-32. Department of Fisheries and Forestry, Victoria, B.C. 392 pp.

Norvell, L.L. 1992. Studying the effects of harvesting on chanterelle productivity in Oregon's Mt. Hood National Forest. In: *Wild Mushroom Harvesting Discussion Session Minutes.* de Geus et al. 1992. B.C. Ministry of Forests, Victoria. p. 9-15.

Oberwinkler, F. and R.J. Bandoni. 1981. Tetragoniomyces gen. nov. and Tetragoniomycetaceae fam. nov. (Tremellales). *Canadian Journal of Botany* 59:1034-1040.

Pomerleau, R. 1980. *Flore des Champignons au Quebec et Regions Limitrophes.* Les editions la presse, Montreal, Ontario.

Redhead, S.A. 1984. Additional Agaricales on wetland monocotyledoneae in Canada. *Canadian Journal of Botany* 62:1844-1851.

Redhead, S.A. 1988. Notes on the genus Xeromphalina (Agaricales, Xerulaceae) In: Canada: Biogeography, Nomenclature, Taxonomy. *Canadian Journal of Botany* 66:479-507.

Redhead, S.A. 1989. A biogeographical overview of the Canadian mushroom flora. *Canadian Journal of Botany* 67:3003-3062.

Redhead, S.A. 1993. Part III: Macrofungi Inventory Requirements. In: *Non-vascular Plants: Inventory Requirements for British Columbia.* Ryan, M.T. Goward and S.A. Redhead. Unpublished report prepared by Arenaria Research and intrepretation for British Columbia Ministry of Environment, Lands and Parks and British Columbia Ministry of Forests.

Redhead, S.A. and L.L. Norvell. 1993. Phaeocollybia in western Canada. *Mycotaxon* 46:343-358.

Redhead, S.A. and D.A. Reid. 1983. Craterellus humphreyi, an unusual Stereopsis from western North America. *Canadian Journal of Botany* 61:3088-3090.

Thomas, G.P. 1958. Studies in forest pathology XVIII. The occurrence of the Indian Paint Fungus, Echinodontium tinctorium E. & E., in British Columbia. *Canadian Department of Agriculture Publication* 1041. Ottawa, Ontario.

de Vries, B.W.L. 1990. On the quantitative analysis of wood-decomposing macrofungi in forests. In: *Forest Components.* 1990. R.A.A. Oldeman et al. (eds.). Wageningen Agricultural University Papers 90-6. I. p. 93-101.

Ziller, W.G. 1974. The tree rusts of western Canada. *Canadian Forestry Service Publication* 1329. Victoria, B.C.

Chapter 10
Rare and Endangered Vascular Plants in Britsh Columbia

Rare Plant Distributions and their Implications for Conservation[1]

Rare, threatened, and endangered vascular plants form an important component of biodiversity in British Columbia. Rarity is now thought to be usually a consequence of several interacting causes, which may be, for example, genetic, historic, geographic, and ecological in origin, and which have differing effects on the different rare taxa (Drury, 1980; Stebbins, 1980; Kruckeberg and Rabinowitz, 1985). It may be expected that the multitude of pathways to rarity is likely connected to a similar multitude of forms of rarity. In this paper, I simplify the topic by concentrating not so much on the causes of rarity as on two different types of rarity and how they affect conservation options.

Hans Roemer
Ministry of Environment, Lands and Parks
800 Johnson Street
Victoria, B.C.
V8V 1X4

Geographical Distribution

The first type of rarity is based upon geographical distribution. The two extremes of distributional rarity are: 1) species that occur in large numbers within a narrow geographic range and 2) species that occur in small numbers over a wide geographic range. Among the British Columbian flora, examples of the first extreme could be the Pacific waterleaf (*Hydrophyllum tenuipes*) and the golden Indian paintbrush (*Castilleja levisecta*). Both are geographically very restricted in the province but, where they do occur, they are concentrated in dense populations, the former in deciduous floodplain forests and the latter in maritime meadows on small offshore islands. Two examples of the second extreme are the white malaxis (*Malaxis monophyllos*) and the lance-leaved grape fern (*Botrychium lanceolatum*) whose occurrences are widely scattered and whose numbers in any one area are very small.

Between these two extremes, there are, of course, a number of intermediate distributions. On the basis of distribution and number of individuals alone, Drury, 1980 (quoting Mayr, 1963), portrays three cases of rarity. The first two are analogous to the two extremes mentioned above and the third is defined as distribution in widely scattered localities, each with a few individuals. For British Columbia, Straley et al. (1985) work with four classes of rarity (R1 to R4) on the basis of distribution, number of individuals, and degree of rarity, as described in the next section of this chapter. In this scheme, R3 is equivalent to Drury's case 2, R2 and R4 are variations on the theme of case 1, and R1 emphasizes mainly degree of rarity.

---

[1] Parts of this article were taken from an earlier publication by the same author in *BioLine*, Vol.9, No.3, 1990.

Field botanists know that there are geographic areas where several to many rare plants are concentrated. Such centres of rarity can be shown by superimposing a number of rare species distribution maps. Figure 10-1 was produced by combining in a single map the localities of all rare taxa reported from three or fewer localities in British Columbia according to Straley et al. (1985). It is true that the pattern emerging from this composite distribution map must be interpreted with caution: to a degree, and especially in the north, the concentrations reflect botanically explored versus unexplored areas. For instance, the "concentration" along the Alaska highway corridor is an artifact of accessibility. Further, while the agglomerations of rare occurrences along the boundaries of the province, especially in the south, are real, they are partly a result of enumerating rarity within too narrow (that is, provincial) boundaries. Nevertheless, geographic "hotspots," such as the Peace River corridor, the southern Rocky Mountain Trench, the South Okanagan, Southern Vancouver Island, and the two areas on the Queen Charlotte Islands, represent real centres of rarity.

Habitat Specificity

The second type of rarity is based upon habitat specificity. Only a few rare species are habitat-vague—that is, distributed over a large number of different habitats. The great majority occur in very specific, well-defined habitats. More often than not, these habitats are also rare (Kruckeberg and Rabinowitz, 1985; Rabinowitz et al., 1986). For instance, the southern maidenhair fern (*Adiantum capillus-veneris*, endangered in Canada) is so habitat-specific in British Columbia that it occurs only in the calcareous tufa deposits of Fairmont Hotsprings. Other rare species occur in unusual bogs, under overhanging cliffs, on very dry, shallow soils, and so forth.

In fact, just as several to many rare plants co-occur in certain geographic hotspots, above-average numbers of rare taxa also co-occur in certain unusual habitats. Calcareous fens and seeps, for example, are unusually rich in native orchids and other rare plants. Interior saline sites contain above-average numbers of rare plants. Ultramafic rocks frequently have several associated rare species (Kruckeberg, 1969). Special limestone floras are well known from Europe and are increasingly recognized in the Pacific Northwest (Bamberg and Major, 1968; Roemer and Ogilvie, 1983). In some cases, geographic concentrations of rarity may actually be due to an unusual diversity of special habitats.

In attempting to define the different forms of rarity, Rabinowitz (1981) looked at both the types of rarity discussed here, as well as at numbers of individuals. She created eight theoretical combinations out of three dichotomies of rarity: geographic range small or large; local population size small or large; and habitat specificity narrow or wide. Since one of the possible combinations (large geographic range, large local population, and wide habitat specificity) is common rather than rare, she arrived at "seven forms of rarity."

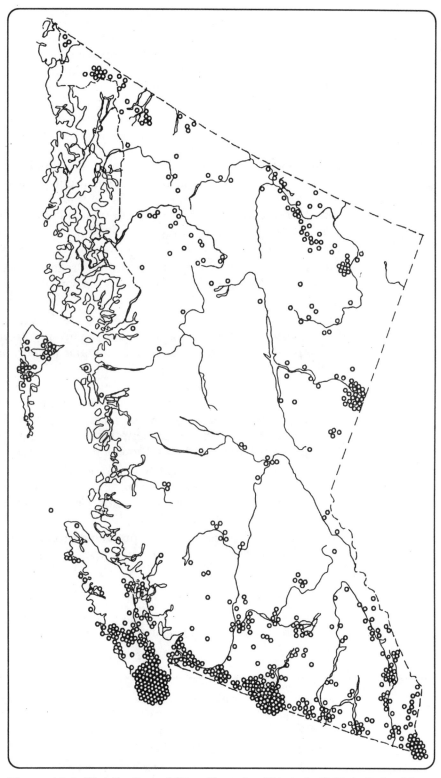

Figure 10-1. Distribution of Rare Vascular Plants in British Columbia.
Circles represent occurrences of those taxa which are known from three or
fewer locations in the province. Source: based on individual distribution
maps for taxa shown in Straley et al., 1985.

Implications for Conservation

The type of rarity—whether it is based upon geographical distribution or habitat specificity—has important implications for rare plant protection in a system of protected areas (Rabinowitz et al., 1986; McCoy and Mushinsky, 1992). Rare plants which are geographically concentrated and/or habitat specific benefit from area-based protection, while those which are not, do not. For example, an above-average density of rare plant reserves in the geographic hotspots shown in Figure 10-1 would efficiently and significantly contribute to the protection of certain species. On the other hand, in any landscape, the smallest number of plant rarities generally occurs in the most average (or zonal/mesic) habitats and plant communities which, in British Columbia, means that few rare, threatened or endangered plants are typical forest species. Rare plant protection efforts, therefore, should generally have little impact on British Columbia's forest industry.

In conclusion, the recommendations regarding rare plant protection which emerge from this discussion of types of rarity are listed below.

1) Protect in reserves those species which have narrow distributions in few localities and/or are tied to very specific habitats of which all or a major proportion may be included in a reserve.

2) Provide alternative means of protection for rare species with diffuse distribution and for species that are habitat-vague. Such species would be particularly dependent on suitable management prescriptions applying outside of protected areas and on a Protected Species Act.

3) Support research on rarity type, including life histories and habitat dependencies, for all provincially listed rare plants (Table 10-1), so that the best mode of protection can be determined for these plants.

4) Allow for greater density of protected areas in centres of rarity (rarity/biodiversity hotspots).

Acknowledgements

The original version of this manuscript was reviewed by Gerald Straley, the Botanical Garden, and George Douglas, Conservation Data Centre.

References

Bamberg, S.A. and Major. 1968. Ecology of the vegetation and soils associated with calcareous parent materials in three alpine regions of Montana. *Ecological Monographs* 38: 127-167.

Drury, W. H. 1980. Rare species of plants. *Rhodora* Vol.82, No.829: 3-48.

Kruckeberg, A. R. 1969. Plant life on serpentine and other ferromagnesian rocks in northwest North America. *Syesis* 2: 15- 114.

Kruckeberg, A. R. and D. Rabinowitz. 1985. Biological aspects of endemism in higher plants. *Annual Review of Ecology and Systematics* 16: 447-479.

McCoy, E.D. and H.R. Mushinsky. 1992. Rarity of organisms in the sand pine scrub habitat of Florida. *Conservation Biology* 6(4):537-548

Rabinowitz, D. 1981. Seven forms of rarity. In: *The Biological Aspects of Rare Plant Conservation*. Ed. H. Synge. John Wiley & Sons. pp. 205-217.

Rabinowitz, D., S. Cairns and T. Dillon. 1986. Seven forms of rarity and their frequency in the flora of the British Isles. In: *Conservation Biology: the Science of Scarcity and Diversity*. Ed. M.E. Soule. Sinauer Associates. Sunderland, Massachusetts. pp. 182-204.

Roemer, H. L. and R. T. Ogilvie. 1983. Additions to the flora of the Queen Charlotte Islands on limestone. *Canadian Journal of Botany* 61: 2577-2580.

Stebbins, G. L. 1980. Rarity of plant species: a synthetic viewpoint. *Rhodora* Vol.82, No.829:77-86.

Straley, G. B., R. L. Taylor and G. W. Douglas. 1985. *The Rare Vascular Plants of British Columbia*. Syllogeus No. 59. National Museums of Canada, Ottawa.

Rare and Endangered Vascular Plants—an Update[2]

In 1985, the first and only list of rare vascular plants of the province was published as part of the Syllogeus series by the National Museums of Canada (Straley et al., 1985). This lengthy list of 816 taxa resulted from more than ten years of concerted efforts by a large committee of interested people, initially organized by Dr. Roy L. Taylor, at that time Director of the University of British Columbia Botanical Garden. The list reflected what was known at that time, and probably more importantly, what was unknown or poorly understood about the status of rare plants in British Columbia. It also reflected the need for further field work on the floristics of the province, at a time when money and interest in such field work was waning. But the list was a starting point for, hopefully, many future refinements and updates, of which this article is the first.

The 1985 list differed from other provincial lists of rare plants in that there were four categories of rarity, R1 to R4, which unfortunately cannot be directly related to endangered, threatened, vulnerable and rare. Plants with a designation of R1 are known from only a single or very few populations, with few individuals in each population. R2's have few to several populations, but usually with larger numbers of individuals in the populations. R3's have a wide geographical range of isolated populations with small numbers of plants. R4's have a very narrow range in the province (often at the extreme limits of their distribution) but may be relatively common locally and much more common outside the province. The presence of a number of the plants on this large list probably reflects the small amount of field work that has been done in the province. Many of these plants, especially those from more northerly parts of the province, may prove to be more widespread than current records indicate.

Since the publication of the list, any new information, including additions, deletions, misidentifications, newly named species, nomenclatural changes and range extensions of rare plants, has been recorded. Herbaria and literature have been rechecked during the preparation of the new four-volume Flora of British Columbia

Gerald B. Straley
The Botanical Garden
6804 S.W. Marine Drive
University of British Columbia
Vancouver, B.C.
V6T 1Z4

George W. Douglas
British Columbia Conservation
Data Centre
British Columbia Ministry of
Environment, Lands and Parks
780 Blanshard St
Victoria, B.C.
V8V 1X5

[2] Originally published in the Summer 1990 (Volume 9, Number 3) issue of *BioLine*, official publication of the Association of Professional Biologists of British Columbia. Revised and updated for this publication.

(Douglas et al., 1989; 1990; 1991; and 1992). The editors and a number of other experts in the field have contributed a great deal of new information to the 1985 rare list.

The following are but a few examples of the new information on rare species in the province that has accumulated since the publication of the rare list.

Plants Now Thought to be Extirpated

Pink sand-verbena (*Abronia umbellata var. acutulata*), coyote tobacco (*Nicotiana attenuata*) and Lobb's water-buttercup (*Ranunculus lobbii*) are definitely all known from historic records in the province. All are obvious enough that they probably have not been just overlooked in the field, but there are no known extant populations in the province, although they do occur beyond our boundaries and they should still be looked for in British Columbia. Pink sand-verbena is also thought to be extirpated from adjacent coastal Washington.

Formerly Rare Plants Re-identified by Recent Researchers

A few specimens of the buckwheat family (*Polygonum*) have been in the past assumed to be erect knotweed (*P. erectum*), but a part of the genus has been studied recently by Dr. Deborah Katz, who finds that none of the specimens from Western Canada are this species. All specimens assigned to leafy bluebells (*Mertensia oblongifolia*) in British Columbia seem to be forms of the widespread long-flowered bluebells (*Mertensia longifolia*). All poverty oatgrass (*Danthonia sericea*) specimens have been determined to be other species. The only specimen of *Claytonia saxosa* on which the British Columbia record was based, has been re-determined by Dr. Kenton Chambers as a variant of the common miner's lettuce (*Claytonia perfoliata*). These formerly rare plants have been deleted from our list of rare flora, although they may yet be found.

Plants More Common than Previously Believed

Plants which are in this category and should, therefore, be removed from the rare list include: floating marsh marigold (*Caltha nutans*) pigmyweed (*Crassula aquatica*), dwarf montia (*Montia dichotoma*), water chickweed (*Montia fontana*), threetip (*Artemisia tripartita*) and probably all of the ladyslipper (*Cypripedium spp.*) species. Ladyslippers are not really very rare in the province, but their flowers are often picked or their plants transplanted to gardens (usually unsuccessfully!) and there are sentimental reasons for leaving them on rare lists. Pigmyweed and the two *Montia* species are so small that they are more overlooked than rare. Exploration in remote areas, such as northwestern B.C., now indicates that species such as nutzotin milk-vetch (*Astragalus nutzotinensis*) and Setchell's willow (*Salix setchelliana*) are actually locally common rather than rare.

New Species Named since the List was Prepared

At least two species, from the Pacific Northwest, including British Columbia, have recently been named: salish daisy (*Erigeron salishii*; Douglas & Packer 1988) and cascade parsley fern (*Cryptogramma cascadensis*; Alverson 1989).

Recorded in Literature but no Herbarium Specimens Extant

Some species, including Pacific hound's tongue (*Cynoglossum grande*), are attributed to our flora based on photographs and notes in Clark (1973), but no herbarium specimens are known. In the case of Pacific hound's tongue, a showy plant, it seems unlikely that it has been overlooked in the field. Clark's photograph may have been taken in Washington.

New Species that Qualify for the Red List

New species that have been recently recorded in the province and qualify for the rare list include: elk thistle (*Cirsium scariosum*), sierra cryptantha (*Cryptantha nubigena*), false mermaid (*Floerkia proserpinacoides*), Okanogan stickseed or hackelia (*Hackelia ciliata*), dwarf hesperochiron (*Hesperochiron pumilus*), monardella (*Monardella odoratissima*), branched phacelia (*Phacelia ramosissima*), common twinpod (*Physaria didymocarpa*), Lemmon's willow (*Salix lemmonii*), Tweedy's willow (*Salix tweedyi*) and Wolf's trisetum (*Trisetum wolfii*).

Nomenclatural Changes

Recent workers have proposed scientific name changes for the following plants that were included in the rare list: wild bergamot (*Monarda menthifolia* = *Monarda fistulosa* var. *menthifolia*), purple blue-eyed-grass (*Sisyrinchium inflatum* = *Olsynium douglasii* var. *inflatum*), and hairy goldfields (*Lasthenia minor* var. *maritima* = *Lasthenia maritima*). The scientific name of the Okanogan fameflower or talinum, which is on the globally rare list, has also been changed: *Talinum okanoganense* = *Talinum sediforme*.

British Columbian Rare Plant Status Reports

Three more status reports for British Columbia's rare plants have recently been prepared for the Committee on the Status of Endangered Wildlife in Canada (COSEWIC). Adolph and Oldriska Ceska prepared the report for the British Columbia endemic, Macoun's meadow-fern (*Limnanthes macounii*), which is now designated by the Committee on the Status of Endangered Wildlife in Canada (COSEWIC) as rare in Canada. The Okanogan fameflower or talinum has just been completed by Trevor Goward and Helen Knight and was given no status by COSEWIC, though our updated list classifies it as globally rare. The phantom orchid (*Cephalanthera austinae*) has been completed by Brian and Rose Klinkenberg and is now designated as vulnerable.

A committee was struck in early 1987 at the request of the Wildlife Branch of the British Columbia Ministry of Environment to look at possible legislation for protecting rare and endangered plants. A major responsibility of the committee was to formulate a short list of 50-70 of the rarest plants in the province, especially those whose habitats are endangered or threatened. Most of these plants were selected from the R1 and R2 categories of the 1985 rare list. A second "monitor list" includes plants that need taxonomic work or much more field study to determine the current status of their populations. The proposed legislation was submitted this spring but was withdrawn by Cabinet Committee, because of other more urgent and essential legislation.

Formulating a short list of plants in need of protection is much like comparing apples to oranges. For every species included on the short list there is one or more other species that could have been included. For many of those plants found in the northern and otherwise inaccessible parts of the province, it is very difficult to assess just how common or rare they may be. Rarity may only reflect where collecting has been done. A number of northern species have been collected only along the Alaska, Haines or Stewart-Watson Lake highways. We can only assume that they may be more widespread, but that their habitats are inaccessible. For a few plants, we have a very good idea of the total distribution and numbers of populations and individuals in populations. These are mostly plants found around the major urban centres, where the greatest amount of collecting and study has been done. For many others we know very little, other than that one or a few herbarium specimens exist.

The committee used a complicated formula in considering the potential species for inclusion in the short list of rare and endangered species. This formula took into account: life forms; reproductive potential of the populations; population size; pressures on habitats from urbanization, agriculture and forestry; ornamental value (showy plants are likely to be dug and moved to gardens or have their flowers and, thus, seed potential removed); and, where known, whether populations have been increasing, decreasing or staying about the same during recent years.

In March of 1991, the B.C. Conservation Data Centre was established (Chapter 2, this volume). Since then, the rare plants of B.C. have been tracked on a computer database through which the categories of rarity have been standardized for most of North America. On the Red List (Table 10-1) produced from this database are: threatened or endangered plants (124 taxa); historic or extirpated plants (29 taxa); globally rare plants (not threatened or endangered; 12 taxa); vulnerable plants (101 taxa); and other rare plants (368 taxa). On the Blue List (or Watch List) are frequent, locally frequent or locally common plants which may become vulnerable in the near future as a consequence of resource development or use. At this time, then, there are 634 plants designated as rare in B.C., of which 124 are considered threatened or endangered and will probably be included in a new legislation upon completion of status reports. Future work by the Conservation Data Centre and other B.C. botanists will include active field tracking (such as determination of exact distribution and population sizes) for all rare plants.

The bottom line is that we need to have so much more done before we can even begin to know our flora. By the time it is done it may be too late for many species or populations in our province.

Acknowledgements

The original manuscript was reviewed by George Douglas of the British Columbia Conservation Data Centre, Victoria, who became a coauthor after providing additional material, including the current rare plant list.

References

Alverson, E.R. 1989. *Cryptogramma cascadensis,* a new parsley-fern from Western North America. *American Fern Journal* 79(3):95-102.

Clark, L.J. 1973. *Wild Flowers of British Columbia.* Gray's Publishing Ltd., Sidney, B.C.

Douglas, G.W. and J.G. Packer. 1988. *Erigeron salishii,* a new *Erigeron* (Asteraceae) from British Columbia and Washington. *Canadian Journal Botany* 66:414-416.

Douglas, G.W., G.B. Straley and D. Meidinger. *The Vascular Plants of British Columbia.* B.C. Ministry of Forests, Victoria. 1989. Part 1—Gymnosperms and Dicotyledons (Aceraceae through Cucurbitaceae). 1990. Part 2—Dicotyledons (Diapensiaceae through Portulacaceae). 1991. Part 3—Dicotyledons (Primulaceae through Zygophyllaceae) and Pteridophytes. 1992 (in press). Part 4—Monocotyledons.

Straley, G.B., R.L. Taylor and G.W. Douglas. 1985. *The Rare Vascular Plants of British Columbia.* National Museums of Canada. Syllogeus 59. Ottawa, Ontario.

THREATENED/ENDANGERED PLANTS

Scientific Name	English Name
Adiantum capillus-veneris	Venus-hair or maidenhair fern
Ammannia coccinea	Scarlet ammannia
Amsinckia retrorsa	Rigid fiddleneck
Artemisia cana ssp. cana	Silver sagebrush
Artemisia longifolia	Long-leaved mugwort
Aster curtus	White-top aster
Aster radulinus	Rough-leaved aster
Astragalus filipes	Threadstalk milk-vetch
Astragalus lentiginosus var. salinus	Freckled milk-vetch
Astragalus spaldingii ssp. spaldingii	Spalding's milk-vetch
Atriplex argentea ssp. argentea	Silvery orache
Atriplex nuttallii	Saltsage or moundscale
Azolla mexicana	Mexican mosquito fern
Balsamorhiza deltoidea	Deltoid balsamroot
Bidens vulgata	Tall beggerticks
Botrychium paradoxum	Two-spiked moonwort
Bouteloua gracilis	Blue grama
Brickellia grandiflora	Large-flowered brickellia or thoroughwort
Brickellia oblongifolia ssp. oblongifolia	Narrow-leaved brickellia
Calochortus lyallii	Lyall's mariposa lily
Camissonia andina	Andean evening-primrose
Camissonia contorta	Contorted-podded evening-primrose
Castilleja exilis	Annual paintbrush
Castilleja levisecta	Golden paintbrush or Indian paintbrush
Castilleja lutescens	Yellowish paintbrush
Castilleja pallescens	Palish paintbrush or Indian paintbrush
Centaurium exaltatum	Western centaury
Centaurium muhlenbergii	Muhlenberg's centaury
Cephalanthera austinae	Phantom orchid
Chenopodium atrovirens	Dark lamb's-quarters
Chenopodium leptophyllum var. oblongifolium	Narrow-leaved goosefoot
Cimicifuga elata	Tall bugbane
Coreopsis atkinsoniana	Atkinson's coreopsis
Crassula erecta	Erect Pigmyweed
Crepis atrabarba ssp. atrabarba	Slender hawksbeard
Crepis modocensis ssp. modocensis	Low hawksbeard
Cryptantha ambigua	Obscure cryptantha
Cryptantha celosioides	Cockscomb cryptantha
Cryptantha watsonii	Watson's cryptantha
Cyperus erythrorhizos	Red-rooted cyperus
Dryopteris marginalis	Marginal wood fern
Eleocharis atropurpurea	Purple spike-rush
Elmera racemosa var. racemosa	Elmera
Epilobium densiflorum	Dense spike-primrose
Epilobium torreyi	Brook spike-primrose
Epipactis gigantea	Giant helleborine
Eragrostis pectinacea	Tufted lovegrass
Eriogonum strictum ssp. proliferum	Strick buckwheat
Euonymus occidentalis	Western wahoo
Gaura coccinea	Scarlet gaura

Table 10-1: Present Status of Rare Native Vascular Plants of British Columbia. Source: B.C. Conservation Data Centre

Scientific Name	English Name
Gayophytum ramosissimum	Hairstem groundsmoke
Githopsis specularioides	Common bluecup
Glycyrrhiza lepidota var. glutinosa	Wild licorice
Hackelia ciliata	Okanogan stickseed or hackelia
Halimolobos whitedii	Whited's halimolobos
Haplopappus bloomeri	Rabbitbrush goldenweed
Haplopappus carthamoides ssp. carthamoides	Columbian goldenweed
Helianthus nuttallii var. nuttallii	Nuttall's sunflower
Hemicarpha micrantha	Small-flowered hemicarpha
Hutchinsia procumbens	Hutchinsia
Idahoa scapigera	Scalepod
Ipomopsis minutiflora	Small-flowered ipomopsis
Juncus kelloggii	Kellogg's rush
Lathyrus bijugatus	Pinewood peavine
Lewisia tweedyii	Tweedy's lewisia
Leymus triticoides	Creeping wildrye
Limnanthes macounii	Macoun's meadow-foam
Linanthus septentrionalis	Northern linanthus or flaxflower
Lindernia anagallidea	False-pimpernel
Liparis loeselii	Loesel's liparis
Lotus formosissimus	Seaside birds-foot trefoil or lotus
Lotus nevadensis var. douglasii	Nevada birds-foot trefoil, or deervetch
Lotus pinnatus	Bog birds-foot trefoil
Lupinus arbustus var. pseudoparviflorus	Montana lupine
Lupinus argenteus var. argenteus	Silvery lupine
Lupinus densiflorus var. scopulorum	Dense-flowered lupine
Lupinus lepidus var. lepidus	Prairie lupine
Lupinus sulphureus ssp. subsaccatus	Sulphur lupine
Marsilea vestita	Hairy water-clover
Meconella oregana	White meconella
Microseris bigelovii	Coast microseris
Microseris lindleyi	Lindley's microseris
Mimulus dentatus	Tooth-leaved monkey-flower
Minuartia pusilla	Dwarf sandwort
Montia howellii	Howell's montia
Muhlenbergia andina	Foxtail muhly
Myosurus aristatus	Bristly or sedge mousetail
Navarretia intertexta	Needle-leaf navarretia
Nemophila breviflora	Great Basin nemophila
Olsynium douglasii var. inflatum	Purple blue-eyed-grass
Ophioglossum pusillum	Northern adder's-tongue
Orobanche corymbosa ssp. mutabilis	Flat-topped broomrape
Orthocarpus bracteosus	Rosy owl-clover
Orthocarpus castillejoides	Paintbrush owl-clover
Orthocarpus faucibarbatus ssp. albidus	Bearded owl-clover
Orthocarpus hispidus	Hairy owl-clover
Pectocarya penicillata	Winged combseed
Phlox speciosa	Showy phlox
Plagiobothrys figuratus	Fragrant popcornflower
Plagiobothrys tenellus	Slender popcornflower
Polygala senega	Seneca-root
Polygonum ramosissimum	Bushy or yellow-flowered knotweed
Potentilla paradoxa	Bushy cinquefoil

Table 10-1: Present Status of Rare Native Vascular Plants of British Columbia (continued).

Scientific Name	English Name
Psilocarphus elatior	Tall woolly-heads
Psilocarphus tenellus var. tenellus	Slender woolly-heads
Ranunculus alismaefolius var. alismaefolius	Water-plantain buttercup
Ranunculus californicus	California buttercup
Ranunculus flabellaris	Yellow water-buttercup or water-crowfoot
Rotala ramosior	Tooth-cup meadow-foam
Salix amygdaloides	Peach-leaf willow
Salix lemmonii	Lemmon's willow
Sanicula arctopoides	Snake-root, or bear's-foot sanicle
Sanicula bipinnatifida	Purple sanicle
Schizachyrium scoparium	False melic
Senecio foetidus var. foetidus	Sweet-marsh butterweed
Senecio foetidus var. hydrophiloides	Sweet-marsh butterweed
Sidalcea hendersonii	Henderson's checker-mallow
Sidalcea oregana var. procera	Oregon checker-mallow
Silene scouleri ssp. grandis	Scouler's campion or catchfly
Sphenopholis obtusata var. obtusata	Prairie wedgegrass
Sporobolus airoides	Hairgrass dropseed
Sporobolus asper	Rough dropseed
Thelypteris nevadensis	Nevada marsh fern
Townsendia exscapa	Easter daisy
Townsendia hookeri	Hooker's townsendia
Townsendia parryi	Parry's townsendia
Triteleia howellii	Howell's tritelia
Veronica catenata	Pink water speedwell
Viola praemorsa ssp. praemorsa	Canary violet or upland yellow violet
Wolffia columbiana	Water-meal
Woodwardia fimbriata	Giant chain fern

HISTORIC (PRE-1950) OR EXTIRPATED (*) PLANTS

Scientific Name	English Name
Abronia umbellata ssp. acutalata	Pink sand-verbena
Agastache foeniculum	Giant-hyssop
Astragalus convallarius	Lesser rusty milk-vetch
Atriplex alaskensis	Alaskan orache
Caltha palustris ssp. asarifolia	Yellow or western marshmarigold
Carex interrupta	Green-fruited sedge
Castilleja fulva	Boreal paintbrush
Catabrosa aquatica	Brookgrass or water hairgrass
Chrysosplenium iowense	Iowa golden-saxifrage or golden carpet
Crepis modocensis ssp. rostrata	Low hawksbeard
Cryptantha fendleri	Fendler's cryptantha
Cryptantha nubigena	Sierra cryptantha
*Downingia elegans**	Common downingia
Epilobium pygmaeum	Smooth spike-primrose
Gilia sinuata	Shy gilia
Hydrocotyle ranunculoides	Floating water pennywort
Hydrocotyle verticillata	Whorled water pennywort
Leucanthemum arcticum	Arctic daisy
Lupinus rivularis	Stream or streambank lupine
Lupinus sulphureus ssp. kincaidii	Sulphur lupine
Montia diffusa	Branching montia
Nicotiana attenuata	Coyote tobacco

Table 10-1: Present Status of Rare Native Vascular Plants of British Columbia (continued).

Scientific Name	English Name
Orobanche pinorum	Pine broomrape
Oryzopsis canadensis	Canada ryegrass
Pleuricospora fimbriolata	Sierra-sap or fringed-pinesap
Poa abbreviata ssp. pattersonii	Patterson's bluegrass
Poa laxa ssp. banffiana	Banff bluegrass
Poa nervosa	Wheeler's bluegrass
Polystichum californicum	California sword-fern
Prenanthes racemosa ssp. multiflora	Purple rattlesnake-root
Ranunculus lobbii	Lobb's water-buttercup
Ranunculus rhomboideus	Prairie buttercup
Senecio hydrophilus	Alkali-marsh butterweed
Senecio integerrimus var. ochroleucus	Western groundsel
Sphaeralcea coccinea	Scarlet globe-mallow
Sphaeralcea munroana	Munro's or white-stemmed globe-mallow
Thellungiella salsuginea	Salt-water cress
Trifolium bifidum	Pinole clover

GLOBALLY RARE PLANTS (NOT THREATENED OR ENDANGERED)

Androsace alaskana	Alaskan fairy-candelabra
Aphragmus eschscholtzianus	Little nightmare
Douglasia nivalis	Snow douglasia
Draba stenopetala	Star-flowered draba or whitlow-grass
Erigeron salishii	Salish daisy
Erigeron trifidus	Three-lobed daisy
Geum schofieldii	Queen Charlotte avens
Phacelia lyallii	Lyall's phacelia
Phacelia mollis	MacBryde's phacelia
Salix raupii	Raup's willow
Salix reticulata ssp. glabellicarpa	Netted willow
Saxifraga nelsoniana ssp. carlottae	Dotted saxifrage
Senecio conterminus	High alpine butterweed
Talinum sediforme	Okanogan fameflower or talinum

VULNERABLE PLANTS

Acorus americanus	American sweet flag
Agoseris lackschewitzii	Pink agoseris
Allium crenulatum	Scalloped onion
Allium validum	Pacific or swamp onion
Androsace chamaejasme	Sweet-flowered androsace
Anemone drummondii var. drummondii	Drummond's anemone
Anemone piperi	Pipers anemone
Anemone riparia	Riverbank anemone
Apocynum sibiricum var. salignum	Clasping-leaved dogbane or Indian hemp
Arctophila fulva	Pendantgrass
Arenaria longipedunculata	Low sandwort
Arnica chamissonis ssp. incana	Leafy or meadow arnica
Arnica louiseana	Lake Louise arnica
Artemisia alaskana	Alaska sagebrush
Artemisia furcata var. heterophylla	Three-forked mugwort

Table 10-1: Present Status of Rare Native Vascular Plants of British Columbia (continued).

Scientific Name	English Name
Artemisia ludoviciana var. incompta	Western mugwort
Asplenium adulterinum	Corrupt spleenwort
Aster ascendens	Long-leaved aster
Aster paucicapitatus	Olympic mountain aster
Astragalus crassicarpus	Ground or buffalo plum
Astragalus drummondii	Drummond's milk-vetch
Astragalus microcystis	Least bladdery milk-vetch
Astragalus umbellatus	Tundra milk-vetch
Besseya wyomingensis	Wyoming kitten-tails
Botrychium ascendens	Upset moonwort
Botrychium hesperium	Western moonwort
Botrychium montanum	Western goblin
Botrychium pedunculosum	Stalked moonwort
Botrychium pinnatum	Northwestern moonwort
Braya purpurascens	Purple braya
Bupleurum americanum	American thorough-wax
Cacaliopsis nardosmia ssp. glabrata	Silvercrown
Callitriche marginata	Winged water-starwort
Calyptridium umbellatum var. caudiciferum	Mount Hood pussypaws
Carex franklinii	Franklin's sedge
Carex pedunculata	Peduncled sedge
Carex petricosa	Rock-dwelling sedge
Carex rugosperma var. tonsa	Bald sedge
Carex tenera	Slender sedge
Chaenactis alpina	Alpine false yarrow
Chaenactis douglasii var. montana	Hoary false yarrow
Chrysosplenium wrightii	Wright's golden saxifrage or golden carpet
Cicuta maculata var. maculata	Spotted cowbane
Claytonia megarhiza var. megarhiza	Alpine springbeauty
Claytonia tuberosa	Tuberous springbeauty
Cnidium cnidiifolium	Hemlock parsley
Coleanthus subtilis	Moss grass
Corydalis scouleri	Scouler's corydalis
Crepis occidentalis ssp. pumila	Western hawksbeard
Cryptantha intermedia var. grandiflora	Large-flowered cryptantha
Cryptogramma cascadensis	Cascade parsley-fern
Cyperus aristatus	Awned flatsedge or cyperus
Dicentra uniflora	Steer's head
Douglasia gormanii	Gorman's douglasia
Douglasia laevigata	Smooth douglasia
Douglasia montana	Rocky Mountain douglasia
Draba palanderiana	Palander's draba or whitlow-grass
Draba porsildii	Porsild's draba or shitlow-grass
Ellisia nyctelea	Ellisia
Erigeron caespitosus	Gray daisy or tufted fleabane
Erigeron leibergii	Leiberg's fleabane or daisy
Erigeron poliospermus var. poliospermus	Cushion fleabane or daisy
Erigeron uniflorus var. eriocephalus	Northern daisy
Eriogonum pyrolifolium var. coryphaeum	Alpine buckwheat
Festuca minutiflora	Little fescue
Floerkea proserpinacoides	False mermaid
Fraxinus latifolia	Oregon ash
Gayophytum humile	Dwarf groundsmoke

Table 10-1: Present Status of Rare Native Vascular Plants of British Columbia (continued).

Scientific Name	English Name
Geum rossii	Ross' avens
Helictotrichon hookeri	Spike-root
Hesperochiron pumilus	Dwarf hesperochiron
Hydrophyllum tenuipes	Pacific waterleaf
Isoetes bolanderi	Bolander's quillwort
Isoetes howellii	Howell's quillwort
Jaumea carnosa	Fleshy jaumea
Lomatium foeniculaceum var. foeniculaceum	Fennel-leaved desert-parsley or lomatium
Lomatium grayi	Gray's desert-parsley or lomatium
Lomatogonium rotatum	Marsh felwort
Lotus purshianus	Spanish-clover
Luzula rufescens	Rusty woodrush
Malaxis monophyllos var. brachypoda	One-leaved malaxis
Malaxis monophyllos var. diphylos	One-leaved malaxis
Marah oreganus	Manroot or Bigroot
Melica bulbosa	Oniongrass
Minuartia austromontana	Rocky Mountain sandword
Minuartia elegans	Northern sandwort
Montia bostockii	Bostock's montia
Myosurus minimus	Tiny or least mousetail
Orthocarpus imbricatus	Mountain owl-clover
Oxytropis arctica	Arctic locoweed or oxytrope
Oxytropis huddelsonii	Huddelson's locoweed or oxytrope
Oxytropis jordalii ssp. jordalii	Jordal's locoweed
Oxytropis scammaniana	Scamman's locoweed or oxytrope
Phacelia ramosissima	Branched phacelia
Pleuropogon refractus	Nodding semaphoregrass
Poa fendleriana ssp. fendleriana	Muttongrass
Poa pseudoabbreviata	Polar bluegrass
Polemonium caeruleum ssp. amygdalinum	Western polemonium
Polystichum lemmonii	Lemmon's holly fern
Polystichum scopulinum	Crag holly fern
Potamogeton oakesianus	Oakes' pondweed
Potamogeton strictifolius	Stiff-leaved pondweed
Primula stricta	Upright primrose
Primula cuneifolia ssp. saxifragifolia	Wedge-leaf primrose
Psoralea physodes	California tea or California tea scurf-pea
Rhododendron macrophyllum	Pacific or California rhododendron
Rhynchospora capillacea	Brown beak-rush
Ribes montigenum	Alpine prickly currant or mountain gooseberry
Salix tweedyi	Tweedy's willow
Samolus valerandi	Brookweed or water pimpernel
Sarracenia purpurea ssp. purpurea	Common pitcher-plant
Saussurea angustifolia var. angustifolia	Northern sawwort
Saxifraga davurica ssp. grandipetala	Lare-petalled saxifrage
Saxifraga hieracifolia	Hawkweed-leaved saxifrage
Scirpus supinus var. saximontanus	European clubrush
Scutellaria angustifolia	Narrow-leaved skullcap
Senecio congestus	Marsh fleabane
Senecio macounii	Macoun's groundsel
Senecio megacephalus	Large-headed groundsel
Senecio ogotorukensis	Ogotoruk Creek butterweed
Senecio serra	Tall butterweed

Table 10-1: Present Status of Rare Native Vascular Plants of British Columbia (continued).

Scientific Name	English Name
Senecio tundricola	Northern groundsel
Silene taimyrensis	Taimyr campion
Solidago gigantea ssp. serotina	Smooth or giant goldenrod
Sparganium glomeratum	Glomerate bur-reed
Stipa spartea	Porcupinegrass
Thalictrum dasycarpum	Purple meadowrue
Thermopsis montana	Mountain thermopsis
Thermopsis rhombifolia	Round-leaved thermopsis
Tonella tenella	Small-flowered tonella
Trifolium cyathiferum	Cup clover
Trifolium macraei var. dichotomum	Macrae's clover
Trisetum wolfii	Wolf's trisetum
Wolffia borealis	

OTHER RARE PLANTS

Scientific Name	English Name
Agastache urticifolia	Nettle-leave giant hyssop
Agrostis pallens	Dune bentgrass
Allium amplectens	Slim-leaf onion
Allium geyeri var. geyeri	Geyer's onion
Allium geyeri var. tenereum	Geyer's onion
Alopecurus alpinus	Alpine foxtail
Alopecurus carolinianus	Carolina foxtail
Anemone canadensis	Canada anemone
Angelica dawsonii	Dawson's angelica
Apocynum medium	Western dogbane
Arabis nuttallii	Nuttall's rockcress
Astragalus bourgovii	Bourgeau's milk-vetch
Astragalus sclerocarpus	The Dalles milk-vetch
Bidens beckii	Water marigold
Botrychium matricariifolium	Chamomile moonwort
Botrychium simplex	Least moonwort
Calamagrostis montanensis	Plains reedgrass
Calamagrostis sesquiflora	One-and-a-half-flowered reedgrass
Calamovilfa longifolia	Prairie sandgrass
Callitriche anceps	Two-edged water-starwort
Camissonia breviflora	Short-flowered evening-primrose
Cardamine angulata	Angled bitter-cress
Cardamine parviflora	Small-flowered bitter-cress
Cardamine umbellata	Siberian or umbellate bitter-cress
Carex amplifolia	Big-leaf sedge or ample-leaved sedge
Carex bicolor	Two-coloured sedge
Carex comosa	Bristly sedge
Carex crawei	Craw's sedge
Carex epapillosa	Blackened sedge
Carex feta	Greensheathed sedge
Carex geyeri	Geyer's sedge
Carex glareosa	Lesser saltmarsh sedge
Carex gmelinii	Gmelin's sedge
Carex heleonastes	Northern clustered sedge
Carex hystricina	Porcupine sedge
Carex krausei	Krause's sedge
Carex lapponica	Lappland sedge

Table 10-1: Present Status of Rare Native Vascular Plants of British Columbia (continued).

Scientific Name	English Name
Carex maritima	Sea sedge
Carex membranacea	Fragile sedge
Carex misandra	Few-flowered sedge
Carex pansa	Sand-dune sedge
Carex paysonis	Payson sedge
Carex rupestris ssp. drummondiana	Drummond's curly sedge
Carex rupestris ssp. rupestris	Curly sedge
Carex saximontana	Black's sedge
Carex scoparia	Pointed broom sedge
Carex scopulorum	Holm's Rocky Mountain sedge
Carex simulata	Short-beaked or analogue sedge
Carex sprengelii	Sprengel's sedge
Carex sychnocephala	Many-headed sedge
Carex vulpinoidea	Fox sedge
Carex xerantica	Dryland sedge
Castilleja gracillima	Slender paintbrush
Castilleja hyperborea	Northern paintbrush or Indian paintbrush
Castilleja rupicola	Cliff paintbrush
Caucalis microcarpa	California hedge-parsley, or false carrot
Cenchrus longispinus	Burgrass
Centunculus minimus	Chaftweed
Cerastium fischerianum	Fisher's chickweed
Chamaerhodos erecta ssp. nuttallii	Nuttall's chamaerhodos
Cheilanthes gracillima	Lace lip-fern
Cicuta virosa	European water-hemlock
Cirsium scariosum	Elk thistle
Clarkia pulchella	Pink fairies
Clarkia rhomboidea	Common clarkia
Cornus suecica	Dwarf bog bunchberry
Cuscuta pentagona	Field or five-angled dodder
Cystopteris montana	Mountain bladder-fern
Delphinium bicolor	Montana larkspur
Delphinium depauperatum	Slim or dwarf larkspur
Descurainia sophioides	Northern tansymustard
Diapensia lapponica	Diapensia
Disporum smithii	Fairy lantern or Smith fairy-bell
Draba alpina	Alpine draba or rock-cress
Draba cinerea	Gray-leaved draba or whitlow-grass
Draba corymbosa	Baffin's Bay draba or whitlow-grass
Draba densifolia	Nuttall's draba or whitlow-grass
Draba fladnizensis	Austrian draba or whitlow-grass
Draba glabella var. glabella	Smooth draba or whitlow-grass
Draba lactea	Milky draba or whitlow-grass
Draba lonchocarpa var. thompsonii	Lance-fruited draba or whitlow-grass
Draba lonchocarpa var. vestita	Lance-fruited draba or whitlow-grass
Draba longipes	Long-stalked draba or whitlow-grass
Draba macounii	Macoun's draba or whitlow-grass
Draba reptans	Carolina draba or whitlow-grass
Draba ruaxes	Coast Mountain draba or whitlow-grass
Draba ventosa	Wind River draba or whitlow-grass
Dryas drummondii var. eglandulosa	Yellow mountain avens
Dryas octopetala ssp. alaskensis	White mountain avens
Dryopteris arguta	Coastal wood fern

Table 10-1: Present Status of Rare Native Vascular Plants of British Columbia (continued).

107

Scientific Name	English Name
Dryopteris cristata	Crested wood fern
Elatine rubella	Three-flowered waterwort
Eleocharis kamtschatica	Kamchatka spike-rush
Eleocharis parvula	Small spike-rush
Eleocharis rostellata	Beaked spike-rush
Elodea nuttallii	Nuttall's spike-rush
Elymus saundersii	Saunder's wildrye
Elymus yukonensis	Yukon wildrye
Epilobium ciliatum ssp. watsonii	Purple-leaved willowherb
Epilobium davuricum	Swamp willowherb
Epilobium foliosum	Foliose willowherb
Epilobium glaberrimum ssp. fastigiatum	Smooth willowherb
Epilobium glareosum	Gravelly willowherb
Epilobium halleanum	Hall's willowherb
Epilobium hornemannii ssp. behringianum	Hornemann's willowherb
Epilobium leptocarpum	Small-flowered willowherb
Epilobium mirabile	Hairy-stemmed willowherb
Epilobium oregonense	Oregon willowherb
Epilobium saximontanum	Rocky Mountain willowherb
Eriogonum pauciflorum var. pauciflorum	Few-flowered buckwheat
Eriophorum vaginatum ssp. spissum	Sheathed cotton-grass
Erysimum arenicola var. torulosum	Sand-dwelling wallflower
Erysimum asperum	Prairie rocket, or western wallflower
Erysimum pallasii	Pallas' wallflower
Erythronium montanum	Alpine fawn-lily, avalanche lily or fawn-lily
Euphorbia serpyllifolia	Thyme-leaved spurge
Euphrasia arctica var. disjuncta	Arctic eyebright
Eutrema edwardsii	Edward's wallflower
Galium labradoricum	Northern bog bedstraw
Galium mexicanum ssp. asperulum	Rough bedstraw
Galium trifidum ssp. trifidum	Small bedstraw
Gentiana affinis	Pleated or Prairie gentian
Gentiana calycosa	Explorer's gentian or mountain bog gentian
Gentianella crinita ssp. macounii	Macoun's fringed gentian
Gentianella tenella ssp. tenella	Slender gentian
Gilia capitata var. capitata	Bluefield or globe gentian
Glyceria leptostachya	Slender-spike mannagrass
Glyceria occidentalis	Western mannagrass
Gratiola neglecta	Common American hedge-hyssop
Gymnocarpium jessoense ssp. parvulum	Nahanni oak fern
Hackelia diffusa var. diffusa	Spreading stickseed or hackelia
Halimolobos mollis	Soft halimolobos
Hammarbya paludosa	Bog adder's-mouth orchid
Hedysarum boreale ssp. boreale	Norhtern hedysarum
Hedysarum occidentale	Western hedysarum
Helenium autumnale var. grandiflorum	Mountain sneezeweed
Heterocodon rariflorum	Heterocodon
Hippuris tetraphylla	Four-leaved mare's-tail
Hydrophyllum fendleri var. albifrons	Fendler's waterleaf
Hypericum scouleri var. nortoniae	Western St. John's wort
Hypericum scouleri var. scouleri	Western St. John's wort
Impatiens aurella	Orange touch-me-not
Impatiens capensis	Spotted touch-me-not

Table 10-1: Present Status of Rare Native Vascular Plants of British Columbia (continued).

Scientific Name	English Name
Impatiens ecalcarata	Spurless or western touch-me-not
Isoetes nuttallii	Nuttall's quillwort
Iva axillaris ssp. robustior	Poverty-weed
Juncus albescens	Three-flowered rush
Juncus arcticus ssp. alaskanus	Arctic rush
Juncus bolanderi	Bolander's rush
Juncus brevicaudatus	Short-tailed rush
Juncus bulbosus	Mud rush
Juncus covillei	Coville's rush
Juncus oxymeris	Pointed rush
Juncus regelii	Regel's rush
Juncus stygius	Bog rush
Kobresia sibirica	Siberian kobresia
Koenigia islandica	Iceland koenigia
Lappula redowskii var. cupulata	Western stickseed
Lasthenia maritima	Hairy goldfields
Lathyrus littoralis	Grey beach peavine
Ledum palustre ssp. decumbens	Northern Labrador tea
Leersia oryzoides	Cutgrass or rice cutgrass
Lepidium densiflorum var. pubicarpum	Prairie pepper-grass
Lesquerella arctica var. arctica	Arctic bladderpod
Leucanthemum integrifolium	Entire-leaved daisy
Lewisia columbiana var. columbiana	Columbia lewisia
Lewisia triphylla	Three-leaved lewisia
Ligusticum verticillatum	Verticillate-umbel lovage
Lilaea scilloides	Flowering quillwort
Linanthus harknessii	Harkness' linanthus or flaxflower
Linnaea borealis ssp. borealis	Twinflower
Lomatium brandegei	Brandegee's lomatium
Lomatium sandbergii	Sandberg's desert-parsley or lomatium
Lupinus arbustus var. neolaxiflorus	Spurred or grassland lupine
Lupinus kuschei	Yukon or Kusche's lupine
Lupinus wyethii	Wyeth's lupine
Luzula arctica	Arctic woodrush
Luzula confusa	Confused woodrush
Luzula groenlandica	Greenland woodrush
Lythrum alatum	Winged loosestrife
Melica smithii	Smith's melica
Melica spectabilis	Showy oniongrass
Mertensia maritima	Sea bluebells or mertensia, sea-lungwort
Mertensia paniculata var. borealis	Tall bluebells, or panicled mertensia
Mimulus breweri	Brewer's monkey-flower
Minuartia biflora	Mountain sandwort
Mirabilis hirsuta	Hairy umbrella wort
Mitella caulescens	Leafy mitrewort
Monardella odoratissima	Monardella
Montia chamissoi	Chamisso's montia
Muhlenbergia racemosa	Satin grass
Myrica californica	California wax-myrtle or bayberry
Myriophyllum quitense	Waterwort water-milfoil
Myriophyllum ussuriense	Ussurian water-milfoil
Nemophila pedunculata	Meadow nemophila
Nothochelone nemorosa	Woodland penstemon

Table 10-1: Present Status of Rare Native Vascular Plants of British Columbia (continued).

Scientific Name	English Name
Nymphaea tetragona	Pygmy waterlily
Oxalis oregana	Redwood sorrel
Oxytropis maydelliana	Maydell's locoweed or oxytrope
Oxytropis podocarpa	Stalked-pod locoweed
Papaver alboroseum	Pale poppy
Papaver alpinum	Dwarf poppy
Papaver macounii	Macoun's poppy
Parietaria pensylvanica	Pennsylvania pellitory
Parrya nudicaulis	Common twinpod
Pedicularis contorta	Coil-beaked lousewort
Pedicularis oederi	Oeder's lousewort
Pedicularis parviflora ssp. parviflora	Small-flowered lousewort
Pedicularis verticillata	Whorled lousewort
Pellaea atropurpurea	Purple cliff-brake
Penstemon albertinus	Alberta penstemon
Penstemon gormanii	Gorman's penstemon
Penstemon gracilis	Slender penstemon
Penstemon lyallii	Lyall's penstemon
Penstemon nitidus	Shining penstemon
Penstemon richardsonii var. richardsonii	Richardson's penstemon
Phlox alyssifolia	Alyssum-leaved phlox
Phlox hoodii	Hood's phlox
Physaria didymocarpa var. didymocarpa	Common twinpod
Pinguicula villosa	Hairy butterwort
Pinus banksiana	Jack pine
Piperia maritima	Seaside rein orchid
Plantago canescens	Arctic or Siberian plantain
Plantago eriopoda	Alkali or saline plantain
Platanthera dilatata var. albiflora	White bog orchid
Poa eminens	Eminent bluegrass
Polemonium boreale	Northern Jacob's-ladder
Polemonium elegans	Elegant Jacob's-ladder or polemonium
Polygonum douglasii ssp. austiniae	Austin's knotweed
Polygonum engelmannii	Engelmann's knotweed
Polygonum hydropiperoides	Water-pepper
Polygonum polygaloides ssp. kelloggii	Kellogg's knotweed
Polygonum prolificum	Proliferous knotweed
Polygonum punctatum	Dotted smartweed
Polypodium virginianum	Virginia or rock polypody
Polystichum kruckebergii	Kruckeberg's holly fern
Polystichum setigerum	Alaska holly fern
Potamogeton perfoliatus	Richardson's pondweed
Potentilla biflora	Two-flowered cinquefoil
Potentilla diversifolia var. perdissecta	Diverse-leaved or blue-leaved cinquefoil
Potentilla elegans	Elegant cinquefoil
Potentilla ovina	Sheep cinquefoil
Potentilla quinquefolia	Five-leaved cinquefoil
Primula mistassinica	Mistassini primrose
Primula sibirica	Siberian primrose
Pteridium aquilinum ssp. latiusculum	Bracken fern
Pyrola dentata	Nootka wintergreen
Pyrola elliptica	White wintergreen or shinleaf
Ranunculus cardiophyllus	Heart-leaved buttercup

Table 10-1: Present Status of Rare Native Vascular Plants of British Columbia (continued).

110

Scientific Name	English Name
Ranunculus eschscholtzii var. suksdorfii	Subalpine, mountain or snowpatch buttercup
Ranunculus macounii var. oreganus	Macoun's buttercup
Ranunculus pedatifidus	Birdfoot buttercup
Ranunculus sulphureus	Sulphur buttercup
Rhus diversiloba	Poison oak or poison ivy
Ribes oxyacanthoides ssp. cognatum	Northern or northern smooth gooseberry
Romanzoffia tracyi	Tracy's romanzoffia or mistmaiden
Rosa arkansana	Arkansas rose
Rubus lasiococcus	Evergreen or cutleaf evergreen blackberry
Rubus nivalis	Snow bramble or dewberry
Rumex arcticus	Atctic dock
Rumex paucifolius	Alpine sorrel
Sagina decumbens ssp. occidentalis	Western pearlwort
Sagina nivalis	Snow pearlwort
Salix boothii	Booth's willow
Salix farriae	Farr's willow
Salix petiolaris	Meadow willow
Salix planifolia ssp. pulchra	Diamond-leaved willow
Salix serissima	Autumn willow
Salix sessilifolia	Soft-leaved or sessile-leaved sandbar willow
Sanguisorba menziesii	Menzies' burnet
Sanguisorba occidentalis	Western burnet
Saxifraga hirculus	Yellow marsh saxifrage
Saxifraga serpyllifolia	Thyme-leaved saxifrage
Scirpus fluviatilis	River bulrush
Scirpus olneyi	Olney's bulrush
Scirpus pallidus	Pale bulrush
Scolochloa festucacea	Fescue scolochloa
Scrophularia lanceolata	Lance-leaved figwort
Selaginella oregana	Oregon selaginella
Selaginella sibirica	Northern selaginella
Senecio atropurpureus	Purple-haired groundsel
Senecio pseudoarnica	Streambank butterweed
Senecio sheldonensis	Mount Sheldon butterweed
Senecio yukonensis	Yukon groundsel
Shepherdia argentea	Thorny buffalo-berry
Silene drummondii var. drummondii	Crummond's campion
Silene involucrata ssp. involucrata	Arctic campion
Silene repens	Pink campion
Smelowskia calycina	Alpine smelowskia
Smelowskia ovalis	Short-fruited smelowskia
Sparganium fluctuans	Water bur-reed
Spergularia macrotheca	Beach sand-spurry
Sphenopholis obtusata var. major	Prairie wedge-grass
Stellaria obtusa	Blunt-sepaled starwort
Stellaria umbellata	Umbellate starwort
Stipa lemmonii	Lemmon's needlegrass
Thelypodium laciniatum var. laciniatum	Thick-leaved thelypody
Thelypodium laciniatum var. milleflorum	Many-flowered thelypody
Thelypteris quelpaertensis	Mountain fern
Tofieldia coccinea	Northern false asphodel
Torreyochloa pallida	Fernald's false-manna

Table 10-1: Present Status of Rare Native Vascular Plants of British Columbia (continued).

Scientific Name	English Name
Trichophorum pumilum	Dwarf clubrush
Trifolium depauperatum	Poverty clover
Triglochin concinnum var. concinnum	Graceful arrow-grass
Utricularia gibba	Humped bladderwort
Valeriana edulis ssp. edulis	Edible valerian
Verbena hastata	Blue vervain
Veronica cusickii	Cusick's speedwell
Viburnum opulus ssp. trilobum	Purslane speedwell
Viola howellii	American bush-cranberry, wild guelder-rose
Viola purpurea var. venosa	Purple-marked yellow violet
Viola septentrionalis	Northern or northern blue violet
Woodsia alpina	Alpine cliff fern, or northern woodsia
Woodsia glabella	Smooth cliff fern or woodsia
Woodsia ilvensis	Rusty cliff fern or woodsia

Table 10-1: Present Status of Rare Native Vascular Plants of British Columbia (continued).

Chapter 11
Benthic Marine Algal Flora (Seaweeds) of British Columbia: Diversity & Conservation Status[1]

Michael W. Hawkes
Department of Botany,
University of British Columbia
#3529 - 6270 University Blvd,
Vancouver, B.C.
V6T 1Z4

More than two-thirds of the Earth's surface is aquatic but, until recently, discussions of biodiversity and conservation have focused on terrestrial rather than marine environments (Eiswerth, 1990; Hawkes, 1990; Earle,1991; Thorne-Miller & Catena, 1991; MacInnis, 1992). Lack of awareness of marine conservation issues is primarily due to the fact that relatively few people have access to the marine environment. It is truly a case of "out-of-sight out-of-mind."

Of the 6,691 plant and animal species that have been designated as threatened or endangered worldwide, only 16 are marine (Thorne-Miller & Catena, 1991). Such low numbers most likely reflect a lack of information on the marine biota rather than any lack of threat.

One must be cautious when looking at biodiversity solely in terms of species numbers. For example, in terms of animal phyla, the marine environment is twice as diverse as the terrestrial environment, as illustrated in Table 11-1 (Chapter 6, this volume; Ray, 1988; Ray & Grassle, 1991). Some marine phyla may not be species rich, but they represent ancient lineages with long evolutionary histories and, in many cases, unique life forms. In addition to species diversity, there is also genetic and ecological diversity to consider.

Terrestrial	Freshwater	Marine
11	14	28
1 Endemic	0 Endemic	13 Endemic

Table 11-1: Distribution of Animal Phyla by Habitat. Source: Ray & Grassle 1991.

Marine Plant Biodiversity

In the article entitled, *Marine biodiversity: The Sleeping Dragon*, Kaufman (1988) discusses many vertebrate and invertebrate marine animals, but makes no mention of marine plants. As in terrestrial ecosystems, plants are at the base of the food web. In the oceans "all flesh is algae." Benthic (bottom-dwelling) marine algae, or seaweeds, occur from the high intertidal region down to depths of 30 metres or more in the subtidal zone. The Canadian west coast is blessed with one of the richest and most diverse marine floras in the world. Scagel et al. (1993) presented a recent synopsis of the biodiversity of the seaweed flora in the Pacific Northwest (Oregon to southeast Alaska) (Table 11-2).

There are 3 main groups of seaweeds: browns (Phaeophyta), greens (Chlorophyta), and reds (Rhodophyta). This colour grouping is based on pigment types, but there are also other biochemical and

[1] Portions of this paper have been adapted from Hawkes (1990, 1991 and 1992).

structural features that characterize these three groups. While these organisms contain chlorophyll and are, therefore, photosynthetic, they are not flowering plants and have very cryptic reproductive structures and frequently complicated life histories. In addition, these marine plants lack true leaves, stems, and roots, although they frequently have parts that resemble these land plant structures. Because seaweeds are suspended in an aquatic environment, nutrients and dissolved gases can be absorbed over the entire plant surface.

Chlorophyta (green algae):	51 genera	117 taxa
Phaeophyta (brown algae):	66 genera	143 taxa
Rhodophyta (red algae):	161 genera	373 taxa
Chrysophyta (golden algae):	3 genera	6 taxa
Anthophyta (seagrasses):	3 genera	6 taxa
Total:	**284 genera**	**645 taxa**

Table 11-2: Diversity of Marine Flora.

Best known of the brown seaweeds are the kelp, and B.C. is the world centre of diversity for this group. Both bull kelp (*Nereocystis leutkeana*) and giant kelp (*Macrocystis integrifolia*) form extensive underwater forests that support diverse communities. Kelp beds are major components of our coastal ecosystem, being more productive than tropical rain forests (Leigh et al., 1987). The rate at which carbon is fixed into organic compounds through photosynthesis is a measure of productivity. Kelp beds on the B.C. coast fix carbon at a rate of about 2.5 kilograms/square metre/year, whereas tropical rain forests fix carbon at 2.0 kilograms/square metre/year. Giant kelp (*Macrocystis*) can grow 35 centimetres a day and reach lengths of up to 25 metres or more. In addition to the kelp, there are many bladed, sac-like, and filamentous brown seaweeds in our flora.

Green seaweed comes in bladed forms (such as the common sea lettuce, *Ulva* spp.), spongy types (such as *Codium* spp.), and many simple to branched filamentous types. It is from the green algal lineage that terrestrial plants evolved.

Red seaweeds are the most numerous in terms of taxonomic diversity and vary in size from microscopic unicells, to filaments, to one metre long blades (such as *Rhodymenia pertusa*). Red seaweeds take on many different forms, including hollow sacs (such as *Halosaccion*), umbrella-like blades (such as *Constantinea*), iridescent fans (such as *Fauchea*) and delicately branched blades (such as *Delesseria*). There are also epiphytic (growing on host plant but drawing nourishment from surroundings rather than from host) and parasitic red seaweeds. Probably the most unplant-like reds are the calcified crustose and articulated coralline algae. Despite their superficial resemblance to some coral-like animal, they actually are plants with photosynthetic tissue under the deposits of calcium carbonate.

Seagrasses, a fourth and smaller group of marine plants, are the whales of the flowering plant world, since they also evolved in a terrestrial environment and then returned to the sea. Marine plants in all four groups are benthic, growing attached to rocks, animals, other plants, or in mud and/or sand in the case of some seagrasses.

Due to a shortage of systematists and the relative inaccessibility (in terms of both geography and weather) of much of the B.C. coast, we are still in the pioneering phase of documenting the west coast seaweed flora. Furthermore, many species are seasonal, often alternating between macroscopic and microscopic life history stages.

So even if climate and remoteness can be overcome, species in the microscopic phase of their life history will be difficult or impossible to find.

Are There Endangered Marine Plant Species in B.C.?

To generate a list of rare or potentially vulnerable marine plant species for the west coast of Canada, I did a search of the benthic marine algal herbarium data base at the University of British Columbia. This computerized data base, which contains 37,374 records for the Pacific Northwest, is by far the largest and most valuable collection of B.C. seaweeds in the world. As a starting point, I chose to limit the search to species with 6 records or less as a measure of their rarity or infrequency of collection. Other criteria I used for the selection of rare species included:

1. small population size; and
2. narrow habitat specificity and/or restricted geographic range.

Many of the species with six or fewer records are small (less than two centimetres in size), poorly known taxonomically, present only seasonally, denizens of the subtidal region, and often from northern localities (both subtidal regions and northern coasts are incompletely surveyed in terms of their marine plant biota). The rarity of such species is probably more apparent than real and much more field data are needed before these species could be recommended for rare or endangered status.

There are several other species that genuinely are rare in B.C. (Table 11-3). The red alga, *Whidbeyella cartilaginea*, is noteworthy as a local endemic (a species not found anywhere else), being known from two locations in B.C. and in the San Juan Islands, Washington State. The green alga, *Codium ritteri*, was reported from Botanical Beach, Vancouver Island, around the turn of the century, but has never been rediscovered at that site. Presently it is known from only one site in northern B.C. (Campania Island) (Hawkes et al., 1978).

Some of the kelps, while relatively abundant, have patchy distributions and/or narrow habitat requirements and would be especially vulnerable to commercial scale harvesting or oil pollution. For example, the sea palm kelp (*Postelsia palmaeformis*) occurs only on extremely exposed sites on the west coast of Vancouver Island. It is absent from comparable sites along the shores of Haida Gwaii (Queen Charlotte Islands) and along the mainland coast of northern B.C. Commercial groups have expressed interest in harvesting this plant both here and in Washington State (Dethier et al., 1989; personal communication on 6 Oct, 1990, with Michael Coon, Ministry of Agriculture and Fisheries). As far as we know, no seaweeds have been extirpated from B.C. waters. If any did go extinct, it is highly likely that the extinction occurred when the sea otter (*Enhydra lutris*) was eliminated from our coast. Until relatively recently sea otters were abundant along the Pacific coast of North America from California to Alaska. By the early 1900's, the sea otter had been hunted to extinction in British Columbia (although some

Reds:
Arthrocardia silvae
Antithamnion kylinii
Bonnemaisonia geniculata
Hollenbergia nigricans
Thuretellopsis peggiana
Whidbeyella cartilaginea
Greens:
Codium ritteri
Browns:
Dictyoneuropsis reticulata
Dictyoneurum californicum
Laminaria farlowii
Laminaria longipes
Laminaria sinclairii

Table 11-3: Rare seaweeds

may have survived in an isolated population around Calvert Island). With the fur trade effectively removing sea otters from the ecosystem, one of the otter's prey species, herbivorous sea urchins, underwent a dramatic population increase, with devastating consequences for kelp beds, other fleshy algae, and many associated organisms. The rocky, nearshore ecosystems that we have in B.C. today are actually significantly disturbed systems which should have much higher seaweed biomass and diversity. Evidence supporting this conclusion comes from studies of subtidal habitats before and after sea otter re-introduction (personal communication 6 Oct, 1990, with Jane Watson, University of California, Santa Cruz; Watson, 1993).

There is renewed interest in harvesting kelps commercially in B.C.. Some of the finest kelp beds in the province are in Queen Charlotte Strait, between Port McNeill and Port Hardy off the northeast coast of Vancouver Island. No part of this kelp forest ecosystem has protected status. It is also worth noting that the State of Washington has declared a moratorium on commercial harvesting of kelp. Mumford (1989) states, "the eventual leasing of kelp beds is problematical—their greatest value may be in habitat and food web support."

Determining whether there are endangered marine plant species in B.C. is complicated by the fact that systematists, those most knowledgeable about biodiversity, are themselves an endangered species. There are few systematists (working on any group of organisms) currently employed in university or museum environments and even fewer students receiving proper systematic training. In the field of phycology (seaweeds and algae), for example, between Alaska and Mexico, there is only one systematist who is permanently employed. After the *Exxon Valdez* oil spill, one of the most difficult tasks was to locate enough people who knew the diversity of organisms (both plant and animal) to take part in the impact assessment. There is an urgent need to support systematic research and researchers.

Acknowledgements

Canadian NSERC Operating Grant 580384 supported this work. Dr. Paul Gabrielson, William Jewell College made several suggestions regarding rare marine algae. Olivia Lee, University of British Columbia herbarium, kindly did the computer search of the marine algal data base. Michael Coon, Ministry of Agriculture & Fisheries, Victoria provided information on kelp-harvesting proposals in B.C.

References

Dethier, M.N., D.O. Duggins and T.F. Mumford, Jr. 1989. Harvesting of non-traditional marine resources in Washington State: Trends and concerns. *The Northwest Environmental Journal* 5:71-87.

Earle, S.A. 1991. Sharks, squids, and horseshoe crabs—the significance of marine biodiversity. *BioScience* 41:506-509.

Eiswerth, M.E. 1990. *Marine Biological Diversity: Report of a Meeting of the Marine Biological Diversity Working Group.* Woods Hole Oceanographic Institution Technical Report 90-13.

Hawkes, M.W. 1990. Benthic marine algal flora (seaweeds) of B.C. Diversity and conservation status. *BioLine* 9:18-21.

Hawkes, M.W. 1991. Seaweeds of British Columbia: Biodiversity, ecology, and conservation status. *Canadian Biodiversity* 1(3):4-11.

Hawkes, M.W. 1992. Conservation status of marine plants & invertebrates in British Columbia: Are our marine parks & reserves adequate? In: *Community Action for Endangered Species. A public symposium on B.C.'s Threatened & Endangered Species and their Habitat, 28-29 September 1991*. S. Rautio (ed.). Federation of B.C. Naturalists & Northwest Wildlife Preservation Society, Vancouver, B.C. pp. 87-101.

Hawkes, M.W., C. Tanner and P.A. Lebednik. 1978 [1979]. The benthic marine algae of northern British Columbia. *Syesis* 11:81-115.

Kaufman, L. 1988. Marine biodiversity: The sleeping dragon. *Conservation Biology* 2:307-308.

Leigh, E.G, Jr., R.T. Paine, J.F. Quinn and T.H. Suchanek. 1987. Wave energy and intertidal productivity. *Proceedings of the National Academy of Science USA* 84:1314-1318.

MacInnis (ed.). 1992. *Saving the Oceans*. Key Porter Books, Toronto, Ontario.

Mumford, T.F. 1989. Seaweed aquaculture...an overview. *NCRI News* 4:8-9.

Ray, G.C. 1988. Ecological diversity in coastal zones and oceans. In: *Biodiversity*. E.O Wilson (ed.). National Academy Press, Washington, D.C. pp. 36-50.

Ray, G.C. and J.F. Grassle. 1991. Marine biological diversity. *BioScience* 41:453-457.

Scagel, R.F., P.W. Gabrielson, D.J. Garbary, L. Golden, M.W. Hawkes, S.C. Lindstrom, J.C. Oliveira and T.B. Widdowson. 1993. *A Synopsis of the Benthic Marine Algae of British Columbia, Southeast Alaska, Washington and Oregon*. Phycological Contribution Number 3, Department of Botany, University of British Columbia, Vancouver, B.C. 532 pp.

Thorne-Miller, B. and J.G. Catena. 1991. *The Living Ocean. Understanding and Protecting Marine Biodiversity*. Island Press, Washington, D.C. and Covelo, California. 180 pp.

Watson, J.C. 1993. *The Effects of Sea Otter (Enhydra lutris) Foraging on Rocky Subtidal Communities off Northwestern Vancouver Island, British Columbia*. Ph.D. thesis. University of California, Santa Cruz.

Chapter 12
Reptiles in British Columbia

The vertebrate class, Reptilia, with more than 6,500 described species (Halliday and Adler, 1986), is represented today by a few, very distantly related lineages. Crocodilians, turtles, lizards, snakes, amphisbaenians and tuataras, though all reptiles, are remarkably dissimilar from one another in their morphology, physiology and behaviour. These evolutionary divergences date back many tens of millions of years and, consequently, have resulted in profound interordinal differences, as well as in great diversity throughout the lower taxonomic ranks. For this reason, general statements about reptile conservation are difficult to make because the survival problems they face are as varied as the species within their class.

Can reptiles live in a world dominated by humans? The historical evidence suggests that many species cannot. Reptiles generally inspire a strong emotional response. But whether we are inspired by reverence or hatred, our attention invariably threatens their survival. We commonly categorize and treat all of the living orders of reptiles as either vermin or as commercially exploitable commodities. Crocodilians, for example, are feared and killed due to an exaggerated perception of their ferocity, but they are also desired for their attractive and durable leather. Turtles, tortoises and some lizards are generally liked by humans but, thereby, suffer from the intense demand for them as pets. Snakes are almost universally regarded with enmity and are, thus, killed on sight, but the beautiful patterns of their skin and symmetry of the scales make snake leather a valuable commercial item.

Thus, like many species of ornamental birds, reptiles are pursued as a pestilence while at the same time sought after as a valued commodity. Their numbers dwindle for the now commonplace reasons of gross habitat loss and human planetary crowding, but the rate of their decline is often accentuated by human persecution and commercial exploitation. Conservation actions have not begun to keep pace with the problems and, hitherto in British Columbia, awareness of the need for such actions has had few positive effects.

While this paper focusses primarily on conserving the diversity of the relatively small reptile fauna of British Columbia (at least 16 species), it is important also to recognize the global context of these problems. Canadians trade commercially in a variety of exotic reptiles and, therefore, our actions contribute to a growing international conservation crisis. Consequently, our political and environmental responsibilities to this class extend well beyond the boundaries of British Columbia.

Stan A. Orchard
Nearctic Zoophiles Incorporated
27-2891 Craigowan Road
Victoria, B.C.
V9B 1M9

Turtles

In British Columbia, the painted turtle (*Chrysemys picta*), is the only native fresh water species. There are, however, two species of sea turtles: the green sea turtle (*Chelonia mydas*; Figure 12-1), which has been found only twice at this latitude; and the leatherback sea turtle (*Dermochelys coriacea*) which is much more commonly reported.

Contrary to appearances, the painted turtle is threatened. Its wide distribution and the fact that it is regularly sighted belie its true provincial population status, which is not at all stable. Turtles are long-lived and, thus, merely sighting adult turtles over many years is no sure indicator of the vitality of the population. Recruitment into the populations must be measured, and natural recruitment rates have, in some cases, been found to be very low (St. Clair, 1989).

Based on my observations, threats to this species range from lakeside settlement and agricultural expansion to their use in target practice. Gravid females are frequently run over by automobiles on lake shore roads while in search of nesting sites, and the hatchlings are similarly threatened as they migrate from nest to water. The developing embryos and hatchlings spend almost a year in the nest. Sitting just a few centimetres below the ground surface, the nest is quite vulnerable to natural predation as well as to trampling by cattle, people and their vehicles. In the past, commercial collectors have been permitted to sell and export painted turtles for the pet trade. Although this practice is no longer permitted, enforcement is difficult, particularly when the base of operation is outside British Columbia. In the province's extreme southwestern corner, including the Lower Mainland, Gulf Islands and southern Vancouver Island, introduced populations of the American bullfrog (*Rana catesbeiana*) have become established and are now abundant. Unfortunately, an adult bullfrog is easily capable of swallowing a turtle in its first year. These are merely a few illustrations of the myriad problems that menace painted turtles.

Both sea turtle species found in British Columbian waters are internationally recognized as endangered. Net fishing takes its toll, but of greater harm is the loss of tropical nesting beaches to human settlement and the killing of nesting females and collection of their eggs for human consumption.

Canada is a signatory to the Convention on International Trade in Endangered Species of Wild Fauna and Flora (CITES). Ironically, this document, which protects sea turtles from

Figure 12-1. Green Sea Turtle (*Chelonia mydas*). Source: Gregory and Campbell, 1984.

being taken across Canada's international border, affords them no protection whatever in Canadian territorial waters, where they may be harried and killed with impunity. Likewise, the leatherback sea turtle is listed as endangered by the Committee on the Status of Endangered Wildlife in Canada (COSEWIC), but this recognition of endangerment confers no correspondent protection.

It may be argued that our encounters with leatherback sea turtles in British Columbian waters are too few in number to affect the overall population. In fact, this issue has not been investigated. It is, however, well documented that, especially where they concentrate to reproduce, the rates of decline are increasing and the threats to survival are mounting for virtually all species of sea turtles. These impediments to survival now so encumber this group that it cannot afford any useless and avoidable casualties (Bjorndal, 1981).

Lizards

Three species of lizard are known to range naturally into British Columbia. The short-horned lizard (*Phrynosoma douglassii*) has been taken only in the southern Okanagan Valley, but the northern alligator lizard (*Gerrhonotus coeruleus*) and the western skink (*Eumeces skiltonianus*) are found across the southern third of this province.

In North America, horned lizards were for many decades one of the reptiles most often sold in pet shops, usually under the common name "horned toads." American states began to prohibit the trade when it became clear that populations were rapidly dwindling from over-collection and habitat loss.

The short-horned lizard was first reported in British Columbia before 1898, when two specimens were collected near Osoyoos. Both specimens now reside in the collections of the Royal British Columbia Museum in Victoria. COSEWIC lists this lizard as extirpated. Recent searches have been conducted through much of the remaining habitat, but no evidence of the species has been found.

Virtually all of the habitat that remains undeveloped in the Osoyoos area is currently in private hands, so the future of any remnant short-horned lizard populations is, at best, uncertain. Reintroduction of the species from northern populations in Washington State might be considered at some point in the future, but only if a sizeable property becomes available for purchase as a wildlife refuge. It should be borne in mind that this creature may have been driven from British Columbia by environmental factors, such as harsh winters, that are independent of human disturbance and, therefore, attempts at reintroduction may be fruitless. Conversely, if populations in Washington State are being seriously curtailed by the effects of human encroachment, the establishment of a protected colony north of the Canada-U.S. Border might be considered a noble effort.

The northern alligator lizard is found in varying densities across the southern third of British Columbia. It generally concentrates in rocky open clearings. In urban areas, it suffers the

121

depradations of domestic cats and children. The recent addition of the European wall lizard (*Podarcis muralis*) to the provincial herpetofauna portends an ominous future for populations of the native northern alligator lizard on southern Vancouver Island. Several European wall lizards were imported from Europe approximately 20 years ago and were released into a garden on southern Vancouver Island. The colony has prospered and is slowly increasing its geographical distribution. European wall lizards are fast, aggressive, and productive. They are sun-loving and live in precisely the same type of rocky, forest clearings as northern alligator lizards. It is reasonable to suppose that European wall lizards will compete with northern alligator lizards and possibly displace them. The real effects will need to be studied over time, but it is interesting to note that, in Europe, where relatives of the alligator lizard are found together with wall lizards, the alligator lizard types have evolved a legless body form and live largely in the shade and underground. The unfolding of this domestic drama in natural selection will make an interesting study, but the fact that this European lizard is now established in North America is yet another lamentable illustration of the need for tighter restrictions on the free trade in non-native species of plants and animals.

Snakes

Nine snake species are indigenous to British Columbia. They are the: western garter snake (*Thamnophis elegans*); northwestern garter snake (*Thamnophis ordinoides*); common garter snake (*Thamnophis sirtalis*); rubber boa (*Charina bottae*); sharp-tailed snake (*Contia tenuis*); night snake (*Hypsiglena torquata*); western yellow-bellied racer (*Coluber mormon*); gopher snake (*Pituophis melanoleucus*); and western rattlesnake (*Crotalus viridis*; Figure 12-2).

Snakes have many survival problems in common. For instance, roadways resemble the rocky clearings to which diurnal species are drawn to warm themselves in the sun. Pavement also collects solar heat during the day and retains it for a time in the evening. Thus, nocturnal species are similarly attracted to these sun-warmed surfaces. After hibernating in talus slopes, many snake species, including rattlesnakes and gopher snakes, move down into riparian areas, often crossing roads en route. Needless to say, many snakes are crushed by automobiles, now a major source of snake mortality. Additionally, automobile drivers often vent their dislike for snakes by intentionally running them over. Snakes that manage to reach riparian areas often find their habitat usurped by

Figure 12-2. Western Rattlesnake (*Crotalus viridus*). Source: Gregory and Campbell, 1984.

122

concentrations of urban and agricultural development, particularly in dry areas such as the Okanagan.

Garter snakes are the most commonly encountered and geographically widespread reptiles in British Columbia and they, too, have declined due to habitat loss. Unlike the strictly terrestrial northwestern garter snake, the western and common garter snakes are primarily aquatic foragers. Their predilection for lakeside and seashore foraging habitat becomes a serious maladaptation when it conflicts directly with our own plans for real estate, recreation, and general civic development in riparian habitats. Nevertheless, garter snakes would survive such development if they could safely traverse roads that intersect traditional migration routes, avoid predation, find food when ground cover is removed, and escape the persecution of dogs, cats, children and ophidiaphobic people (those having a fear of snakes). If the needs of garter snakes were considered at the outset of development planning, then a degree of compatibility might be achieved and extirpation might be avoidable. Otherwise, though the decline in garter snake numbers may be gradual and, thus, not readily noticeable, our population expansion will inevitably overwhelm even the western and common garter snakes.

The distribution of rubber boas in British Columbia extends from the Lower Mainland, across the southern third of the province, reaching greatest density in the drier valleys of the Southern Interior (Gregory and Campbell, 1984). They are especially sought after for the pet trade because they are harmless, attractive and easy to maintain in captivity. There is also, among reptile keepers, a mystique that surrounds the members of the family Boidae. It may relate to the rubber boa's familial connection with the primitive and exotic giants of the snake world or the popular notion that boa constrictors "crush" their prey. In any case, the tendency toward overcollection of this species and destruction of its habitat has led to its being protected throughout most of its range in western North America. In British Columbia, it is nowhere abundant, though it is occasionally killed on roads. Little is known of its biology.

The sharp-tailed snake is also a mystery species. It is too small (less than 50 centimetres) and too secretive to be easily spotted or recognized. Only a few specimens have been taken from the Gulf Islands in the Strait of Georgia and from the southern tip of Vancouver Island. Consequently, this species' population status is unknown. However, large tracts of fairly undisturbed habitat still remain on the Gulf Islands, and this fact offers some hope that sharp-tailed snakes, though rarely encountered, may be holding their own. Widespread human settlement of the islands is possibly their greatest current threat. A long-term intensive study would be required to uncover the secrets of this inscrutable reptile.

All of the snake species that are found only in the Thompson and Okanagan Valleys, or the dry interior, warrant attention. The mildly venomous night snake is the least common of these, having been found at only two locations in the southern Okanagan Valley. Both sites are in talus on opposite sides of the valley—one being 30 kilometres north of the Canada-US Border and the other 15 kilometres further still. One might postulate that night snakes are more common than they appear, and that their manifest rarity is an illusion based on too few encounters with a species that ventures infrequently from cover, and then only at night. Conversely, the recently identified populations may be remnants of a formerly contiguous distribution stretching south that is presently reduced due to factors unknown. Until proven otherwise, it would be prudent to assume the latter and protect both sites.

Western yellow-bellied racers and gopher snakes, colloquially referred to as "blue racers" and "bullsnakes," are still numerous in many locations in the dry interior. However, on a provincial scale, their numbers appear to be dwindling. It is of mounting concern that the rates of increase in automobile traffic, human settlement and habitat loss to agriculture are taking a heavy toll on snake populations in the long, but very narrow, Thompson and Okanagan valleys where snakes are often killed on sight.

Western rattlesnakes are subject to the same perils as racers and gopher snakes, but to an even greater degree because they have been persecuted for decades as a public health threat. Their habit of overwintering in communal dens is a boon to snake collectors and recreational exterminators because, when the local population emerges in spring, it is concentrated and exposed to wholesale slaughter. There is demand for rattlesnake leather for such uses as hat bands, wallets, and cowboy boots, and the rattles are taken as prized trinkets. Though rattlesnakes are protected in British Columbia under the Wildlife Act, there is strong anecdotal evidence that British Columbian rattlesnakes are being skinned for this trade and the leather taken out of the province for processing. Recently, rattlesnake meat, apparently imported from the United States, has become a popular offering at restaurants in the Vancouver area. Here, at the northern limits of their distribution, western rattlesnakes reproduce only once in three to four years (Macartney, 1985). Thus, these populations will not support the growing demands of a needless and ecologically damaging industry that cannot fail to encourage private collecting for culinary experimentation.

Talus slopes are extremely important to snake survival in the dry interior (Herrington, 1988). The extensive rock piles above the valleys and along their edges have thus far offered refuge and security. The rocks provide cover and are good solar collectors that warm quickly and retain the heat. Unfortunately, some talus slopes are now being quarried and their stones used for dykes, land fill and highway construction. The chances of conserving rubber boas, night snakes, racers, gopher snakes and western rattlesnakes, as well as western skinks, alligator lizards and a host of other species would be significantly improved if the faunal importance of talus sites were investigated thoroughly before excavations are permitted.

Conclusion

Comparatively little is know about our few reptile species, except that one has already been extirpated and others may follow if conservation measures are not taken. Basic information needs include population and distributional data, as well as biological data on, for example, requirements and life cycles. Habitat destruction, particularly of riparian areas, but also of specialized habitat components such as talus slopes, is a major threat to reptile conservation. Introduction of exotic species of reptile predators, such as bull frogs, and competitors, like the European wall lizard, threaten native reptiles, and the level of this threat seems sure to increase. Wilful persecution and collection are also detrimental to some species, such as rattlesnakes. Conservation measures should, therefore, include performing biological surveys that inventory populations; identifying and protecting important habitats and habitat components; improving enforcement of the <u>Wildlife Act</u>; taking measures that limit introduction and spreading of exotic species; applying and enforcing CITES; and improving public education about the importance of conserving reptiles.

Acknowledgements

Patrick Gregory, University of Victoria, reviewed the manuscript.

References

Bjorndal, Karen A. 1981. Biology and conservation of sea turtles. In: *Proceedings of the World Conference on Sea Turtle Conservation, Washington, D.C. 26-30 November 1979.* Smithsonian Institution Press, Washington, D.C. in cooperation with World Wildlife Fund Inc., Washington, D.C.

Gregory, Patrick T. and R. Wayne Campbell. 1984. *The Reptiles of British Columbia.* British Columbia Provincial Museum. Victoria, British Columbia.

Halliday, T.R. and K. Adler. 1986. *The Encyclopedia of Reptiles and Amphibians.* Facts on File Inc., New York, New York.

Herrington, R.E. 1988. Talus use by amphibians and reptiles in the Pacific Northwest. In: *Management of Amphibians, Reptiles, and Small Mammals in North America.* R.C. Szaro, K.E. Severson and D.R. Patton (Technical co-ordinators). Proceedings of the symposium held July 19-20. 1988 in Flagstaff, Arizona. USDA Forest Service General Technical Report RM-166. November, 1988.

Macartney, J.M. 1985. *The Ecology of the Northern Pacific Rattlesnake, Crotalus viridis oreganus, in British Columbia.* Unpublished M.Sc. thesis, University of Victoria, B.C.

St. Clair, Robert C. 1989. *The Natural History of a northern turtle, Chrysemys picta bellii (Gray).* Master's thesis. University of Victoria. Victoria, British Columbia.

Chapter 13
Amphibians in British Columbia: Forestalling Endangerment[1]

Since the first symposium on the subject in 1990, numerous cases from all parts of the globe have brought to light the inexplicable decline and extinction of amphibian populations (Blaustein and Wake, 1990). Most unsettling is the fact that so many of these reports have to do with sites that are largely undamaged and unused by humans. Many theories have been proposed but, to date, the underlying cause or causes have not been satisfactorily identified. To biologists, the broader ecological ramifications of such a widespread phenomenon are most worrisome because amphibians, abundant and low on the food chain, are currently an important link in the energy flow of wetland ecosystems almost everywhere (Bishop and Petit, 1992; Blaustein and Wake, 1990).

In British Columbia, nearly all cases of amphibian population decline and local extinction are readily explained and can be directly connected to our activities. Our numbers are still relatively low in this province, so our impacts are too, but elsewhere, for example in Europe and eastern North America, British Columbia's potential environmental future has already been played out. There, poorly planned and mushrooming settlements have detrimentally affected amphibian populations. Are we wise enough to critically examine habitat degradation in other jurisdictions and to plan novel approaches to forestalling (or even preventing) the further loss of species and habitats here at home?

We usually displace amphibian populations quite unconsciously. Forests are cleared, swamps are drained, ponds are filled in, and the rate of habitat destruction intensifies, all without overt signs of the cost in amphibian diversity. We tend not to notice this cost because amphibians are cryptically coloured, small, only periodically vocal and tend to be most active at night and especially during rainy periods. Also, many of us find the sight of amphibians to be fearsome or loathsome and, thus, consciously avoid them. In any case, the appearance and behaviour of amphibians seem to preclude our attention and concern.

Stan A. Orchard
Nearctic Zoophiles Incorporated
27-2891 Craigowan Road
Victoria, B.C.
V9B 1M9

Threatened Species and Habitats

The most abundant amphibians in British Columbia are the habitat generalists, such as the long-toed salamander (*Ambystoma macrodactylum*), spotted frog (*Rana pretiosa*) and western toad (*Bufo boreas*). They are adaptable to a variety of conditions and are, therefore, widely distributed. Conversely, British Columbia also has an

[1] Originally published in the Spring 1990 (Volume 9, Number 2) edition of *BioLine*, official publication of the Association of Professional Biologists of British Columbia. Revised and updated for this publication.

assortment of species with restricted ranges that require very particular sets of environmental conditions. As habitat specialists, they are much more vulnerable to environmental change.

Eighteen species of amphibians are known to occur naturally in British Columbia, and an additional two introduced species have become established (Orchard, 1990). Five native species are threatened by human activity and warrant immediate conservation attention. They are:

o   northern Pacific giant salamander (*Dicamptodon tenebrosus*)

o   tiger salamander (*Ambystoma tigrinum*)

o   coeur d'alene salamander (*Plethodon idahoensis*)

o   tailed frog (*Ascaphus truei*)

o   Great Basin spadefoot toad (*Scaphiopus intermontanus*)

These five species are found in two habitat types, both of which are threatened: alkali lakes, ponds and ephemeral pools; and cascading streams.

Alkali Lakes, Ponds, and Ephemeral Pools

The hottest and driest parts of British Columbia's dry interior would seem to be the least likely place in which to find amphibians. In fact, this area is home to North America's most nearly desert-adapted salamander, the tiger salamander, and the remarkably xerophilic (likes dry conditions) spadefoot toad (figure 13-1). Uncharacteristically, for amphibians, both species are able to spawn where water conditions can be thermally unstable and highly saline and alkaline; a set of conditions quite common to the pot-hole lakes and ephemeral pools of the dry interior.

Tiger salamanders exhibit two possible developmental states at sexual maturity: one is semi-terrestrial (metamorphic) and the other is persistently or permanently aquatic (neotenous). Metamorphic salamanders reproduce in both permanent and ephemeral water bodies, whereas the neotenous form requires permanent water. Spadefoot toads spawn primarily in ephemeral pools.

The alkali lakes, ponds and ephemeral pools so critical to the survival of these amphibians are threatened by multiple forms of human disturbance. Range cattle often congregate around the edges of these lakes and wade through the shallower ephemeral pools. Trampling muddies the water by churning up the bottom, and this muddied water can clog the gills of salamander larvae and tadpoles. Manure from the cattle can become very concentrated, especially in the smaller water bodies, and can form toxic nitrites and culture harmful bacteria (Larkin, 1974). In addition, off-road vehicles are now commonly run through the shallow basins, stirring up mud and killing toads and tadpoles. Spadefoot

Figure 13-1. Great Basin Spadefoot Toad (*Scaphiopus intermontanus*). Source: Green, 1984.

toads will continue be drawn to these sites to spawn, but the tadpoles may not survive to metamorphosis.

Normally, tiger salamanders colonize small lakes that are not populated with predatory fish but, in recent times in the Okanagan Valley, game fish have been released into most of the permanent lakes. The introduction of game fish, such as trout, can set off a vicious cycle that may reduce salamander density, shrink the gene pool and eventually wipe out neotenous subpopulations. The fish displace the salamanders by preying upon their larvae (Matthews, 1986) and competing with them for food. Game fish naturally attract anglers, some of whom use live fish as bait. This practice is now prohibited in British Columbia, but because the law is virtually unenforceable, species such as shiners and sunfish have become secondarily established in lakes stocked with game species. Sunfish are more voracious competitors and predators than trout, and spawn freely in lakes that lack the inflow and/or outflow streams which trout need to successfully reproduce. The population density of sunfish can rise quickly and, within a span of a decade, they can render the lake ecologically inhospitable for trout. Many lakes, such as Kilpoola Lake in the Okanagan, have been poisoned to remove the unwanted fish species and reinstate the game fish. This periodic lake poisoning removes the entire neotenous subpopulation of tiger salamanders. Significantly, tiger salamander survival may pivot on the success of its neotenic form during periods of protracted drought. Extremely dry conditions are unsuitable for a primarily terrestrial salamander. Thus, in lakes that retain some water throughout a drought, the neotenous form has the adaptive advantage.

Cascading Streams

One of the most interesting and least discussed habitats is found on the steep slopes of the Cascade Mountains in British Columbia's southwest corner, on the north coast, and on the slopes of the Flathead River Valley in the extreme southeast corner of the province. Here, glacier-fed streams plunge down the mountain sides. Their waters, insulated from the sun by the shade of mature forests, remain cold year round. Even on the hottest days, the refrigerating effect that these streams have on their surroundings is immediately noticeable. Within this habitat lives the northern Pacific giant salamander (figure 13-2) and the tailed frog. All age classes of both species live right in the stream, but the metamorphosed adults and juveniles emerge periodically to forage on the forest floor in the damp, cool shade of the adjacent woods.

Traditionally in British Columbia, streams have been managed only if they supported game fish but, generally speaking, game fish do not live in the cascading streams to which tailed frogs and northern Pacific giant salamanders are especially adapted. The water descends at such a steep angle that it forms a series of log-jammed splash

Figure 13-2. Pacific Giant Salamander (Dicamptodon tenebrosus). Source: Green, 1984.

129

pools connected by churning rapids and vertical falls, which are simply too turbulent and obstructed for game fish. Though the Fisheries Act ostensibly protects such upstream habitats, the reality has been that many of them have been drastically and irreparably altered by the removal of the riparian forest and by intrusive logging activity. When the riparian forest is removed through logging or other development, silt washes freely into the stream, detrimentally affecting tailed frog tadpoles, in particular. These tadpoles normally avoid being washed away by firmly attaching their large, inverted-saucer mouths to the undersurface of rocks and boulders that have been scoured clean by the force of the rushing water. However, after a clear-cut, these rocks and boulders can become half submerged in a layer of heavy sand and silt, and otherwise coated in fine particulates. The stream is now bathed in sunlight, which raises its summer temperature and also heats the adjacent surroundings. In addition, a thick layer of algae grows on the now sun-exposed rock surfaces. When the rocks are contaminated in this way by algae and silt, tailed frog larvae are unable to hang on and are easily washed away. Northern Pacific giant salamanders avoid warm or cloudy water; so their survival is likewise jeopardized.

A recent discovery in British Columbia is the coeur d'alene salamander, which is associated with cool and cascading streams in the southern Kootenay Valley. Like the tailed frog and the northern Pacific giant salamander, this species lives right in the stream and along its edge, especially at the base of waterfalls. Consequently, it too is highly vulnerable to the effects of logging. Up to now, this salamander has been found in only a few places, but the entire region has yet to be systematically searched. Regardless of whether this species proves to have a larger distribution in British Columbia, the wet, cool, shaded habitats that it appears to favour are isolated from one another by tracts of unsuitable, relatively dry and well-drained terrain, which seem to form an effective barrier to dispersal. Thus, there is good reason to monitor the known sites closely and to provide strict habitat protection.

Conclusion

The survival of amphibians is dependent upon protection of their habitat and reduction of other causes of population decline. Many of the negative effects of single species management techniques or industrial activities on local amphibian fauna are predictable and should be routinely taken into account in the planning stage. Projects that alter habitat should require amphibian population assessments before, during, and for several years after implementation. If amphibian conservation and habitat protection are not given higher priority than in the past, those species that I have listed above will be the first in their class to be eradicated from British Columbia.

Acknowledgements

Patrick Gregory, University of Victoria, reviewed the manuscript.

Bishop, C.A. and K.E. Petit. 1992. *Declines in Canadian Amphibian Populations: Designing a National Monitoring Strategy.* Canadian Wildlife Service, Environment Canada. Occasional Paper #76.

Blaustein, A.A. and D.B. Wake. 1990. Declining amphibian populations: a global phenomenon. *Trends in Ecology and Evolution* 5:203-204.

Green, David M. 1984. *The Amphibians of British Columbia.* British Columbia Provincial Museum. Victoria, B.C.

Larkin, P.A. 1974. *Freshwater Pollution, Canadian Style.* Sponsored by the Canadian Society of Zoologists. Environmental Damage and Control in Canada Series. McGill-Queen's University Press, Montreal and London.

Matthews, S. 1986. *A Proposal to Chemically Rehabilitate Burnell Lake (South Okanagan) For Fisheries Enhancement.* Unpublished Report. Ministry of Environment, Fish and Wildlife, Penticton, British Columbia. January, 1986.

Orchard, Stan A. 1990. Amphibians. In: *The Vertebrates of British Columbia: Scientific and English Names.* R.A. Cannings and A.P. Harcombe (eds.). Royal British Columbia Museum Heritage Record No. 20; Wildlife Report No. R24. Ministry of Municipal Affairs, Recreation and Culture and Ministry of Environment, Victoria, British Columbia.

Chapter 14
Threats to Fish Diversity in the Fresh Waters of British Columbia[1]

"Biodiversity crisis" is the current term for catastrophic species loss, such as that now occurring in Amazonian rain forests. A similar crisis is evident for fishes indigenous to western North America (Deacon et al., 1979), where economic development (in particular, agriculture and sport fishing) conflicts with protection of lesser known fishes. Decisions affecting economic development in river basins are typically made in the absence of information on the life-forms being affected, and few scientists speak out to encourage public concern for non-economic fishes.

Alex Peden
Royal British Columbia Museum
675 Belleville Street
Victoria, B.C.
V8V 1X4

Why is Fish Diversity Important?

Why should we be concerned about the diversity of fish? In contrast to plants, fishes are not known to be storehouses of undiscovered chemicals or pharmaceutical drugs. On the other hand, the genetic potential of fish stocks has barely been studied. We do know that genetic diversity among salmonids contributes to the health of salmonid stocks and, therefore, to the viability of the fishing industry that is dependent upon them. Fish are also known to be good bioindicators. In addition, preservation of the diversity of fish may be seen as a moral imperative.

Loss of genetic variation threatens species by limiting their ability to adapt to changing conditions (Walters, 1988; Goodman, 1990). In the case of salmonids, for example, intraspecific genetic diversity enables local stocks to survive better in their home environment than introduced stocks. Yet major losses of this diversity have been documented. Nehlson et al. (1991) compiled a list of 106 genetically distinct stocks of salmon and steelhead on the American west coast that have become extinct, including stocks from transboundary rivers originating in British Columbia (Okanagan, Kettle, Kootenay and Columbia Rivers); and a further 214 stocks (8 in B.C. border rivers) that are currently threatened with extinction. They note that a third of salmon and steelhead habitat in the Columbia River basin has been lost as a result of impassable dams (Northwest Power Planning Council, 1986, referenced in Nehlson et al. 1991) and estimate that natural production in the Columbia River basin is now about 4-7% of pre-development levels. "This," they say, "likely represents a significant loss of genetic diversity." No such inventory has been completed for British Columbia. Nor do we know whether most other native fishes show similar genetic diversity, since the genetic characteristics of their populations have not been surveyed.

[1]Originally published in the Spring 1990 (Volume 9, Number 2) issue of *BioLine*, official publication of the Association of Professional Biologists of British Columbia. Revised and updated for this publication.

Various species of fish can be used as indicators of ecosystem health. For example, an increase in the population of a species that is highly resistant to pollution or habitat alteration may signal environmental degradation. Monitoring of other species can help in the identification of pollution sources. For instance, fresh water sculpins (*Cottus*) in the Columbia River are occasionally sampled for heavy metals. In such sampling, it is important that the vagaries of the different species' life histories be taken into account lest the results be misleading, as they may be in the case of the sculpins, where it is assumed that all those sampled share the wide-ranging habits of the prickly sculpin *(Cottus asper)*. If the different territorial habits of the shorthead *(C. confusus)* or mottled sculpin *(C. bairdi)*, were taken into account, interpretations of pollution sources and dispersal characteristics might be expected to differ (Peden and Clermont, 1989). This example of the diversity among sculpins and their use as bioindicators illustrates both the complexity and importance of developing an understanding of and maintaining fish diversity.

And then there is the moral argument. Whatever we may think of them, species have an intrinsic value and right to exist, as was recognized in the Biodiversity Convention, which was signed by Canada at the 1992 United Nations Conference on Environment and Development in Rio de Janeiro, Brazil. We are morally bound to reorganize our priorities and to be willing to make economic sacrifices for the welfare of the flora and fauna of this province.

Measures of Endangeredness

Table 14-1 provides a list of B.C.'s endangered fish species. Determination of endangeredness is invariably arbitrary because inadequate data are available for most non-economic fishes. Information on localized genetic adaptations of populations reaching the limits of their known range in Canada is needed for complete analysis of the diversity of most of the province's fish species. In the absence of such data, I look for populations of an identifiable taxon in separate water drainages to get an idea of the status of some of the lesser known fishes.

Speciation in B.C.

British Columbia's glacial periods probably eliminated many endemic species. Following glaciation, fish populations reinvaded the province from five refugia (McPhail and Lindsey, 1986) and have morphologically differentiated over the last 13,000 years. Invasion of under-utilized habitats and niches continues today—though some of these invasions have occurred at an accelerated rate due to unplanned introductions of foreign fish species. Nevertheless, there are still examples of unique populations diverging into distinct taxa. Sticklebacks (*Gasterosteus*) found in Enos Lake, north of Nanaimo, are both the most noteworthy and the most threatened in the province. These fish have evolved into two distinct populations, the plankton-feeding and the benthic-feeding, which go through very different life cycles. Dr. J.D. McPhail (1984, and personal communi-

Species		Distribution	Status	Notes	Authority
Scientific Name	English Name				
Lampetra macrostoma	Vancouver lamprey	Cowichan & Mesachie lakes	Rare* Special Concern*	Limited distribution	Campbell 1987:166 Beamish 1982,1987 Williams et al 1989:3
Acipenser medirostris	green sturgeon	Anadromous, coastal waters, Fraser River	Rare*	Few published records	Campbell 1988:82 Houston 1988:286
Acipenser transmontanus	white sturgeon	Marine & freshwater Columbia & Fraser rivers	Unknown	Needs study	Campbell 1988:83
Sardinops sagax	Pacific sardine, pilchard	Coastal, marine	Rare*	Almost extirpated from BC	Campbell 1988:82 Schweigert 1988:296
Spirinchus thaleichthys	longfin smelt	Anadromous, land locked in Harrison & Pitt lakes, BC	Possibly rare	Harrison Lake population alleged to be unique	Campbell 1988:84
Coregonus sp.	Dragon Lake whitefish	Dragon Lake near Quesnel	Extinct	Eliminated during lake rehabilitation	Lindsey et al 1970
Coregonus artedi	cisco, lake herring	Maxahamish Lake, common outside BC	Rare in BC	Only known in one BC lake	Specimens in RBCM collections
Prosopium coulteri	pygmy whitefish	Giant race found in Maclure and McLeese lakes	Unknown	Alleged to be rare relict populations in BC	Campbell 1988:83
Catostomus sp.	Salish sucker	Lower Fraser Valley populations in Campbell and Fraser Rivers reduced or extinct.	Endangered; Special concern	Deterioration of restricted habitat in BC; taxonomy needs verification	McPhail 1980,1987 Campbell 1988:82 Williams et al 1989:5; Inglis & Rosenau 1993
Catostomus platyrhynchus	mountain sucker	Mountain streams of Fraser and Columbia rivers	Possibly rare	Alleged to be rare in BC	Campbell 1988:84
Couesius plumbius ssp.	Liard Hot Springs chub	Liard Hot Springs	Rare	Restricted to hot spring	Unusual specimens at RBCM

Table 14-1: Unique Fish Populations in British Columbia
The status is the author's suggestion except where an asterisk indicates approval by the Committee on the Status of Endangered Wildlife in Canada. "Threatened" has been reinterpreted as "vulnerable." More information is needed to clarify the status of many of the species on this list.
Key: RBCM = Royal British Columbia Museum

135

Species		Distribution	Status	Reason	Authority
Scientific Name	English Name				
Notropis atherinoides	emerald shiner	Fort Nelson River, common outside BC	Rare	Rarely sampled in BC	Scott and Crossman 1973:440
Notropis hudsonius	spottail shiner	Maxahamish Lake area, common outside BC	Not known	Survey needed of total range in BC	In RBCM collections
Pimephales promelas	fathead minnow	One-island Lake and Fraser River System, common outside BC	Not known	May have been introduced	In RBCM collections Smith & Lamb 1976:188
Rhinichthys sp.	Nooksack dace	Bertrand Creek, Nooksack River drainage	Endangered, Vulnerable	Urbanization of habitat; taxonomy needs verification	Campbell 1988:84 McPhail 1967 1980 Williams et al 1989:5
Rhinichthys falcatus	leopard dace	Fraser and Columbia region	Vulnerable in Columbia, common in Fraser River	Dams and/or eutrophication in Columbia drainage	Peden 1989 Campbell 1988:84
Rhinichthys umatilla	Umatilla dace	Columbia,Kettle,and Similkameen rivers	Rare; Vulnerable in Columbia	Proposed dam in Columbia; rare elsewhere	Peden and Hughes 1988 Campbell1988 Hughes & Peden 1985
Gasterosteus sp.	giant Mayer stickleback	Mayer Lake, Queen Charlotte Islands	Rare Vulnerable*	Confined to one lake; taxonomy needs verification	Moodie 1972 Campbell 1987:166 Williams et al 1989:11
Gasterosteus sp.	unarmed Boulton stickleback	Boulton Lake, Queen Charlotte Islands	Probably vulnerable; Special concern	Confined to one lake by highway	Campbell 1987:166 Moodie & Reimchen 1973 Williams et al 1989:11 Reimchen 1982
Gasterosteus sp.	black Drizzle stickleback	Drizzle Lake, Queen Charlotte Islands	Rare	Confined to one lake	Moodie & Reimchen 1973

Table 14-1: Unique Fish Populations in British Columbia (Continued)
Key: RBCM = Royal British Columbia Museum

Species		Distribution	Status	Reason	Authority
Scientific Name	English Name				
Gasterosteus sp.	Enos sticklebacks	Enos Lake, near Nanoose Bay	Two forms endangered, Vulnerable	Urbanization	McPhail 1984 Campbell 1988:83 Williams et al 1989:11
Gasterosteus sp.	Heisholt sticklebacks	Paxton, Priest, Emily, and Balkwill Lakes, Texada Island	Apparently rare	Confined to one lake	MacLean 1980 Campbell 1988:84
Pungitius pungitius	ninespine stickleback	Liard River area	Rare, common outside British Columbia	Only one British Columbia record	Scott & Crossman 1973
Asemichthys = Radulinus taylori	spinynose sculpin	Marine	Thought to be rare	Common subtidally	Peden observations Campbell 1988:85
Cottus aleuticus ssp.	Cultus Lake sculpin	Cultus Lake	Rare midwater population	Status needs study; taxonomy needs verification	Campbell 1988:85
Cottus bairdi ssp.	Flathead sculpin	Flathead River	Vulnerable*	Proposed coal mine; taxonomy needs verification	Peden & Hughes 1984 Campbell 1987:166 Peden 1991
Cottus confusus ssp.	shorthead sculpin	Kettle and Columbia rivers	Rare	Some affected by proposed dam; taxonomy needs verification	Peden & Clermont 1989 Peden 1991
Cottus bairdi	mottled sculpin	Kettle, Columbia & Similkameen rivers	Common in Similkameen; vulnerable to rare elsewhere	Limited range in Kettle; proposed & present dams in Columbia	Peden 1991
Ocella inpi	pixie poacher	Marine, Queen Charlotte Islands	Only one specimen	Taxonomy needs clarification	Peden's opinion Campbell 1988:85
Allolumpenus hypochromus	Y-prickleback	Marine	Possibly rare	Few observed	Campbell 1988:85

Table 14-1: Unique Fish Populations in British Columbia (Continued)
Key: RBCM = Royal British Columbia Museum

cation) argues that each group presently evolving into a plankton-feeding population is a more noteworthy evolutionary example than Darwin's famous examples from the Galapagos Islands, since this speciation is occurring now. Other examples of stickleback speciation in process are known in Paxton, Priest, Emily, and Balkwill Lakes on Texada Island, all the lakes on Lasqueti Island, and Drizzle and Mayer Lakes on the Queen Charlotte Islands (Moodie and Reimchen, 1976). According to McPhail, introductions of exotic pumpkin seed sunfish (*Lepomis gibossus*), have eliminated native sticklebacks in southeastern British Columbia and, consequently, we do not know how many other unique populations may have been destroyed.

Whitefishes (*Coregonus*), which are found across northern Canada, have also been evolving since glacial times. One example of speciation in process within British Columbia probably occurred at Dragon Lake near Quesnel, but management favouring sport fishes destroyed this example before it was properly studied (Lindsey et al. 1970).

Threats to Fish Diversity

Threats to fish diversity in British Columbia fall into three main, though somewhat overlapping, categories: fishery management, urbanization, and resource exploitation.

Fishery Management

Fishery management has tended to focus on preservation of the economically profitable species, to the detriment of the flora and fauna thought to lack economic value. Lake poisoning, which was the management action of choice at Dragon Lake, is one common way of eradicating unwanted species (Orchard, 1988). Ironically, often the unwanted species itself has been introduced to the lake. Carp (*Cyprinus*), goldfish (*Carassius*), catfish (*Ictalurus*), and the exotic pumpkin seeds that precipitated the elimination of native sticklebacks are some species that have been introduced and whose presence is generally judged to be deleterious. However, the toxicants traditionally used to eliminate introduced species can be devastating to indigenous fauna. If this practise is unavoidable, more effort should be allocated to rescuing and reintroducing stocks of indigenous species to their original habitat after lake poisoning.

Introductions can also be used as a fishery management technique. For example, the Fisheries Branch has planned to introduce Enos sticklebacks into remote, publicly inaccessible lakes (personal communication with J.D. McPhail, University of British Columbia). Very careful study is required to minimize displacement and/or predation of indigenous populations. Such introductions could be disastrous for Umatilla dace (*Rhinichthys umatilla*) and morphologically distinct shorthead sculpins (*Cottus confusus*), which are considered rare or threatened in the Columbia River. Since waterfalls can prevent upstream dispersal of unwanted species, rare species can be introduced above the falls to remove them from downstream threats, like alien species or dams. But there are always impacts on the local ecosystem. For instance, introduction of mottled sculpins or Umatilla dace above the falls in the Kettle River near Christina Lake could cause competitive displacement of local slimy sculpins (*Cottus cognatus*) *or* the rare speckled dace (*Rhinichthys osculus*) (Peden and Hughes, 1981a and 1981b; Peden and Clermont, 1989; Peden et al., 1989).

Hybridization is a further potential effect of using introductions as a management technique. It is well known that introductions of foreign stocks, whether as bait fish or through hatchery programs, can cause indigenous stocks to lose their genetic distinctiveness. For example, as a consequence of introductions, identification of the natural genetic stocks of golden trout. Gila trout and Rio Grande trout in the mountains of the American Southwest have required great effort.

Urbanization

Urbanization still seems to be synonymous with habitat degradation. Among the effects of development are pollutants in the streams or water table, clearing of the vegetation which shades the water, and removal of protective aquatic weeds from the shallows. Consequently, fish are poisoned, overheated, or eaten in greater numbers than ever before. The insidious impacts of urbanization are compounded by the fact that it tends to concentrate in lowland areas—the same areas critical to the survival of some of the unique fishes having the most restricted habitat in British Columbia. McPhail (1967 and 1980) and McPhail and Lindsey (1986) have described unique populations which are related to longnose dace (*Rhinichthys cataractae*) and northern sucker (*Catostomus catostomus*) in drainages south of the Lower Fraser River, where they are threatened by, among other things, removal of shore vegetation and siltation from gravel pits. Enos sticklebacks are threatened by a major retirement housing development designed to appeal to those who seek the luxuries of golfing, yachting and a semi-rural lifestyle.

Resource Exploitation

There is not enough space in this short article to discuss the impacts of logging, mining and other forms of resource exploitation on river basins and the fish within their waters, but the effects of hydroelectric dams on the highly modified Columbia River Basin deserve special comment (Peden and Clermont, 1989; Peden et al., 1989). Only a few vestiges of natural, large river habitat remain in the main stem of the Columbia River. Various storage dams, which raise river water levels at times when they are naturally low, affect water temperature and sedimentation and eliminate the river currents required by river-adapted species. Consequently, conservation of several fishes, such as speckled dace, Umatilla dace, leopard dace (*Rhinichthys falcatus*), shorthead sculpins and, possibly, mottled sculpins, found within the basin has become a matter of some concern. Some of these species are recent natural immigrants to the province and have not dispersed past natural barriers to other river basins (McPhail and Lindsey, 1986).

Minimizing the Threat: the Ecosystem Approach

Preservation of the diversity of fishes requires an ecosystem approach, which encompasses all organisms within, for example, a river basin like the Columbia. Presumably, all the species in an ecosystem will have evolved together, developing genetic traits unique to that region. Management plans designed for the entire biota in an ecosystem are, therefore, far more effective and ecologically sound than those designed to support only one or two economically profitable species. In this regard, I am pleased to see the many recent ecosystem-based government and citizens' initiatives.

Acknowledgements

Dr. J.D. McPhail, University of British Columbia, reviewed the manuscript.

References

Beamish, R.J. 1982. *Lampetra macrostoma*, a new species of freshwater parasitic lamprey from the west coast of Canada. *Canadian Journal of Fisheries and Aquatic Sciences* 39(5):736-747.

Beamish, R.J. 1987. Status of lake lamprey, *Lampetra macrostoma*, in Canada. *Canadian Field-Naturalist* 101(2):186-189.

Campbell, R.R. 1987. Rare and endangered fishes and marine mammals of Canada: COSEWIC fish and marine mammal subcommittee status reports: III. *Canadian Field-Naturalist* 101(2):165-309.

Campbell, R.R. 1988. Rare and endangered fishes and marine mammals of Canada: COSEWIC fish and marine mammal subcommittee status reports: IC. *Canadian Field-Naturalist* 102(1):81-86.

Deacon, J.E., G. Kebetich, J.D. Williams, S. Contreras, and other members (A.E. Peden) of the Endangered Species Committee of the American Fisheries Society. 1979. Fishes of North America endangered, threatened, or of special concern: 1979. *Fisheries* 4:29-44.

Goodman, M.L. 1990. Preserving the genetic diversity of salmonid stocks: a call for federal regulation of hatchery programs. *Environmental Law* 20(111):110-165.

Houston, J.J. 1988. Status of the green sturgeon, *Acipenser medirostris*, in Canada. *Canadian Field-Naturalist* 102(2):286-290.

Hughes, G.W. and A.E. Peden. 1985. *The Canadian Status of the Umatilla dace, Rhinichthys Umatilla (Pisces: Cyprinidae)*. Unpublished report submitted to COSEWIC. December, 1985.

Inglis, S.D. and M.L. Rosenau. 1993. *The Distribution, Abundance and Habitat Requirements of Rare and Endangered Fishes in an Expanding Urban Setting*. Watchable Wildlife and Nongame Symposium. Oct. 13-15, 1993. Ministry of Environment. Victoria, B.C.

Lindsey, C.C., J.W. Clayton and W.C. Franzin. 1970. Zoological problems and protein variation in the *Coregonus clupeaformis* whitefish species complex. In: *Biology of Coregonid Fishes*. C.C. Lindsey and C.S. Woods (eds.). University of Manitoba Press, Winnipeg. pp. 127-146.

MacLean, J. 1980. Ecological genetics of threespine sticklebacks in Heisholt Lake. *Canadian Journal of Zoology* 58:2026-2039.

McPhail, J.D. 1967. Distribution of freshwater fishes in western Washington. *Northwest Science* 41(1):1-10.

McPhail, J.D. 1980. Distribution and status of freshwater fishes in British Columbia. In: *Threatened and Endangered Species and Habitats in British Columbia and the Yukon*. R. Stace-Smith, L. Johns and P. Joslin (eds.). 1980. Proceedings of the Symposium co-sponsored by the Federation of British Columbia Naturalists;

Institute of Environmental Studies, Douglas College; British Columbia Ministry of Environment, Fish and Wildlife Branch. Richmond, British Columbia March 8-9, 1980.

McPhail, J.D. 1984. Ecology and evolution of sympatric sticklebacks (*Gasterosteus*): morphological and genetic evidence for a species pair in Enos Lake, British Columbia. *Canadian Journal of Zoology* 62(7):1402-1408.

McPhail, J.D. 1987. Status of the Salish sucker, *Catostomus sp.*, in Canada. *Canadian Field Naturalist* 101(2):231 - 236.

McPhail, J.D. and C.C. Lindsey. 1986. Zoogeography of freshwater fishes of Cascadia (the Columbia System and rivers north to the Stikine). In: *The Zoogeography of North American Freshwater Fishes*. C.H. Hocutt and E.O. Wiley (eds.). John Wiley and Sons, Toronto, pp. 615-636.

Moodie, G.E.E. 1972. Morphology, life history, and ecology of an unusual stickleback (*Gasterosteus aculeatus*) in the Queen Charlotte Islands, Canada. *Canadian Journal of Zoology* 50(6):721-732.

Moodie, G.E.E. and T.E. Reimchen. 1973. Endemism and conservation of sticklebacks in the Queen Charlotte Islands. *Canadian Field-Naturalist* 87(2):173-175.

Moodie, G.E.E. and T.E. Reimchen. 1976. Glacial refugia, endemism and stickleback populations of the Queen Charlotte Islands, British Columbia. *Canadian Field-Naturalist* 90:471-474.

Nehlson, W., J.E. Williams and J.A. Lichatowich. 1991. Pacific salmon at the crossroads: stocks at risk from California, Oregon Idaho and Washington. *Fisheries* 16(2): 4-24.

Orchard, S.A. 1988. Lake poisoning: its effects on amphibians and reptiles. *BioLine* 7(2):12-13.

Peden, A.E. 1989. *Status of Leopard dace, Rhinichthys falcatus, in Canada*. Draft report submitted to COSEWIC, Subcommittee on Fish and Marine Mammals.

Peden, A.E. 1991. *Updated report on Status of Shorthead Sculpin (Cottus confusus) in Canada*. Unpublished report submitted to COSEWIC: Subcommittee and Fish and Marine Mammals, Dec. 1991.

Peden, A.E. and T. Clermont. 1989. *Updated report on Status of Shorthead Sculpin (Cottus confusus) in Canada*. Unpublished report submitted to COSEWIC: Subcommittee and Fish and Marine Mammals, June 1989.

Peden, A.E. and G.W. Hughes. 1981a. Life history notes relevant to the Canadian status of the speckled dace (*Rhinichthys osculus*). *Syesis* 14:21-31.

Peden, A.E. and G.W. Hughes. 1981b. *Status of the Speckled Dace, Rhinichthys osculus, in Canada during 1980*. Unpublished report submitted to COSEWIC.

Peden, A.E. and G.W. Hughes. 1984. Status of the shorthead sculpin, *Cottus confusus*, in the Flathead River, British Columbia. *Canadian Field-Naturalist* 98:127-133.

Peden, A.E. and G.W. Hughes. 1988. Sympatry in four species of *Rhinichthys* (Pisces), including first known occurrences of *R. Umatilla* in the Canadian drainages of the Columbia River. *Canadian Journal of Zoology* 66:1846-1856.

Peden, A.E., G.W. Hughes and W.E. Roberts. 1989. Morphologically distinct populations of the shorthead sculpin, *Cottus confusus*, and mottled sculpin, *Cottus bairdi* (Pisces, Cottidae) near the western border of Canada and the United States. *Canadian Journal of Zoology* 67:2711-2720.

Peden, A.E. and T. Clermont. 1989. *Updated report on Status of Shorthead Sculpin (Cottus confusus) in Canada*. Unpublished report submitted to COSEWIC: Subcommittee and Fish and Marine Mammals. June 1989.

Reimchen, T.E. 1982. *Status Report on Unarmored and Spine-Deficient Populations of Threespine Stickleback (Gasterosteus aculeatus) on the Queen Charlotte Islands, British Columbia*. Unpublished report submitted to COSEWIC.

Schweigert, J.F. 1988. Status of the Pacific Sardine, *Sardinops sagax*, in Canada. *Canadian Field-Naturalist* 102(2):296-303.

Scott, W.B. and E.J. Crossman. 1973. Freshwater Fishes of Canada. *Bulletin of the Fisheries Research Board of Canada* 184-1-966.

Smith, J.F. and A. Lamb. 1976. Fathead minnows (*Pimephales promelas*) in northeastern British Columbia. *Canadian Field-Naturalist* 90(2):188.

Walters, C.J. 1988. Mixed stock fisheries and the sustainability of enhancement production for chinook and coho salmon. In: *Proceedings of the World Salmon Conference*. W.T. McNeill (ed.). Oregon State University Press.

Williams, J.E., J.E. Johnson, D.A. Hendrickson, S. Contreras-Balderas, J.D. Williams, M. Navarro-Mendoza, D.E. McAllister and J.E. Deacon. 1989. Fishes of North America endangered, threatened, or of special concern: 1989. *Fisheries* 14(6):2-20.

Chapter 15
Endangered Mammals in British Columbia[1]

For a northern temperate region, British Columbia supports a remarkably rich mammalian fauna. The province is home to 105 native terrestrial and 29 native marine species (Nagorsen, 1990). No other Canadian province or territory has such a diverse and varied mammalian fauna. In comparison to more populated regions in eastern Canada, British Columbia's mammals survived European settlement relatively intact. For example, the province still supports large populations of caribou *(Rangifer tarandus)*, elk *(Cervus elaphus)*, grizzly bear *(Ursus arctos)*, cougar *(Felis concolor)*, and wolverine *(Gulo gulo)*, species that have long disappeared from some eastern provinces. Nonetheless, the sea otter *(Enhydra lutris)*, bison *(Bison bison)*, and white-tailed jackrabbit *(Lepus townsendii)* were extirpated from the province and several marine mammals were exploited to near extinction by the late 19th century. Moreover, there is growing concern about habitat loss, particularly in populated southern areas of British Columbia, and its impact on mammalian diversity.

David Nagorsen
Royal British Columbia Museum
675 Belleville Street
Victoria, B.C.
V8V 1X4

Designation of Endangered Mammals

The various systems for categorizing species at risk and assigning them to appropriate categories are described in Chapter 2. Such lists are useful for setting priorities and promoting awareness among governmental agencies and the general public. The Committee on the Status of Endangered Wildlife in Canada (COSEWIC) has now prepared status reports for a number of the province's marine mammals, ungulates, and carnivores. However, the small mammals have received less attention. Management of endangered mammals is under the jurisdiction of federal and provincial agencies, with the federal Department of Fisheries and Oceans (DFO) taking responsibility for marine mammals and the provincial Ministry of Environment managing all terrestrial species. For terrestrial mammals, a committee comprised of directors from various federal and provincial wildlife agencies (RENEW) is responsible for appointing a recovery team for each species listed as endangered or threatened by COSEWIC. The recovery team's task is to develop a technical plan to rehabilitate the species.

Since the 1970's, the Wildlife Branch of the provincial Ministry of Environment has also taken an active role in designating endangered mammals that are under provincial jurisdiction. In addition to the Red and Blue listings discussed in Chapter 2, species can be

---

[1] Originally published in the Spring 1990 (Volume 9, Number 2) issue of *BioLine*, official publication of the Association of Professional Biologists of British Columbia. Revised and updated for this publication.

declared threatened or endangered under the provincial Wildlife Act. The focus of the Wildlife Branch is provincial, but there is close coordination with COSEWIC and the RENEW strategy.

There are problems with any attempt to designate threatened or endangered mammals. All classification systems for categorizing species at risk have a degree of subjectiveness. The definition of the threatened, endangered, and vulnerable categories of COSEWIC is somewhat arbitrary and assigning a species to one of these may be a judgement call. Although the new ranking system developed by the Nature Conservancy and the B.C. Conservation Data Centre is more objective than the COSEWIC system, it also has its problems. Because of sampling bias, the number of occurrences may be a poor indicator of population size or rarity for some mammals. Also, some mammals are naturally rare in communities and rarity per se is not always an indicator of extinction risk. A promising approach recently developed for the International Union for the Conservation of Nature (IUCN) uses threat categories that are based on extinction probabilities derived from population biology (Mace and Lande, 1991).

Another difficulty is determining the taxonomic level at which mammals are designated. To maintain biological diversity, we should be concerned about protecting subspecies and geographically isolated populations as well as species. Because of its diverse topography and environments, British Columbia supports a rich assemblage of mammalian subspecies. Some 238 subspecies (or races) are currently recognized for the 105 terrestrial mammals; 61% of these races occur nowhere else in Canada and 21% are endemic to the province (Nagorsen, 1990). The validity of many of these subspecies is questionable because they were described decades ago, often from small samples and rather subjective methodology. Nonetheless, they are a rough indicator of phenotypic (physical appearance) and genetic variation at the population level. Maintaining this variation is an important dimension of preserving biodiversity in the province. COSEWIC has adopted a liberal definition of species that includes "species, subspecies, or geographically separate populations." However, mammals designated by COSEWIC at the population level have only included cetaceans, ungulates, and carnivores; threatened populations of small mammals have not been addressed. Subspecies, however, are now included in the Wildlife Branch's revised Red and Blue lists. Rare, isolated and insular subspecies such as the Queen Charlotte Islands' ermine *(Mustela erminea haidarum)*, Vancouver Island wolverine *(Gulo gulo vancouverensis)*, Vancouver Island water shrew *(Sorex palustris brooksi)*, and races now isolated by habitat loss such as the Lower Mainland population of southern red-backed vole, *(Clethrionomys gapperi occidentalis)* are obvious potential candidates for threatened or endangered status.

The issue of how to deal with peripheral mammals has also been contentious. From a biogeographical perspective, British Columbia occupies a unique location straddling five terrestrial biomes in western North America. A number of terrestrial mammals reach the limits of their geographic ranges in the province. These peripheral species often occur in small, isolated populations, making them particularly vulnerable to local extinction. Because they may be common elsewhere and their populations would be expected to be unstable in marginal areas, some biologists have argued that peripheral species should be excluded from endangered status. The majority of mammals now under review for provincial designation could all be conveniently dismissed as peripheral. Yet, to ignore these species would be an abdication of our responsibility to conserve populations. Many of these peripheral mammals occur nowhere else in Canada. Furthermore, because these populations have adapted to marginal conditions, they may be differentiated genetically from populations that live at the core of a species range.

Despite the utility of threatened and endangered species lists, there is a risk that they can become an end in themselves. Bureaucratic quibbles (see O'Brien and Mayr, 1991, for examples) over appropriate designations for species can divert attention from more urgent conservation issues. Governments may be tempted to use these lists as a stalling tactic to avoid direct action. There is a general perception among the public that if a species is "officially" designated, it is somehow safe and protected. The COSEWIC designations, however, carry no legal status and a jurisdiction could choose to ignore the recommendations of COSEWIC. The provincial Red List comprises species that are "candidates for designation as endangered or threatened." Yet, to date, no mammalian species on the Red List has been "uplisted" and legislated by the province as threatened or endangered.

Who is Endangered and Why

Three of the province's mammals, the sea otter, right whale *(Balaena glacialis)*, and Vancouver Island marmot *(Marmota vancouverensis)*, are officially designated as endangered. Two species, humpback whale *(Megaptera novaeanglia)* and wood bison *(Bison bison athabascae)*, are listed as threatened by COSEWIC. The revised provincial Red and Blue Lists include 49 other mammalian taxa. There are many causes for mammalian extinctions and population declines (Hayes, 1991), but exploitation by humans and habitat loss probably have had the greatest impact on British Columbian mammals. Large mammals and some of the commercially valuable furbearers have been most affected by over harvesting in historical time. With their protection and regulation under various wildlife regulations, this threat has been largely removed, although a few species (such as bison and baleen whales) have never recovered from these historical declines. Except for bats, where large numbers are occasionally destroyed in human dwellings, the smaller mammals have generally not been impacted by direct killing and their most immediate threat is habitat loss.

Marine Mammals

The history of marine mammals in the north Pacific follows a common pattern: heavy exploitation in historical times, severe reductions in populations, protection, and the subsequent recovery of some populations. Although most marine mammals are no longer threatened by over-killing, they face a number of potential threats, including incidental captures in fishing gear, marine pollution, disturbance on their breeding grounds, and depletion of their prey species.

Many of the large baleen whales were seriously depleted by pelagic (open ocean) and coastal whaling. The eastern north Pacific stocks of large whales are now protected by national and international bans on commercial whaling and some species show evidence of recovery. The right whale and humpback whale, however, are still at risk. A slow swimmer, the right whale was easily exploited by early whalers and it was hunted to near extinction. According to Gaskin (1991), the north Pacific population is estimated at no more than 120 animals and, despite its protection since the 1940's, it has demonstrated no evidence of recovery. Because it migrates in coastal waters, the humpback whale was also vulnerable to whaling. The eastern north Pacific population of about 2,000 represents only a small proportion of its pre-whaling numbers (Whitehead, 1987). The smaller cetaceans escaped intensive commercial hunting. Except for the killer whale *(Orcinus orca)*, however, we have no idea of the number of distinct stocks or populations, let alone reliable estimates of their numbers or population trends.

The sea otter is another marine mammal depleted by overhunting. It has generally been assumed that it was extirpated from the British Columbia coast by the 1920's, although a small colony discovered in 1990 near the Hunter Islands may be a relict population that escaped detection. The reintroduction of the sea otter to the west coast of Vancouver Island is a conservation success story. The greatest threat to this species is a catastrophic oil spill, which could destroy the entire Vancouver Island population. Establishing additional populations on other parts of the British Columbian coast, such as the Queen Charlotte Islands, should be a high priority.

Because they can be studied at their land-based breeding rookeries, we have more reliable data on long term population trends for pinnipeds (seals and sea lions) than other marine mammals. The pinnipeds were also heavily exploited but, with national and international protection, their populations in the eastern north Pacific recovered from low numbers at the turn of the century. In most of the eastern north Pacific, populations of the northern sea lion *(Eumetopias jubatus)* are stable and numbers of California sea lions *(Zalophus californianus)*, northern elephant seals *(Mirounga angustirostris)*, and harbour seals *(Phoca vitulina)* are increasing. Nonetheless, in the past decade pinniped populations in the Bering Sea and Alaska's Aleutian Archipelago have undergone disturbing declines that may be related to food supply (Trites, 1992). Declines in Alaskan populations resulted in the northern sea lion being listed as threatened under the U.S. Endangered Species Act in 1990.

Terrestrial Mammals

British Columbia's only endangered terrestrial mammal is the Vancouver Island marmot. It is considered vulnerable because the population, estimated at 200 to 400, is confined to a small area (Figure 15-1) on south-central Vancouver Island (Nagorsen, 1985;

146

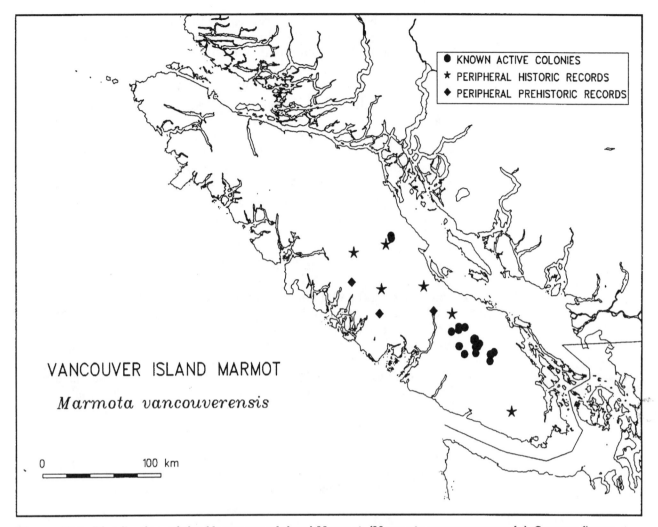

Figure 15-1. Distribution of the Vancouver Island Marmot *(Marmota vancouverensis).* Source: figure provided by author.

Janz et al., 1993). Bones recovered from prehistoric archaeological sites and historical records suggest that this marmot may have been more widespread on Vancouver Island in the past. Nonetheless, this species has not been thoroughly inventoried and there is an urgent need for systematic surveys of all potential habitats on Vancouver Island to determine the distribution and numbers of extant colonies. Attempts to develop a recovery plan are complicated by a lack of data and clear evidence of any human threat. The most contentious issue is the impact of high elevation logging, a relatively recent phenomenon on Vancouver Island. Marmots are living in logging slash, they are breeding successfully there, and at some logged sites have increased in numbers. Debates about overwinter survival and dispersal in logged habitats cannot be resolved until detailed field studies are done.

The wood bison disappeared from northern British Columbia by the late 1800's. Recent sightings in the northeast are probably from the introduced herds in Alberta and the Northwest Territories. Most of the conservation effort (Wood Bison Recovery Team, 1987) has focused on maintaining the integrity of this subspecies and preventing hybridization with the Plains bison race *(Bison bison*

bison). This issue is particularly relevant to British Columbia because the population of the Pink Mountain herd of Plains bison is growing. Although it is justifiable to conserve the taxonomic diversity of bison, I wonder if as much effort would be put into conserving a threatened subspecies of a small furbearer or shrew.

Besides these mammals officially recognized as threatened or endangered, 49 other mammalian taxa are being considered for the Wildlife Branch's Red and Blue Lists. Many of these mammals have not been designated by COSEWIC. The mammals listed are diverse, ranging from caribou and grizzly bears to shrews, but are dominated by rodents, bats, and insectivores. Most of the small mammals of concern represent two faunal groups: a Great Basin group that extends into the grasslands of the southern Interior and a Pacific coast group that occupies the lower Fraser Valley and Cascades. The Great Basin mammals, such as white-tailed jackrabbit, Nuttall's cottontail (Sylvilagus nuttalli), Great Basin pocket mouse (Perognathus parvus), western harvest mouse (Reithrodontomys megalotis), and pallid bat (Antrozous pallidus) are typically associated with arid grassland and shrub steppe habitat. In the Okanagan Valley, irrigated orchards and vineyards and spreading urban centres, such as Kelowna, have destroyed much of this original habitat. Among the Pacific coast fauna, species such as Townsend's mole (Scapanus townsendii), Trowbridge's shrew (Sorex trowbridgii), and the Pacific marsh shrew (Sorex bendirii) that are restricted to low elevation habitats in the Fraser Valley are especially vulnerable to habitat loss through agriculture, forestry, and the urban sprawl of Vancouver. Although hunting pressure, natural predation, and human activities have all reduced populations of ungulates and large carnivores, habitat loss and land use are the most important issues for wildlife biologists managing these species.

What is to be Done

British Columbia's marine mammals present some unique problems. Many are migratory and their movements traverse international and various national waters and their breeding areas may be beyond Canadian jurisdiction. More research certainly could be done on some of the Canadian populations of dolphins and porpoises to determine local movements and population structure. The recent unexplained declines in Alaskan pinniped populations demonstrates the need for continued monitoring of population trends, even for species that appear stable, and more research on changes in food abundance. Marine pollution and over-fishing of prey species are global environmental issues; clearly the conservation of these mammals has to be a cooperative international effort.

Terrestrial mammals now classified as threatened or endangered by COSEWIC are being addressed by the RENEW strategy. Recovery teams have been appointed for the wood bison and Vancouver Island marmot and recovery plans are either completed or in progress. The plans will identify threats, areas for concentrating research and conservation effort, and assist in soliciting funds from sources such as the World Wildlife Fund.

Other mammals of concern will have to be dealt with largely by the provincial Wildlife Branch. Their proposed Red and Blue Lists comprise an ambitious scheme that will require an intensive cooperative research effort among various governmental agencies and universities. Much of the effort should be directed toward the small mammals. The distribution, population numbers and trends, and habitat requirements of ungulates and large carnivores in the province are relatively well known, and entire workshops have been dedicated to single species (for an example, see Page, 1988). I suspect that enough is known to at least develop management plans for most of these species. In contrast, even basic inventory data are lacking for many of the small mammal species. Our information on Townsend's mole in British Columbia, for example, comes largely from museum specimens collected nearly half a century ago.

Although detailed ecological studies of these species would be invaluable, habitat is disappearing quickly and there may not be enough time to carry out single species studies for every mammal on the Red and Blue Lists. An endangered spaces or ecosystem approach would be most effective. I suggest that the lower Fraser Valley and Okanagan Valley are two regions of high priority for field work and research. They support the highest diversity of mammalian species, the greatest number of species of concern (that is, those on the Red or Blue List), and they are the areas of greatest human population growth and habitat destruction. Intensive inventories would yield invaluable data on distributions and habitat requirements. For some species, it may be possible to obtain ecological and genetic data to determine minimum viable population and area sizes (Koenig, 1988). Precise distribution maps should be generated from recent and historical records. Overlaying habitat maps, forestry and urban development plans, and park and ecoreserve boundaries would assist in predicting range declines, habitat fragmentation, and potential areas for protection. As an example, results from a three year study on the western harvest mouse that I have just completed (Figure 15-2), indicate that the northern Okanagan populations of this mouse may be isolated from populations in the southern Okanagan by habitat loss in the Kelowna area. The northern populations are obviously vulnerable to extinction. They occupy a relatively small area outside Vernon, where the native grassland habitats are disappearing due to development. Other mammals, such as the Great Basin pocket mouse, may demonstrate a similar distribution pattern.

In the Okanagan, a start has been made with the South Okanagan Critical Areas Program (SOCAP) (Hlady, 1990). Such work, though, has to be broadened to include the entire Okanagan and possibly adjacent grasslands in the Similkameen, Thompson, and Fraser river valleys. The Lower Mainland area, which supports as many threatened mammals as the Okanagan, certainly has received less attention and it too, could benefit from a SOCAP approach.

Finally, as a museum biologist, I think it is appropriate to conclude with a comment on systematics and taxonomy, which has been discussed in Chapter 3. Expanding the provincial Red and Blue Lists to include endangered subspecies and populations underscores the important role of systematics in preserving biodiversity

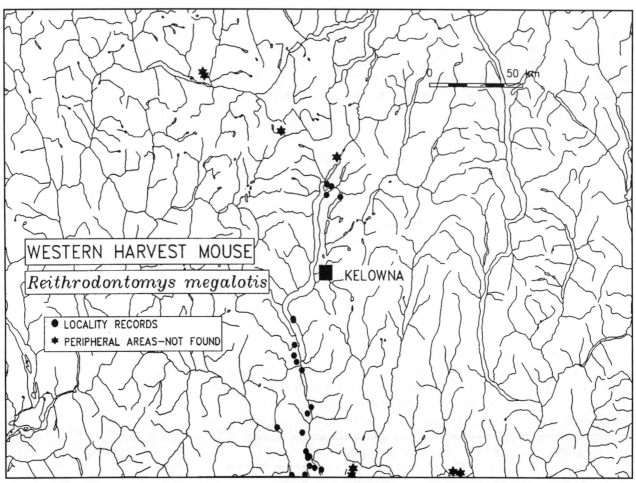

Figure 15-2. Distribution of the Western Harvest Mouse *(Reithrodontomys megalotis)* **in British Columbia.** Records of occurrence are based on all known historical museum records and surveys from 1990-1992. Source: figure provided by author.

and endangered species. The taxonomy of most British Columbian mammals is in need of modern research. Even "high profile" forms such as the kermode and glacier bear races of black bear *(Ursus americanus)*, are based on outdated taxonomic studies done half a century ago. Powerful tools, including multivariate statistical analyses of traditional morphological traits, chromosomes, and molecular genetics (proteins, DNA) are available to the mammalian systematist. These techniques could be profitably applied to a number of marine and terrestrial species to delimit the patterns of geographic variation and identify unique races or populations in the province. Unfortunately, there is only a handful of mammalian systematists in all of Canada and none are on staff at any of the British Columbia universities. Museums, the traditional centres for systematic research, struggle with tight budgets and limited resources. This situation will have to be rectified if we hope to document and maintain genetic diversity.

Acknowledgements

Dr. Art Martell, Canadian Wildlife Service, reviewed the manuscript.

References

Gaskin, D.E. 1991. An update on the Right Whale, *Eubalaena glacialis*, in Canada. *Canadian Field-Naturalist* 105(2):198-205.

Hlady, D.A. 1990. *South Okanagan Conservation Strategy 1990-1995.* Unpublished report, B.C. Ministry of Environment.

Hayes, J.P. 1991. How mammals become endangered. *Wilderness Society Bulletin* 19:210-215.

Janz, D., J.C. Blumensat, N.K. Dawe, B. Harper, S. Leigh-Spencer, W.T. Munro and D.W. Nagorsen. *1993 National Recovery Plan for the Vancouver Island Marmot.* Final draft submitted to RENEW for final approval.

Koenig, W.D. 1988. On determination of viable population size in birds and mammals. *Wilderness Society Bulletin* 16:230-234.

Mace, G.M. and R. Lande. 1991. Assessing extinction threats: towards a reevaluation of the IUCN threatened species categories. *Conservation Biology* 5(2)148-157.

Nagorsen, D.W. 1990. The mammals of British Columbia. A taxonomic catalogue. *Royal British Columbia Museum Memoir* No. 4:1-140.

Nagorsen, D.W. 1985. *Marmota vancouverensis. Mammalian Species* 270:1-5.

O'Brien, S.J. and E. Mayr. 1991. Bureaucratic mischief: recognizing endangered species and subspecies. *Science* 251:1187-1188.

Page, R. 1988. (ed.) *Caribou Research and Management in British Columbia: Proceedings of a Workshop.* B.C. Ministry of Forests, Research Branch WHR-27.

Trites, A.W. 1992. Northern fur seals: why they have declined. *Aquatic Mammals* 18 (1):3-18.

Whitehead, H. 1987. Updated status of the Humpback Whale, *Megaptera novaeangliae,* in Canada. *Canadian Field-Naturalist* 101(2):284-294.

Wood Bison Recovery Team 1987. *Status report on Wood Bison (Bison bison athabascae) in Canada, 1987.* Unpublished COSEWIC Status report.

Chapter 16
Threatened and Endangered Birds in British Columbia[1]

British Columbia's diversity of topography, climate, and habitats is admirably reflected in its avifauna. Our province is home to 452 bird species, 297 of which breed here (Campbell, 1989; Campbell et al., 1990), representing about three-quarters of the total species in Canada on both counts and by far the highest proportion in any one province (Bunnell and Williams, 1980). British Columbia is, therefore, entrusted with a remarkably diverse list of bird species, and the populations of several of these species are of great concern at the moment. Unfortunately, we know very little about the bird populations of our province, especially those of forest birds, which are obviously being affected by our number one industry, forestry. Two species have already disappeared from British Columbia since the turn of the century: Sage Grouse (*Centrocercus urophasianus*), a species associated with large tracts of sagebrush in the Okanagan Valley, and Yellow-billed Cuckoo (*Coccyzus americanus*), a species formerly found in dense riparian thickets on the southern coast.

Several lists have been devised outlining the bird species whose populations are threatened or endangered in British Columbia (such as Weber, 1980) and Canada (such as Munro, 1990). As well, the British Columbia Wildlife Branch has noted such species in its Red and Blue lists (Munro, 1993).

As is the case with most animal groups around the world, the major threat to bird populations in British Columbia is loss of habitat. Four areas of special concern in British Columbia are wetlands, grasslands, riparian woodland, and old growth forests.

Richard Cannings
Department of Zoology, University of British Columbia
6270 University Boulevard
Vancouver, B.C.
V6T 1Z4

Wetlands

As many of the country's wetlands are drained or filled, birds dependent on those marshes, swamps, and estuaries inevitably decline as well. An important wetland in British Columbia in terms of bird populations is the Fraser delta, which faces an onslaught of development proposals, including golf courses, marinas, and housing. The delta, especially the Boundary Bay area, is an essential site for migratory and wintering waterfowl, raptors and shorebirds—essentially the entire world population of Western Sandpipers (*Calidris mauri*) use it twice a year as a refuelling stop between the central United States and Alaska (Butler and Campbell, 1987; Butler and Cannings, 1989). Although most of these populations seem to be stable right now, continued loss of habitat in the delta would result in significant population declines of most species—there simply is nothing like this delta within flying distance.

[1] Revised and updated. Original published in *BioLine*, official publication of the Association of Professional Biologists of British Columbia.

One of the two birds classified as endangered by the Wildlife Branch is the American White Pelican (*Pelecanus erythrorhynchos*), a species which feeds on fish in the shallow waters of large marshy lakes. The only colony in the province is at Stum Lake (White Pelican Provincial Park) in the Chilcotin, and usually numbers about 150 pairs, though over 400 pairs were present in 1993 (personal communication in August, 1993, with R.W. Campbell, B.C. Wildlife Branch) The overall prognosis for White Pelicans is quite good; the North American population is expanding, and the species has been recently delisted by the Committee on the Status of Endangered Wildlife in Canada (COSEWIC).

The interior race of the Peregrine Falcon (*Falco peregrinus anatum*) is listed as endangered by COSEWIC and is on the provincial Red list. This falcon is largely dependent on small waterfowl and shorebirds and, thus, its welfare is closely tied to that of wetlands. The British Columbian interior population of Peregrine Falcons is very low, probably around a dozen pairs (Hodson, 1980), slowly recovering from persecution in the first half of the century and pesticides in the second half. There are indications that some peregrines are now taking large numbers of starlings and pigeons, thereby adapting somewhat to the increasing urbanization of British Columbia (Hodson, 1980).

The Forster's Tern *(Sterna forsteri)* has established itself as a breeding species in British Columbia recently at Creston (Goossen et al., 1982). It is on the Red list due to the vulnerability of its single provincial colony, but is otherwise not threatened. The Caspian Tern (*Sterna caspia*) is considered vulnerable by COSEWIC, but is expanding its range in Washington State and has recently been suspected of breeding on the Roberts Bank jetty on the Fraser delta (Campbell et al., 1990). The extensive marshes at Creston are also home to a major colony of Western Grebes (*Aechmophorus occidentalis*), one of only three colonies in British Columbia. This species requires large lakes with marshy margins; other breeding sites in British Columbia are the north end of Okanagan Lake, and Shuswap Lake at Salmon Arm. A colony at Williams Lake was apparently abandoned in the 1970s. Recently, one or two pairs even nested in the Westham Island marshes on the Fraser delta. The Purple Martin (*Progne subis*), famous for its apartment-style nestboxes in eastern North America, has declined drastically on the west coast. The British Columbian population now numbers perhaps only 30 pairs, most found in estuarine habitats in southeastern Vancouver Island.

Grasslands

The dry intermontane grasslands of British Columbia have been decimated in the last hundred years—most were overgrazed by the turn of the century and since then hundreds of hectares have been turned into alfalfa, orchards, grapes, or houses. The grasslands of the Okanagan Valley, particularly interesting in terms of bird populations, have been very hard hit. The Burrowing Owl (*Speotyto cunicularia*), once a common sight in the Thompson-Okanagan, was becoming rare by the turn of the century (Cannings et al., 1987); by the early sixties only a few pairs were left, and by

1972 the species had essentially disappeared from the province as a breeding species. The Wildlife Branch has carried out an apparently successful reintroduction programme at the north end of Osoyoos Lake in recent years, and about 10 pairs of Burrowing Owls have returned each spring from their southern wintering grounds to nest in burrows made of plastic drainage pipes. In 1993, however, only two pairs were observed returning to nest. Careful monitoring and reintroductions of captive-bred birds will hopefully continue until other sites in the Southern Interior have been repopulated by this fascinating species. Burrowing Owls are considered threatened by COSEWIC and endangered by the Wildlife Branch.

The Prairie Falcon (*Falco mexicanus*) is strongly associated with dry grasslands in southern British Columbia, nesting in spring on rock cliffs at lower elevations and often moving to alpine grasslands in late summer. Populations across the western provinces have apparently declined by 35% or more (Fyfe et al., 1969). This species is on the provincial Red list, but at present is not listed by COSEWIC. The Ferruginous Hawk (*Buteo regalis*), a large grassland Hawk specializing in ground squirrels as prey, is very rare as a breeding species in British Columbia. It is considered threatened by COSEWIC and is on the Wildlife Branch Red list.

The Sage Thrasher (*Oreoscoptes montanus*), as its name suggests, is restricted to sagebrush (*Artemisia tridentata*) grasslands, and is found regularly in Canada only in the south Okanagan. Local numbers fluctuate widely from year to year, perhaps influenced largely by breeding conditions and populations in the Columbia Basin. It prefers stands of large sagebrush, and is quickly affected by range management techniques such as mowing and burning designed to reduce sagebrush cover in favour of grasses. Only a dozen or so pairs are present each year (often at only one site) and its habitat is under constant threat (Cannings, 1992). The Sage Thrasher is considered endangered in Canada by COSEWIC and is on the provincial Red list. Another species closely tied to sagebrush is the Brewer's Sparrow (*Spizella breweri*). It is more common than the Thrasher, however, being found wherever there are significant areas of sagebrush in the southern Okanagan, and also has a sub-species found in northern subalpine shrub habitats.

The Grasshopper Sparrow (*Ammodramus savannarum*) is a red–listed species restricted in British Columbia essentially to the Okanagan Valley; a recent survey tallied only eight singing males. It prefers ungrazed or lightly grazed grasslands with a minimum of sagebrush. This species is found most often around Penticton and Vernon, so management plans for it and the Sage Thrasher need not conflict, since the thrasher is found south of Penticton.

The Sharp-tailed Grouse (*Tympanuchus phasianellus*) is found in grasslands throughout the province, but has disappeared from much of its former range and the remaining populations are scattered and relatively small. This grouse is quite susceptible to human disturbance, particularly over-hunting, and has declined throughout its range in western North America. The subspecies occupying the southern half of British Columbia, *T.p. columbianus*, is on the provincial Blue list. Another blue–listed grassland species with low but stable numbers is the Long-billed Curlew (*Numenius americanus*). Although heavy use of its breeding habitat by cattle may reduce its nesting success, this species seems to be adapting fairly well to some agricultural practices, such as the conversion of dry grassland to irrigated alfalfa fields.

Riparian Woodland

Deciduous woodlands along rivers and lakeshores have suffered disproportionate losses in British Columbia, since they are on prime real estate in a province with very little flat land. Yellow-breasted Chats (*Icteria virens*) and Lewis' Woodpeckers (*Melanerpes lewis*) are found in riparian habitats in the Thompson-Okanagan area. The red-listed Chat breeds in dense thickets, and is locally common only south of Penticton. Lewis' Woodpeckers nest in large cottonwoods in valley bottoms, but also use large ponderosa pine (*Pinus ponderosa*) and Douglas-fir (*Pseudotsuga menziesii*) snags in adjacent grasslands. Lewis' Woodpeckers were formerly found on the south coast as well, nesting on snag-covered, logged slopes which grew over with second-growth forest in the last 50 years. Apparently, early logging left enough unmerchantable timber to provide habitat for the woodpeckers, while current methods of clearcutting do not. It is probably this habitat change, rather than competition with starlings, that resulted in the disappearance of the species from the coast. These Woodpeckers seem to do well in the face of starling competition in the Southern Interior as long as suitable nest-trees are available, but populations are declining there as well as more large trees are cut, so the species has been put on the provincial Blue list. The Vaux's Swift (*Chaetura vauxi)* also uses large, hollow cottonwoods for nesting in the Interior, as well as hollow cedars on the coast, and has elicited concern in Washington State because of the drastic decline in both types of trees. It is unlisted in British Columbia, but deserves blue listing.

Old Growth Forests

The Spotted Owl (*Strix occidentalis*) has become famous throughout North America as the centre of a very political controversy around the cutting of the last old growth temperate rain forests in the Pacific Northwest. In British Columbia, Spotted Owls are found scattered throughout their historic range, from the North Shore watersheds near Vancouver north to Pemberton and south to Manning Provincial Park. Very few birds remain, however; recent Wildlife Branch surveys estimate about 50 birds left in the province. Spotted Owls are listed as endangered by COSEWIC and are on the provincial Red list, but as yet have no official endangered status in British Columbia, although a recovery plan was drawn up recently by the Wildlife Branch. In our increasingly fragmented forests, Spotted Owls are assaulted on several fronts; juveniles starve or are killed by Great Horned Owls (*Bubo virginianus*) when they disperse into cut-over forests and adults are often displaced by the larger, more aggressive, Barred Owls (*Strix varia*) which are more tolerant of second growth forests (Campbell and Campbell, 1984). The Spotted Owl is, of course, only a symbol of a much more extensive problem with the destruction of old growth forests on which several other bird species rely as well.

156

Recently, Marbled Murrelets (*Brachyramphus marmoratus*) have become media stars in their own right. They are a fairly common seabird found along the length of the British Columbian coast, presumably nesting high in old, large conifers. No nest had been found in the province (and only about 10 elsewhere) until 1990, when a recently-fledged nest was discovered on southern Vancouver Island. A nest with a single nestling was also found in 1993 in old growth forest on the Caren Range, Sechelt Peninsula (personal communication in August, 1993, with Paul Jones, Friends of Caren). The British Columbian population still numbers in the thousands, but evidence indicates that murrelet numbers are declining as old growth forest is eliminated along the coast. Marbled Murrelets are on the Blue list and surveys are under way in British Columbia to provide baseline data for much-needed population monitoring.

In the old growth forests of the Interior, other bird species are affected. The White-headed Woodpecker (*Picoides albolarvatus*) is found in ponderosa pine forests, and reaches the northern end of its range in the Okanagan Valley. Its provincial population is unknown, but only two or three sightings are reported annually. It is considered threatened by COSEWIC and is on the Blue list. The tiny Flammulated Owl (*Otus flammeolus*) is found at somewhat higher elevations and latitudes, associated with mature Ponderosa Pine/Douglas-Fir forests. It is considered vulnerable by COSEWIC and is on the Red list. A recent study by Astrid van Woudenberg and Fred Bunnell of UBC Forestry Wildlife investigated the habitat requirements and breeding biology of this little-known species in the Kamloops area, where a relatively dense local population is threatened with logging. Among other things, they found that the Owls seemed to be found most often in areas of spruce budworm outbreaks. The Williamson's Sapsucker (*Sphyrapicus thyroideus*) is on the provincial Blue list because of its apparent dependence on mature western larch (*Larix occidentalis*) forests. That habitat is endangered in the south-central Interior, the only part of Canada which is home to this beautiful Woodpecker.

In the northeastern corner of the province, the red-listed Connecticut Warbler (*Oporornis agilis*) is found in mature aspen (*Populus tremuloides*) forests, which are also slated for logging. The Bay-breasted Warbler (*Dendroica castanea*), also on the Red list, is associated with old-growth white spruce (*Picea alba*) forests in the same area. The Canada Warbler (*Wilsonia canadensis*) is also on the Blue list; its mature spruce forests are threatened by both logging and flooding.

Conclusion

In this brief discussion of the threatened birds of British Columbia, I haven't mentioned Barn Owls (*Tyto alba*) on Fraser Valley farmlands, Great Blue Herons (*Ardea herodias*) and dioxins, nor our vast populations of seabirds threatened with oil spills. Nonetheless, I think if immediate measures are taken to ensure the protection of the habitats mentioned above, we will have gone a long way to maintaining viable populations of our diverse bird fauna.

Acknowledgements

Dr. Rob Butler, Canadian Wildlife Service, reviewed the manuscript.

References

Bunnell, F.L. and R.G. Williams. 1980. Subspecies and diversity—the spice of life or prophet of doom. In: *Threatened and Endangered Species and Habitats in British Columbia and the Yukon.* R. Stace-Smith, L. Johns, and P. Joslin (eds.). B.C. Ministry of Environment, Victoria, B.C. pp. 246-259.

Butler, R.W. and R.W. Campbell. 1987. *The Birds of the Fraser River Delta: Populations, Ecology and International Significance.* Occasional Paper No. 65, Canadian Wildlife Service, Ottawa, Ontario.

Butler, R.W. and R.J. Cannings. 1989. *Distribution of Birds in the Intertidal Portion of the Fraser River Delta, British Columbia.* Technical Report No. 93, Canadian Wildlife Service, Delta, B.C.

Campbell, E.C. and R.W. Campbell. 1984. *Status Report on the Spotted Owl (Strix occidentalis caurina) in Canada—1983.* COSEWIC report, Canadian Wildlife Service. Ottawa, Ontario.

Campbell, R.W. 1989. *Checklist of British Columbia Birds.* Federation of British Columbia Naturalists and Royal British Columbia Museum.

Campbell, R.W., N.K. Dawe, I. McTaggart-Cowan, J.M. Cooper, G.W. Kaiser and M.C.E. McNall. 1990. *The Birds of British Columbia, Vol. 1: Nonpasserines, introduction, and loons through waterfowl;* and *Vol. 2: Nonpasserines, diurnal birds of prey through woodpeckers.* Royal British Columbia Museum in association with Environment Canada, Canadian Wildlife Service, Victoria, B.C.

Cannings, R.A., R.J. Cannings and S.G. Cannings. 1987. *Birds of the Okanagan Valley, British Columbia.* Royal British Columbia Museum, Victoria, B.C.

Cannings, R.J. 1992. *Status Report on the Sage Thrasher (Oreoscoptes montanus) in Canada.* COSEWIC report, Canadian Wildlife Service. Ottawa, Ontario.

Fyfe, R.W., J. Campbell, B. Hayson and K. Hodson. 1969. Regional population declines and organochlorine insecticides in Canadian Prairie Falcons. *Canadian Field-Naturalist* 83:191-200.

Goossen, J.P., R.W. Butler, B. Stushnoff and D. Stirling. 1982. Distribution and breeding status of Forster's Tern, *Sterna forsteri,* in British Columbia. *Canadian Field Naturalist* 96:345-346.

Hodson, K. 1980. Peregrine Falcons in British Columbia. In: *Threatened and Endangered Species and Habitats in British Columbia and the Yukon.* R. Stace-Smith, L. Johns, and P. Joslin (eds.). B.C. Ministry of Environment, Victoria, B.C. pp. 85-87.

Munro, W.T. 1990. Committee on the Status of Endangered Wildlife in Canada. *Bioline* 9:10-12.

Munro, W.T. 1993. Designation of endangered species, subspecies and populations by COSEWIC. In: *Our Living Legacy: Proceedings of a Symposium on Biological Diversity.* M.A. Fenger, E.H. Miller, J.F. Johnson and E.J.R. Williams (eds.). Royal British Columbia Museum, Victoria, B.C. pp. 213-227.

Weber, W.C. 1980. A proposed list of rare and endangered bird species for British Columbia. In: *Threatened and Endangered Species and Habitats in British Columbia and the Yukon.* R. Stace-Smith, L. Johns, and P. Joslin (eds.). B.C. Ministry of Environment, Victoria, B.C. pp. 160-182.

Chapter 17
Exotic Species in British Columbia

Introducing Aliens

Alien (or exotic) species are those that have recently come to British Columbia. As a consequence of the number of recent immigrants, species diversity in British Columbia has been rising. Table 17-1 shows that there are more introduced than extinct vertebrate species (see Chapter 2), and the lists of introduced species of plants and invertebrates provided in the subsequent sections of this chapter are far longer than the number of extinct native species mentioned in previous chapters. Table 17-2 shows the proportion of alien species in various taxonomic groups in British Columbia.

A commonly held view of ecosystem stability has been that each species in its proper proportion is important in maintaining a more or less stable climax plant community (Clements, 1916 and 1920). Although the idea of stasis in species diversity has been useful in such fields as animal population ecology and range management (Johnson, 1984), ecologists currently tend towards a more dynamic view of species diversity (Gleason, 1926; Miles, 1979; McIntosh, 1980; Vale, 1982). In this view, ecosystem functions (such as primary production, cycling of nutrients, and progression to climax) may be maintained, even though biological structure, including species diversity, may change (Johnson, 1984). Indeed, Heady (1973) notes that, because many organisms have effective dispersal mechanisms, immigrations are natural events which differ more in degree than in principle because of human activities. He showed that many newcomers quickly become naturalized and a part of the local succession and climax. However, the organisms that have effective dispersal mechanisms tend to be those that can occupy habitats characteristic of the early stages of succession following disturbance (Sorensen, 1984). Often lacking natural predators, parasites, and other natural controls (Goeden, 1971 and 1974; Maw, 1976), and finding plenty of disturbed habitat to occupy, alien species may increase to capture enough of the available resources to limit or even replace populations of native species (Salisbury, 1961; Harris, 1984). The literature is filled with accounts of alein species altering physical and biological structures and functions of ecosystems. An example close to home is given by Dale (1992) of an exotic species of legume, purposely seeded to assist in restoring a disturbed, formerly forested environment, that prevented re-establishment of native conifers.

Despite the ecological hazards, many alien species such as those sought by hunters or used in biological control of pests are useful. Economically and aesthetically speaking, introductions are only perceived to be negative when they directly affect ecosystem products (such as cattle grazing) or services (such as water flow regulation) that we value. The problem is that ecosystems are so

Lee E. Harding
Canadian Wildlife Service
P.O. Box 340
Delta, B.C.
V4K 3Y3

Scientific Name	English Name
Birds	
Sturnus vulgaris	European Starlings
Passer domesticus	House Sparrow
Acridotheres cristatellus	Crested Myna
Colinus virginianus	Northern Bobwhite
Callipepla californica	California Quail
Oreortyx pictus	Mountain Quail
Phasianus colchicus	Ring-necked Pheasant
Alectoris chukar	Chukar
Melegris gallopavo	Wild Turkey
Columbia livia	Rock Dove
Cygnus olor	Mute Swan
Perdix perdix	Gray Partridge
Alauda arvensis	Eurasian Skylark
Anas rubripes	American Black Duck
Fish	
Alosa sapidissima	American Shad
Salvelinus fontinalis	Brook Trout
Salmo trutta	Brown Trout
Carassius auratus	Goldfish
Cyprinus carpio	Common Carp
Ictalurus nebulosus	Brown Bullhead
Ictalurus melas	Black Bullhead
Lepomis gibbosus	Pumpkinseed
Micropterus dolomieui	Smallmouth Bass
Micropterus salmoides	Largemouth Bass
Pomoxis nigromaculatus	Black Crappie
Mammals	
Rattus norvegicus	Norway Rat
Mus musculus	House Mouse
Myocastor coypus	Nutria
Canis familiaris	Dog
Felis domesticus	Domestic Cat
Didelphis virginiana	Opossum
Oryctolagus cuniculus	European Rabbit
Sylvilagus floridanus	Eastern Cottontail
Sciurus carolinensis	Gray Squirrel
Sciurus niger	Fox Squirrel
Cervus dama	Fallow Deer
Amphibians	
Rana catesbeiana	American Bullfrog
Rana clamitans	Green Frog
Reptiles	
Chelydra serpentina	Snapping Turtle
Chineymys reevesii	Reeve's Turtle
Clemmys marmorata	Western Pond Turtle
Podarcis muralis	European Wall Lizard

Table 17-1: Introduced Vertebrate Species. Sources: Cannings and Harcombe 1990; chapters 12 and 13, this volume.

Group	Alien	Native	%	Authority
Vascular plants	662	2475	21.1	Taylor & MacBryde, 1977
Aquatic macrophytes	11	156	7.1	Newroth, this volume
Wildflowers	114	951	9.3	Clark, 1976
All beetles	248	3378	6.8	Smith, this volume
Ladybird beetles	5	89	5.3	Smith, this volume
Aquatic beetles	1	289	0.3	Smith, this volume
Cerambycid (forest wood decomposing) beetles	0	145	0	Smith, this volume
Predacious beetles	91	1212	7.0	Smith, this volume
Salmonids	2	6	25	Cannings & Harcombe, 1990
Marine fish	3	362	0.8	Cannings & Harcombe, 1990
Freshwater fish	11	60	15.5	Cannings & Harcombe, 1990
Amphibians	2	18	10	Cannings & Harcombe, 1990
Mammals	12	131	8.4	Cannings & Harcombe, 1990
Birds	14	434	3.1	Cannings & Harcombe, 1990
Reptiles	4	15	26.7	Cannings & Harcombe, 1990

Table 17-2: Numbers of Alien and Native Species of Selected Taxonomic Groups in British Columbia.

complex that we cannot accurately predict the impacts of alien species. The following sections give some examples of the costs, as well as benefits, of newly introduced species in British Columbia.

References

Cannings, R.A. and A.P. Harcombe. 1990. *The Vertebrates of British Columbia, Scientific and English Names.* Royal British Columbia Museum, Victoria, B.C.

Clark, L.J. 1976. *Wildflowers of the Pacific Northwest.* Gray's Publishing Ltd., Sidney, B.C.

Clements, F.E. 1916. *Plant Succesion: an Analysis of the Development of Vegetation.* Carnegie Inst. Wash. Pub. No. 242.

Clements, F.E. 1920. *Plant Indicators. The Relation of Plant Communities to Process and Practice.* Carnegie Inst. Wash. Pub. No. 290.

Dale, V.H. 1992. Exotic seeds reduce biodiversity on the Mount St. Helens debris avalanche. The Northwest Environmental Journal 8(1): 183-185.

Gleason, H.A. 1926. The individualistic concept of the plant association. *Torrey Bot. Club Bull.* 53: 7-26.

Goeden, R.D. 1971. The phytophagous insect fauna of milk thistle in southern California. *J. Econ. Ent.* 64: 1101-1104.

Goeden, R.D. 1974. Comparative survey of the phytophagous insect faunas to Italian thstle, *Carduus pycnocephalus,* in southern California and southern Europe relative to biological weed control. *Env. Ent.* 3: 464-474.

Harris, P. 1984. Biocontrol of weeds: bureaucrats, botanists, beekeepers and other bottlenecks. In: *Proc. VI Int. Symp. on Biological Control of Weeds.* E.S. Delfosse (ed.). Univ. of British Columbia. pp. 2-12.

Heady, H.F. 1973. Arid shrublands. In: *Proceedings of the Third Workshop of the United States/Australia Rangelands Panel.* D.N. Hyder (ed.). 1973. Tucson, Arizona.

Johnson, H.B. 1984. Consequences of species introductions and removals on ecosystem function - implications for applied ecology. In: *Proc. VI Int. Symp. on Biological Control of Weeds.* E.S. Delfosse (ed.). Univ. of British Columbia. pp. 27-56.

Maw, M.G. 1976. An annotated list of insects associated with Canada thistle, *Cirsium arvense)* in Canada. Can. Ent. 108: 235-244.

McIntosh, R.P. 1980. The relationship between succession and the recovery process. In: *The Recovery Process in Damaged Ecosystems.* Cairns, J.Jr.(ed.) Ann Arbor Science, Ann Arbor MI, pp. 11-62.

Salisbury, E., 1961. *Weeds and Aliens.* Colins, London.

Sorensen, A.E. 1984. Seed dispersal and the spread of weeds. In: *Proc. VI Int. Symp. on Biological Control of Weeds.* E.S. Delfosse (ed.). Univ. of British Columbia. pp. 121-132.

Taylor, R.L. and B. MacBryde. 1977. Vascular plants of British Columbia: a descriptive inventory. *Botanical Garden Technical Bulletin* No. 4. UBC Press.

Vale, T.R. 1982. Plants and people vegetation change in North America. *Assoc. Amer. Geogr.,* Wash. D.C.

Introduced Wildflowers and Range and Agricultural Weeds in British Columbia

Lee E. Harding
Canadian Wildlife Service
P.O. Box 340
Delta, B.C.
V4K 3Y3

Introduced wildflowers and range and agricultural weeds include a variety of plants with environmental, economic and aesthetic costs, and some benefits. Wildflowers are important to a discussion of biodiversity because they are part of complex ecological interactions, being pollinated by insects and used for food and other purposes by birds, mammals, and other organisms. Introduced range and agricultural weeds, including widlflowers and grasses, will be discussed here because of their economic implications.

Wildflowers

Wildflowers include representatives of both the Monocotyledoneae (irises, *Iridaceae*, and lilies, *Liliaceae*) and Dicotyledoneae (most other flowers) of the angiosperm subdivision of vascular plants. Taylor and McBride (1977) found that 21.1% of all vascular plants in British Columbia were exotics. However, of the 1,065 species of British Columbia wildflowers listed by Clark (1976), only 9.3% were introduced. These figures are an indication of the extent to which our natural environments have been altered. Table 17-3 lists the introduced flowers (not including trees, shrubs, and garden herbs) categorized by Douglas et al. (1991 and 1992).

Alien wildflowers, like many other introduced plants, are strongly associated with disturbed environments (Salisbury, 1961; Montgomery, 1964; Mulligan, 1976; Clark, 1976). Plants that colonize disturbed habitats must be able to disperse widely because, as the ecosystem recovers with the growth of other vegetation following disturbance, the environment becomes unsuitable for these species.

Scientific Name	English name
FAMILY APIACEAE	
Antheum graveolens	Common dill
Anthriscus caucalis	Bur chervil
Carum carvi	Caraway
Conium maculatum	Poison hemlock
Daucus carota	Wild carrot
Eryngium planum	Plains eryngo
Foeniculum vulgare	Sweet fennel
Heracleum mantegazzianum	Giant cow-parsnip
Lactuca biennis	Tall blue lettuce
Lactuca canadensis	Canadian wild lettuce
Lactuca muralis	Wild Lettuce
Pastinaca sativa	Common parsnip
Scandix pecten-veneris	Venus'-comb/Shepherd's needle
Torilis japonica	Upright hedge-parsely
FAMILY APOCYNACEAE	
Vinca major	Large perwinkle
Vinca minor	Common perwinkle
FAMILY ASTERACEAE	
Ambrosia psilostachya	Western ragweed
Anthemis arvensis	Field chamomile
Anthemis tinctoria	Yellow chamomile
Arctium lappa	Great burdock
Arctium minius	Lesser burdock
Artemisia absinthium	Wormwood
Artemisia vulgaris	Common mugwort
Bellis perennis	English daisy
Bidens frondosa	Common beggarticks
Carduus acanthoides	Plumeless thistle
Carduus crispus	Curled thistle
Carduus nutans	Nodding thistle
Carthamus lanatus	Distaff thistle
Centaurea cyanus	Cornflower
Centaurea diffusa	Diffuse knapweed
Centaurea maculosa	Spotted knapweed
Centaurea militensis	Maltese star-thistle
Centaurea montana	Mountain bluet
Centaurea nigrescens	Short-fringed knotweed
Centaurea paniculata	Jersey knapweed
Centaurea pratensis	Meadow knapweed
Centaurea repens	Russian knapweed
Cichorium intybus	Chicoree
Cirsium arvense	Canada thistle
Cirsium palustre	Marsh thistle
Cirsium vulgare	Bull thistle
Conyza canadensis	Horseweed
Coryopsis lanceolata	Garden coryopsis
Cotula coronopifolia	Brass buttons
Crepis capillaris	Smooth hawksbeard
Crepis nicaeenis	French hawksbeard
Crepis tectorus	Annual hawksbeard
Crepis vesicaria	Weedy hawksbeard
Doronicum pardalianches	Great leopard's-bane
Filago arvensis	Field filago
Filago vulgaris	Common filago
Gallinsoga cilliata	Shaggy gallinsoga

Table 17-3: Introduced Flowers of British Columbia. Source: Douglas et al., 1991 and 1992.

163

Scientific Name	English Name
Gnaphalium purpureum	Purple cudweed
Gnaphallum sylvaticum	Woodland cudweed
Gnaphallum uliginosum	Marsh cudweed
Helianthus annuus	Sunflower
Helianthus nuttalli	Nutall's sunflower
Helianthus rigidus	Rigid sunflower
Hieracium aurantiacum	Orange hawkweed
Hieracium murorum	Wall hawkweed
Hieracium pilosella	Mouse-ear hawkweed
Hieracium piloselloides	Tall hawkweed
Hypochoeris glabra	Smooth cat's ear
Hypochoeris radicata	Cat's ear
Inula helenium	Elecampane
Iva xanthifolia	Marsh-elder
Krigia virginica	Virginia dwarf dandelion
Lactuca canadensis	Canadian wild lettuce
Lactuca muralis	Wall lettuce
Lactuca serriola	Prickly lettuce
Lapsana communis	Nipplewort
Leontodon autumnalis	Fall dandelion
Leontodon taraxacoides	Hairy hawkbit
Leucanthemum vulgare	Ox-eye daisy
Matricaria perforata	Scentless mayweed
Matricaria recutita	Wild chamomile
Onopordum acanthium	Scotch thistle
Petasites japonicus	Japanese butterbur
Rudbeckia hirta	Black-eyed susan
Senecio jacobaea	Tansy ragwort
Senecio sylvaticus	Wood groundsel
Senecio viscosus	Sticky ragwort
Senecio vulgaris	Common groundsel
Silybum marianum	Milk thistle
Sonchus arvensis	Sow thistle
Sonchus arvensis	Perennial sow-thistle
Sonchus asper	Prickly sow thistle
Sonchus oleraceus	Common sow thistle
Tanacetum parthenium	Feverfew
Tanacetum vulgare	Common tansy
Tanacetum vulgare	Tansy
Taraxacum laevigatum	Red-seeded dandelion
Taraxacum officinale	Common dandelion
Tragopogon dubius	Yellow salsify
Tragopogon porrifolius	Oyster plant
Tragopogon praetensis	Meadow salsify
Trapopogon dubius	Yellow salsify
Trapopogon porrifolius	Common salsify, oyster plant
Trapopogon pratensis	Meadow salsify
Tussilago farfara	Coltsfoot
Xanthium strumarium	Common cocklebur

FAMILY BALSAMINACEAE

Impatiens glandulifera	Policeman's helmet
Impatiens parviflora	Small touch-me-not/Balsam

FAMILY BERBERIDACEAE

Berberis vulgaris	Common barberry

FAMILY BORAGINACEAE

Anchusa officinalis	Common anchusa

Table 17-3: Introduced Flowers of British Columbia (Continued).

Scientific Name	English Name
Asperugo procumbens	Madwort/Catchweed
Borago officinalis	Common borage
Cynoglossum officinale	Hound's tongue
Echium vulgare	Blueweed
Lappula echinata	Common stickseed
Lithospermum arvense	Corn gromwell
Myosotis arvense	Forget-me-not
Myosotis discolor	
Myosotis scorpioides	Common forget-me-not
Myosotis stricta	Blue forget-me-not
Myosotis sylvatica	Wood forget-me-not
Symphytum asperum	Rough comfrey
Symphytum officinale	Common comfrey

FAMILY BRASSICACEAE

Scientific Name	English Name
Alyssum alyssoides	Pale alyssum
Alyssum desertorum	Desert alyssum
Alyssum murale	Wall alyssum
Arabidopsis thallana	Mouse-ear
Arabis glabra	Tower mustard
Armoracia rusticana	Common horseradish
Barbarea verna	Early winter cress
Barbarea vulgaris	Yellow rocket
Berteroa incana	Hoary alyssum
Brassica campestris	Rape/Field mustard
Brassica hirta	White mustard
Brassica juncea	Indian mustard
Brassica kaber	Charlock, Wild mustard
Brassica napus	Turnip/Winter rape
Brassica nigra	Black mustard
Cakile maritima	European searocket
Camelina microcarpa	Littlepod/Hairy flax
Camelina sativa	Falseflax/Gold-of-pleasure
Capsella bursa-pastoris	Shepherd's purse
Cardaria chalepensis	Chalapa hoary-cress
Cardaria draba	Heart-podded hoary-cress
Cardaria pubescens	Globe-pod hoary-cress
Cardimine hirsuta	Hairy bitter-cress
Chorispora tenella	Blue mustard
Conringla orientalis	Hare's-ear mustard
Coronopus didymus	Lesser swine-cress
Descurainia sophia	Felix weed
Draba verna	Vernal whitlow-grass
Erucastrum gallicum	Dog mustard
Erysimum cheiri	Common wallflower
Hesperis matronalis	Dame's violet
Isatis tinctoria	Dyers woad/Asp-of-Jerusalem
Lepidium campestre	Field pepper-grass
Lepidium heterophyllum	Smith's pepper-grass
Lepidium perfollatum	Clasping pepper-grass
Lepidium sativum	Garden cress
Lobularia maritima	Sweet alyssum
Lunaria annual	Honesty
Nasturtium microphyllum	One-rowed water cress
Nasturtium officinale	Water cress
Neslia paniculata	Ball mustard
Rorippa sylvestris	Creeping yellow cress
Sisymbrium altissimum	Tall tumblemustard
Sisymbrium loeselii	Loesel's tumblemustard

Table 17-3: Introduced Flowers of British Columbia (Continued).

Scientific Name	English Name
Sisymbrium officinale	Hedge mustard
Teesdalia nudicaulis	Shepherd's cress
Thiaspi arvense	Field pennycress
FAMILY CAMPANULACEAE	
Campanula medium	Centerbury-bells
Campanula rapunculoides	Creeping bellflower
Lobelia inflata	Indian tobacco
FAMILY CANNABACEAE	
Humulus lupulus	Common hop
FAMILY CAPRIFOLIACEAE	
Lonicera etrusca	Etruscan honeysuckle
FAMILY CARYOPHYLLACEAE	
Agrostemma githago	Corn cockle
Arenaria serpyllifolia	Thyme-leaved sandwort
Cerastium fontanum	Mouse-ear chickweed
Cerastium semidecandrum	Little chickweed
Cerastium tomentosum	Snow-in-summer
Corrigiola litoralis	Strapwort
Dianthus armeria	Deptford pink
Dianthus barbatus	Sweet William
Dianthus deltoides	Marden pink
Gypsophila paniculata	Baby's breath
Holosteum unbellatum	Umbellate chickweed
Lychnis coronaria	Mullien pink
Moencia erecta	Upright chickweed
Myosoton aquaticum	Water chickweed
Petrohagia saxifraga	Tunic flower
Polycarpon tetraphyllum	Four-leaved all-seed
Sagina japonica	Japanese pearlwort
Saponaria officinalis	Bouncing-bet/soapwort
Scieranthus annuus	Annual knawel
Silene alba	White cockle/Campion
Silene armeria	Sweet William catchfly
Silene cserei	Biennial campion
Silene dichotoma	Forked catchfly
Silene doica	Red campion
Silene gallica	Small-flowered catchfly
Silene noctiflora	Night-flowering catchfly
Silene vulgaris	Bladder campion
Spergularia rubra	Pink sand spurry
Spurgula arvensis	Com-spurry/Stickwort
Spurgularia marina	Salt-marsh sand-spurry
Stellaria alsine	Bog starwort
Stellaria graminea	Grass-leaved starwort
Stellaria media	Chickweed/Sommon starwort
Vaccaria pyramidata	Cow-basil
FAMILY CHENOPODIACEAE	
Atriplex heterosperma	Russian orache
Atriplex hortensis	Garden orache
Atriplex oblongifolia	Oblong-leaved orache
Atriplex patula	Common orache
Atriplex rosea	Red/Tumbling orache
Axyris amaranthoides	Russian pigweed
Bassia hyssopifolia	Five-hooked bassia

Table 17-3: Introduced Flowers of British Columbia (Continued).

Scientific Name	English Name
Chenopodium album	Lamb's quarters
Chenopodium botrys	Jerusalem oak/Feather geranium
Chenopodium capitatum	Strawberry blight
Chenopodium glaucum	Oak-leaved goosefoot
Chenopodium hypridum	Maple-leaved goosefoot
Chenopodium urbicum	Upright goosefoot
Kochia scoparia	Summer-cypress
Salsola kali	Tumbleweed/Russian thistle
FAMILY CONVOLVULACEAE	
Cuscuta cephalanthii	Button-bush dodder
Cuscuta epithymum	Common or Thyme dodder
Convolvulus arvensis	Field bindweed
Convolvulus sepum	Hedge bindweed/Morning glory
Cuscuta approximata	Clustered or Alfalfa dodder
FAMILY CRASSULACEAE	
Sedum acre	Goldmoss stonecrop
Sedum album	White stonecrop
FAMILY CUCURBITACEAE	
Echinocystis lobata	Wild/Prickly cucumber
FAMILY DIPSACACEAE	
Dipsacus sylvestris	Teasel
Knautia arvensis	Wild scabious
FAMILY EUPHORBIACEAE	
Euphorba helioscopia	Summer spurge
Euphorba lathyris	Caper spurge
Euphorba supina	Milk spurge
Euphorbia cyparissias	Cypress spurge
Euphorbia escula	Leafy spurge
Euphorbia exigua	Dwarf spurge
Euphorbia peplus	Petty spurge
FAMILY FABACEAE	
Astragalus cicer	Chick-pea milk vetch
Coronilla varia	Common crown vetch
Cytisus scoparius	Scotch broom
Lathyrus latifolius	Perennial pea
Lathyrus praetensis	Meadow peavine
Lathyrus sphaesicus	Grass peavine
Lathyrus sylvestris	Narrow-leaved peavine
Lotus corniculatus	Bird's foot trefoil
Lotus pedunculatus	Pedunculate bird's-foot trefoil
Lotus tenuis	Narrow-leaved bird's-foot trefoil
Lupinus arborus	Tree lupine
Lupinus densiflorus	Dense-flowered lupine
Medicago arabica	Spotted medic
Medicago falcata	Sickle medic
Medicago lupulina	Black medicago
Medicago polymprpha	Bur-clover
Medicago sativa	Alfalfa
Meliotus alba	White sweet-clover
Meliotus officinalis	Yellow sweet-clover
Onobrychis viciaefolia	Sandfain
Trifolium arvense	Haresfoot clover
Trifolium aureum	Yellow clover

Table 17-3: Introduced Flowers of British Columbia (Continued).

Scientific Name	English Name
Trifolium campestre	Low hop-clover
Trifolium dubium	Small hop-clover
Trifolium fragiferum	Strawberry clover
Trifolium hybridum	Alsike clover
Trifolium incarnatum	Crimson clover
Trifolium pratense	Red clover
Trifolium repens	White clover
Trifolium subterraneum	Burrowing clover
Vicia craccia	Cow vetch
Vicia hirsuta	Hairy vetch
Vicia sativa	Common vetch
Vicia tetrasperma	Slender vetch
Vicia villosa	Hairy vetch
FAMILY FUMARIACEAE	
Fumeria bastardii	Bastard fumatory
Fumeria officinalis	Common fumatory
FAMILY GENTIANACEAE	
Centaurium erythracaea	Common centaury
FAMILY GERANIACEAE	
Erodium cicutarium	Stork's bill
Genanium dissectum	Cut-leaved geranium
Geranium molle	Crane's bill
Geranium pusillium	Small-flowered geranium
FAMILY HYPERICACEAE	
Hypericum perforatum	St. John's wort
FAMILY IRIDACEAE	
Iris psuedacorus	Yellow flag
FAMILY LAMIACEAE	
Ajuga reptans	Bugle-weed
Dracocephalum thymiflorum	Eurasian dragonhead
Geleopsis tetrahit	Mint
Glecoma hederacea	Ground ivy
Marrubium vulgare	Common horehound
Melissa officianlis	Lemon balm
Mentha arvensis	Fielf mint
Mentha pulegium	Pennyroyal
Mentha spicata	Spearmint
Mentha suaveolens	Applemint
Mentha x gentillis	Red mint
Mentha x piperita [citrata]	Peppermint
Nepeta cataria	Catnip
Stachys arvensis	Field hedge-nettle
Stachys byzantina	Cooley's hedge-nettle
FAMILY LILIACEAE	
Asperagus officinalis	Asperagus
FAMILY LINACEAE	
Linum bienne	Pale flax
Linum usitatissimum	Common flax

Table 17-3: Introduced Flowers of British Columbia (Continued).

Scientific Name	English Name
FAMILY LYTHRACEAE	
Lythrum salicaria	Purple loose-strife
Peplis portula	Water-purslane
FAMILY MALVACEAE	
Malva moschata	Musk mallow
Malva neglecta	Dwarf mallow
Malva parviflora	Small-flowered mallow
Malva pusilla	Small mallow
Malva sylvestris	Common mallow
FAMILY NYCTAGINACEAE	
Mirabillis nyctaginea	Umbrellawort
FAMILY NYMPHAEA	
Nymphaea alba	European white waterlily
Nymphaea mexicana	Yellow waterlily
Nymphaea odorata	Fragrant water lily
FAMILY ONAGRACEAE	
Epilobium hirsutum	Hairy willowherb
Oenothera biennis	Common evening primrose
Oenothera glazioviana	Red-sepaled evening-primrose
Oenothera perennis	Perennial sundrops
FAMILY ORCHIDACEAE	
Epipactis helleborine	Helleborine
FAMILY OXALIDACEAE	
Oxalis corniculata	Yellow oxalis
Oxalis stricta	Upright yellow oxalis
FAMILY PAPAVERACEAE	
Chelidonium majus	Celandine
Escholtzia californiana	California poppy
Papaver nudicaule	Iceland poppy
Papaver somniferum	Opium poppy
FAMILY PLANTAGINACEAE	
Plantago coronopus	Buckshorn plantain
Plantago lancolata	English/Narrow-leaved plantain
Plantago major	Common/Broad-leaved plantain
Plantago media	Hoary plantain
Plantago psyllium	Whorled plantain
FAMILY POLYGONACEAE	
Fagopyrum esculentum	Buckwheat
Polygonum arenastrum	Oval-leaved knotweed
Polygonum aviculare	Common knotweed
Polygonum convolvulus	Bindweed
Polygonum cuspidatum	Japanese knotweed
Polygonum hydropiper	Marshpepper smartweed
Polygonum nepalense	Nepalese knotweed
Polygonum persicaria	Lady's thumb
Polygonum polystachyum	Himalayan knotweed
Polygonum sachalinense	Sachalene knotweed

Table 17-3: Introduced Flowers of British Columbia (Continued).

Scientific Name	English Name
Rumex acetosella	Sheep sorrel
Rumex conglomeratus	Clustered dock
Rumex obtusifolius	Bitter dock

FAMILY PORTULACACEAE
Portulaca oleracea	Purslane

FAMILY PRIMULACEAE
Anagallis arvensis	Scarlet pimpernel
Lysimachia nummularia	Creeping jenny
Lysimachia punctata	Spotted loosestrife
Lysimachia terrestris	Bog loosestrife
Lysimachia vulgaris	Yellow loosestrife

FAMILY RANUNCULACEAE
Clematis tangutica	Golden clematis
Clematis vitalba	Traveler's joy/Old man's beard
Consolida ambigua	Rocket larkspur
Ranunculaceae testiculatus	Hornseed buttercup
Ranunculus acris	Tall buttercup
Ranunculus ficaria	Lesser celandine
Ranunculus repens	Creeping buttercup
Ranunculus sardous	Hairy buttercup
Reseda alba	White mignonette
Reseda lutea	Yellow mignonette

FAMILY ROSACEAE
Alchemilla subcrenata	Lady's mantle
Aphanes arvensis	Field parsely-piert
Aphanes microcarpa	Small-fruited parsley-piert
Duchesnea indica	Indian strawberry
Potentilla argentea	Silvery cinquefoil
Potentilla recta	Sulphur cinquefoil
Rosa canina	Dog rose
Rosa eglanteria	Sweetbrier
Rubus allegheniensis	Allegheny blackberry
Rubus discolor	Himalayan blackberry
Rubus laciniatus	Evergreen blackberry
Santuisorba minor	Salad burnet

FAMILY RUBIACEAE
Galium aparine	Bedstraw
Galium mollugo	White bedstraw
Gallium odoratum	Sweet woodruff
Gallium vernum	Yellow bedstraw
Rubus procerus	Himalayan blackberry
Sherardia arvensis	Field madder

FAMILY SAXIFRAGACEAE
Saxifraga tridactylites	Rue-leaved saxifrage

FAMILY SCROPHULARIACEAE
Chaenorrhinum minus	Common dwarf snapdragon
Cympalaria muralis	Ivy-leaved toadflax
Digitalis purpurea	Common foxglove
Euphrasia nemorosa	Eastern eyebright
Kickxia elatine	Sharp-leaved fluellen
Linaria purpurea	Purple toadflax
Linaria vulgaris	Butter-and-eggs/Common toadflax

Table 17-3: Introduced Flowers of British Columbia (Continued).

Scientific Name	English Name
Parentucellia viscosa	Yellow bartsia
Verbascum blattaria	Moth mullien
Verbascum phlomoides	Woolly mullien
Verbascum thapsus	Common mullien
Veronica anagallis-aquatica	Water speedwell
Veronica arvensis	Wall speedwell
Veronica chamaedrys	Germander speedwell
Veronica filiformis	Slender speedwell
Veronica hederaefolia	Ivy-leaved speedwell
Veronica officinalis	Common speedwell
Veronica persica	Bird's-eye speedwell
Veronica verna	Spring speedwell
FAMILY SOLANACEAE	
Datura stramonium	Jimsonweed
Lycium halimifolium	Matrimony vine
Solanum americanum	Black nightshade
Solanum dulcamara	Bittersweet
Solanum rostratum	Buffalo-bur
Solanum sarrachoides	Hairy nightshade
FAMILY SCROPHULARIACEAE	
Linaria genistifolia	Dalmatian toadflax
FAMILY THYMELAEACEAE	
Daphne laureola	Spurge-laurel
FAMILY URTICACEAE	
Urtica urens	Dog nettle
FAMILY VALERIANACEAE	
Valeriana officinalis	Garden heliotrope
Valerianella locusta	Cornsalad
FAMILY VIOLACEAE	
Viola arvensis	European field pansy
Viola lanceolate	Lance-leaved violet
Viola odorata	Sweet violet
Viola tricolor	European wild pansy/Johnny-jump-up
FAMILY ZYGOPHYLLACEAE	
Tribulus terrestris	Puncture vine

Table 17-3: Introduced Flowers of British Columbia (Continued).

Hence, their progeny must be able to find new, disturbed environments, and therefore have evolved mechanisms for wide dispersal, such as wind-borne seeds and animal-borne burrs (Sorensen, 1984). Many arrived in the grain or hay of livestock feed (Burcham, 1957), in ballast from ships, in nursery or agricultural shipments and on equipment such as tractors, hay mowers and harvesters (Montgomery, 1964). Because so many of these products move by rail, and the railroad beds provide ideal habitats for disturbed-site species, railroad rights of way have been major routes of introduction (Montgomery, 1964). Some species, such as burdock (*Arctium lappa, A. minius*) and hound's tongue (*Cynoglossum officinale*) are spread by furred wildlife, livestock and people. A few alien flowers,

171

such as black-eyed susan (*Rudbeckia hirta*) are native to elsewhere in North America and have recently migrated to British Columbia on their own. Of course, many alien wildflowers escaped from cultivation in domestic gardens.

Many introduced plant species serve useful functions, or are aesthetically pleasing. English daisies (*Bellis perenis*) add interest to a lawn, for example, and many of the alien mints (*Mentha citrata, M. piperite, M. spicata, M. rotundafolia*) seem to improve a meadow or roadside wet spot with their blue or purple flowers and fresh scent. Even the much maligned dandelions (*Taraxacum officinale, T. laevigatum* and other species) are pretty on someone else's lawn. Lupines (*Lupine arborus, L. densiflorus*), clovers (*Trifolium arvense, T. pratense and T. subterraneum*) and a number of other wildflowers are routinely seeded to add nitrogen and prevent erosion on rights of way. Several attractive, introduced blue-flowered vetches (*Vicia craccia, V. hirsuta, V. sativa*), now wild, also perform this function. Some flowers, like the bright gold California poppy (*Escholtzia californiana*) are seeded for colour along urban roadsides. The invasive bush Scotch broom (*Cytisus scoparius*) gilds vacant lots, roadside embankments and other waste spaces along the coast with brilliant gold blooms in the spring, while adding nitrogen to the soil. Ox-eye daisies (*Chrysanthemum leucanthemum*), are widely naturalized along roadsides and in meadows and fields. Other flowers, such as chickory (*Cichorium intybus*) which is especially abundant along roadsides in the dry interior, helleborine (*Epipactus helleborine*), crane's bill (*Geranium molle*), herb robert (*G. robertanium*), baby's breath (*Gypsophila paniculata*), dame's violet (or rocket) (*Hesperis matronalis*), yellow flag Iris (*Iris pseudacorus*), linaria (*Linaria dalmatica*), toadflax (*L. vulgaris*), mullien pink (*Lychnis coronaria*), purslane (*Portulaca oleracea*), have simply escaped from flower or herb gardens. As well as contributing to soil stability and development, introduced wildflowers provide seeds for birds and small mammals and nectar for hummingbirds and insects. However, these benefits are often a mixed blessing.

Notwithstanding the aesthetic, economic (such as slope stabilization and right-of-way maintenance) and even environmental benefits of alien wildflowers, they do take the place of, and in many cases directly displace, native flora. Moreover, they often alter the biological structure, and possibly the functions, of some ecosystems. For example, purple loose-strife *(Lythrum salicaria)* (see facing page) can be so invasive in undisturbed riparian habitats that it can destroy wetlands fish and wildlife habitat. Similarly, Scotch broom has invaded several environments, including the endangered garry oak-Douglas-fir communities on Vancouver Island, where it converts wildflower meadows to shrub-dominated communities.

Range and Agricultural Weeds

Alien range and agricultural weeds of economic importance (those listed under the <u>Weed Control Act</u>) are listed in Table 17-4. The distributions of several important species are shown in Figure 17-1.

Purple Loosestrife in British Columbia

Peter R. Newroth
Water Quality Branch
Ministry of Environment, Land and Parks
765 Broughton Street
Victoria, B. C.
V8V 1X5

Purple loosestrife (*Lythrum salicaria*) is an herbaceous perennial plant which grows in wetland areas. Probably introduced to the United States in the early 1800's, it is believed to have been brought to Canada in about 1930 (Thompson et al., 1987).

The biological characteristics of Purple Loosestrife include:

o   preference for moist organic soils, but tolerance for a wide range of textures;

o   preference for full sun, but tolerance for up to 50% shade;

o   production of up to 2.7 million seeds per plant per year;

o   seed germination in a range of conditions, though several years' delay possible;

o   seed dispersal by wind and water, with possible assistance from bird and mammal movement; and

o   ability to establish itself in disturbed areas.

Because of its physical beauty, purple loosestrife is dispersed through sale by nurseries.

Populations in British Columbia that have been known for over ten years are spreading and enlarging (an estimated 208 hectares were affected in 1991) and threaten major impacts on the wetlands where they are expected to thrive. The main adverse impact of purple loosestrife is its propensity to grow in dense stands and displace native species. Densities of plant stems may reach 200,000 stalks per hectare. The species displaced may include plants, waterfowl and mammals (White and Haber, 1992).

Only limited research and control of purple loosestrife have occurred in British Columbia to date. There is growing concern, expressed by the public and naturalist groups, about the spread of this plant. An inventory is being performed and a database on loosestrife in British Columbia has been established. In the Okanagan Valley, 70 loosestrife colonies were documented in 1991. This information will provide evidence of distribution and changes in areal coverage (Enns and Grainger, 1992).

Some experimental controls have been implemented in British Columbia, but effective control is difficult and may require several years work at individual sites. Eradication appears to be impossible, except in the smallest populations. Mechanical, chemical and biological control approaches have been considered but all have limitations (White and Haber, 1992), though several biological control agents were introduced in 1993. Integrated management of selected sites, together with public information campaigns and restriction of further nursery sales, appears to be the most promising management strategy.

Acknowledgements

V. Bartnik, Environment Canada, reviewed the manuscript.

References

Enns, K.A. and K.L. Grainger. 1992. *Distribution and Abundance of Purple Loosestrife in the Okanagan Valley.* Report prepared for Ducks Unlimited Canada and Habitat Conservation Fund.

Thompson, D.Q., R.L. Stuckey and E.B. Thompson. 1987. *Spread, Impact and Control of Purple Loosestrife (Lythrum salicaria) in North American Wetlands.* Fish and Wildlife Research 2. U. S. Department of the Interior, Fish and Wildlife Service, Washington, D. C.

White, D.J. and E. Haber. 1992. *Invasive Plants of Natural Habitats in Canada. Part 1: Wetland Species.* Report prepared for Canadian Wildlife Service.

English Name	Scientific Name	Location
Diffuse Knapweed	*Centaurea diffusa*	Southern Interior
Spotted Knapweed	*Centaurea maculosa*	Southern Interior
Russian Knapweed	*Acroptilon repens*	Southern & Central Interior
Canada Thistle	*Cirsium arvense*	Widespread
Nodding Thistle	*Carduus nutans*	Okanagan, Cariboo, Kootenay
Leafy Spurge	*Euphorbia esula*	Widespread
Toadflax	*Linaria vulgaris*	Okanagan, Thompson & Cariboo
Heart-podded Hoary Cress	*Cardaria draba*	Okanagan-Shuswap, Nicola & Vancouver Island
Lens-podded Hoary Cress	*Cardaria chalepensis*	Thompson & Okanagan
Globe-podded Hoary Cress	*Cardaria pubescens*	Thompson, Nicola, Kootenay & Peace River
Sulphur Cinquefoil	*Potentilla recta*	Thompson, Okanagan, & eastern Kootenay
Common Burdock	*Arcticum minus*	Widespread
Great Burdock	*Arcticum lappa*	Lower Mainland & Peace River
Common Tansy	*Tanacetum vulgare*	Widespread
Tansy Ragwort	*Senecio jacobeae*	Lower Mainland, Okanagan & Vancouver Island
Common Tansy	*Tanacetum vulgare*	Widespread
Blueweed	*Echium vulgare*	East Kootenay, Nicola, Clinton & Christina Lake
Scentless Mayweed	*Matricaria maritima*	Widespread, especially in Peace River area
Ox-eye Daisy	*Chrysanthemum leucanthemum*	Widespread
Hound's Tongue	*Cynoglossum officinale*	Thompson, Okanagan, Kootenay & Cariboo
Common Toadflax	*Linaria vulgaris*	Widespread
Dalmatioan Toadflax	*Linaria dalmatica*	Southern Interior
White Cockle	*Silene alba*	Peace River & Thompson-Okanagan
Morning Glory	*Convolvulus arvensis*	Peace River
Cleavers (Bedstraw)	*Galium aparine*	Peace River
Night-flowering Catchfly	*Silene noctiflora*	Peace River
Wild Mustard	*Brassica kaber*	Peace River
Scentless Camomile	*Matricaria maritima*	Cariboo, Peace River
Tartary Buckwheat	*Fagopyrum tataricum*	Peace River
Clustered Dodder	*Cucuta approximata*	Widespread
Common Dodder	*Cuscuta epithymum*	Widespread
Quackgrass	*Agropyron repens*	Peace River
Wild Oats	*Avena fatua*	Widespread

Table 17-4: Introduced Range and Pasture Weeds of Economic Importance. Source: <u>Weed Control Act</u>.

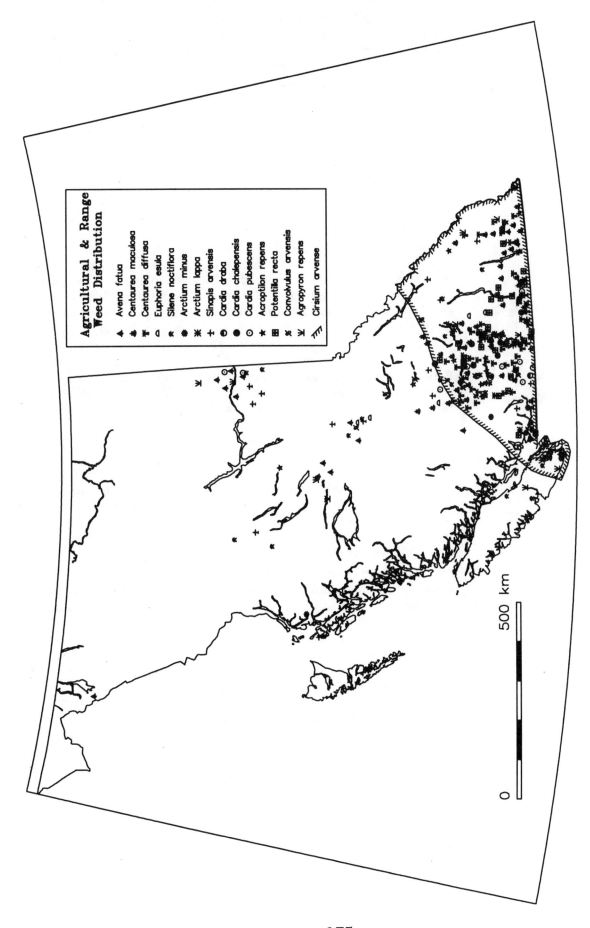

Figure 17-1. Agricultural & Range Weed Distribution. Sources: Best et al., 1980; Gross et al., 1980; McNeill, 1980; Moore, 1975; Mulligan and Bailey, 1975; Mulligan and Findlay, 1974; Mulligan and Munro, 1981; Sharma and Vanden Born, 1978; Watson, 1980; Weaver and Riley, 1982; Werner and Rioux, 1977; Werner and Soule, 1976.

Agricultural & Range
Weed Distribution

Avena fatua
Centaurea maculosa
Centaurea diffusa
Euphoria esula
Silene noctiflora
Arctium minus
Arctium lappa
Sinapis arvensis
Cardia draba
Cardia chalepensis
Cardia pubescens
Acroptilon repens
Potentilla recta
Convolvulus arvensis
Agropyron repens
Cirsium arvense

500 km

0

175

Some wildflowers, such as ox-eye daisy and night-flowering catchfly (*Silene noctiflora*), contaminate seed, hay or grain crops, reducing their value. Field bindweed (*Convolvulus arvensis*) causes mechanical damage to harvesting equipment. Canada thistle (*Cirsium arvense*), also alien, is one of the most troublesome agricultural weeds in British Columbia. A perennial, it can reduce grain yields by over 60% and may release into the soil toxic chemicals that inhibit other plants (undated brochure, Ministry of Agriculture and Food). Other alien wildflowers that cause agricultural damage include wild mustard *(Sinapsis arvensis)*, quackgrass *(Agropyron repens)* and nodding thistle *(Carduus nutans)*.

Range weeds that affect grassland ecosystems reduce forage for livestock and negatively affect wildlife. Many range weeds make poor livestock and wildlife forage and degrade pasture and rangelands by competing with native grasses and forbs for space, soil moisture and nutrients. Disturbance is clearly associated with some aspects of the initial infestation and later spread in the overwhelming majority of cases, although, once present, many weeds can compete with native species and infest healthy grasslands. However, other environmental factors, such as prolonged cool, wet periods or droughts, can also favour weed species (Sturko and Wikeem, 1990). Conversely, healthy, undisturbed grasslands can better withstand weed infestations without being dominated by them, and some native grassland communities are more resistant than others (personal communication in September, 1993 with B. Wikeem, Ministry of Forests). For example, grassland communities dominated by rough fescue (*Festuca campestris [F. scabrella]*) seem more resistant to knapweeds than nearby blue-bunch wheatgrass (*Elymus spicatus [Agropyron spicatum]*) communities (Sturko and Wikeem, 1990). Once an infestation is established, the alien weeds can maintain environmental conditions suitable for themselves, with the possible result that the original climax ecosystem may never be attained and productivity may be permanently lowered (personal communication in September, 1993 with B. Wikeem, Ministry of Forests).

Among the worst weeds for rangelands are knapweeds (especially the spotted knapweed *(Centaurea maculosa)* and the diffuse knapweed *(C. diffusa)*. They are primarily weeds of grasslands in the Southern Interior, with major concentrations in the Okanagan, Kootenay and Thompson regions, and smaller concentrations throughout southern, central and eastern British Columbia (Watson and Renney, 1974). In 1988, knapweeds were found in the Peace River district. According to the Ministry of Agriculture, Fisheries and Food (undated brochure), diffuse and spotted knapweeds have caused major environmental deterioration and loss of beef production, as well as undetermined losses in the value of grassland for recreational and wildlife use. Over 40,000 hectares of grazing lands are affected, with forage reductions of up to 88%. There is the potential in British Columbia for knapweed to spread to one million hectares of grassland range and an additional undetermined area of transitional forest. If knapweeds spread to their current ecological limits, the total susceptible grassland in Western Canada would be 8 to 10 million hectares.

Canada thistle (*Cirsium arvense*), also alien, is widespread in pastures. It does not limit forage production except in local situations, such as around ephemeral ponds and wet meadows that dry in summer and receive heavy grazing, where it can form dense stands. It is also found in logged areas, where the wind-borne seeds germinate on disturbed soils not seeded to a competitive perennial cover.

Other potential new problem weeds include wild four o'clock (*Mirabilis nyctajinea*), recently found at Spences Bridge; yellow starthistle (*Centaurea solstitialis*), not yet in British Columbia but identified in adjacent Washington and Idaho; and rush skeleton weed (*Chondrilla juncea*) which was recently found in the Kootenays and the North Okanagan (personal communication in September, 1993 with D. Blumenauer, Ministry of Agriculture, Fisheries, and Food).

A discussion of introduced rangeland plants would be incomplete without mention of alien grasses. Many different seed mixtures of several dozen species of grasses and other plants have been used in British Columbia by different agencies for a variety of purposes, such as seeding rights of way. These domestic species only become a problem if they invade natural habitats to the detriment of native species. Provincial government agencies and private ranchers have seeded crested wheatgrass *(Agropyron cristatum)* and other grasses and legumes to improve forage production and for rangeland rehabilitation. These mixes are increasingly seeded on clear-cut areas in the interior, to provide livestock forage during the early reforestation period. Cheatgrass *(Bromus tectorum)*, a very invasive annual grass, may have spread north from California, where it was introduced by early Spanish settlers with hay for their cattle, and has been associated with widespread rangeland degradation. It sprouts from foxtail-type seeds dispersed by furred wildlife, grows quickly, sets seed in spring and then dies. Under heavy grazing pressure on the preferred perennial grasses, annuals such as this can gain a competitive advantage (Dasman, 1965). Although now widespread in the Southern Interior, cheatgrass is not yet a serious problem for grazing in British Columbia and will not likely become one so as long as rangelands are maintained in good condition (personal communication in September, 1993 with D. Blumenauer, Ministry of Agriculture, Fisheries, and Food). Taken together, whether a cause of ecosystem degradation or merely a consequence of other disturbance, these alien weeds considerably degrade the land's ability to support livestock and wildlife.

Costs of Control

No one can pretend that it would be possible to eradicate the long established and widely dispersed alien wildflowers and weeds from the province, but they are the subject of intensive control efforts. For example, control of knapweeds on rangelands costs the province about $1,000,000 per year (Environment Canada, 1991). Various crop/range management, mechanical and biological control programs have limited—but not stopped—the spread and reduced the damage caused by these weeds. This control is accomplished through considerable interagency dialogue and cooperative programs between federal, provincial and regional district governments as well as non-government groups, such as cattlemen's associations. The emphasis of current weed control programs is on containing existing infestations and preventing new ones. However, controls are not completely without impacts of their own. Herbicides such as Picloram, used to control broad-leaved weeds and to encourage grasses, may affect biodiversity of native plant communities within the treatment areas (Sturko and Wikeem, 1990). Pesticides also enter stormwater systems in urban and suburban environments and, thence, streams and rivers, where the "first flush" of rain after a long dry spell in urban environments often produces a "slug" of runoff water with high levels of pesticide residues that are toxic to aquatic plants, insects and fish (Whitfield and Wade, 1992). Hence, herbicides applied to control alien and other weeds may temporarily reduce biodiversity of non-target organisms.

A ray of hope in the control of introduced wildflowers in urban areas is the increasing awareness of the importance of biodiversity (see Chapter 22, this volume). Many municipalities across North America are increasingly using native species in parks and other municipal landscaping. This change, to some degree, begins a shift from urban environments being a source of the problem to being part of the solution, providing that the native species are obtained from seed or vegetative propagation, rather than digging up wild stock. Landscaping with native species is not only good for the environment, but may also reduce landscape maintenance budgets and conserve water. A more proactive solution would be to legislate, by provincial law or municipal ordinance, against importation or release of potentially damaging flowers, such as purple loosestrife, much as is done for range weeds of economic importance.

Acknowledgements

Roy Cranston, Don Blumenauer and Alan Sturko of theBritish Columbia Ministry of Agriculture, Fisheries and Food, and Brian Wikeem of the British Columbia Ministry of Forests, reviewed the manuscript and provided valuble information on range and agricultural weeds.

References

Best, K.F., G.G. Bowes, A.G. Thomas and M.G. Maw. 1980. The biology of Canadian weeds 39: *Euphorbia esula*. L. *Canadian Journal of Plant Science* 60:651-663.

Burcham, L.T. 1957. *California Range Land: an Historico-ecological Study of the Range Resource of California.* Department of Natural Resources, Sacramento, California.

Clark, L.J. 1976. *Wildflowers of the Pacific Northwest.* Gray's Publishing Ltd., Sidney, British Columbia.

Dasman, R.F. 1965. *The Destruction of California.* Macmillan Co., New York, New York.

Douglas, G.W., G.B. Straley and D. Meidinger, 1989-91. The Vascular Plants of British Columbia Part I Gymnosperms and Dicotyledons (*Aceraceae* through *Cucurbitaceae*), April 1989; Part II Dicotyledons (*Diapensiaceae* through *Portulacaceae*), April 1990; and Part III Dicotyledons (*Primulaceae* through *Zygophyllaceae* and *Pteridophytes*). Ministry of Forests.

Environment Canada. 1991. *The State of Canada's Environment.*

Gross, R.S., P.A. Werner and W.R. Hawthorn. 1980. The biology of Canadian weeds 38: *Arctium minus* (Hil) Bernth and *A. Lappa L. Canadian Journal of Plant Science* 60:621-634.

Montgomery, F.H. 1964. *Weeds of Canada and the northern United States.* Ryerson Press, Toronto

Moore, R.J. 1975. The biology of Canadian weeds 13: *Cirsium arvense L. Canadian Journal of Plant Science* 55:1033-1048.

Mulligan, G.A. 1976. *Common weeds of Canada.* McClelland and Stewart Ltd.

Mulligan, G.A. and L.G. Bailey. 1975. The biology of Canadian weeds 8: *Sinapis arvensis L. Canadian Journal of Plant Science* 55:171-183.

Mulligan, G.A. and J.N. Findlay. 1974. The biology of Canadian weeds 3: *Cardaria draba, C. chalepensis,* and *C. pubsecens. Canadian Journal of Plant Science* 54:149-160.

Mulligan, G.A. and D.B. Munro. 1981. The biology of Canadian weeds 48: *Cicuta maculata L., C. douglasii* (DC.) Coult. & Rose and *C. virosa L. Canadian Journal of Plant Science* 61:93-105.

McNeill, J. 1980. The biology of Canadian weeds 46: *Silene noctiflora L. Canadian Journal of Plant Science* 60:1243-1253.

Salisbury, E., 1961. Weeds and aliens. Colins, London.

Sharma, M.P. and H. Vanden Born. 1978. The biology of Canadian weeds 27: *Avena fatua L. Canadian Journal of Plant Science* 58:141-157.

Sorensen, A.E. 1984. Seed dispersal and the spread of weeds. In: Proc. VI *Int. Symp. on Biological control of Weeds.* E.S. Delfosse (ed.). Univ. of British Columbia. pp. 121-132.

Sturko, A. and B.M. Wikeem. 1990. *The Effects of Weeds on Grassland Biodiversity.* Paper for 1990 Section Meeting, Society for Range Management, Pacific Northwest Section, Vernon, B.C. October 26-27, 1990.

Taylor, R.L. and B. MacBryde, 1977. *Vascular plants of British Columbia. A descriptive resource inventory.* U.B.C. Botanical Garden Tech. Bull. No. 4.

Watson, A.K. 1980. The biology of Canadian weeds 43: *Acroptilon (Centurea) repens* (L.)DC. *Canadian Journal of Plant Science* 60:993-1004.

Watson, A.K. and A.J. Renney. 1974. The biology of Canadian weeds 6: *Centurea diffusa* and *C. maculosa. L. Canadian Journal of Plant Science* 54:687-701.

Weaver, S.E. and W.R. Riley. 1982. The biology of Canadian weeds 53: *Convolvulus arvensis L. Canadian Journal of Plant Science* 62:461-472.

Werner, P.A. and R. Rioux. 1977. The biology of Canadian weeds 24: *Agropyron repens* (L.)Beauv. *Canadian Journal of Plant Science* 57:905-919.

Werner, P.A. and J.D. Soule. 1976. The biology of Canadian weeds 18: *Potentilla recta L., P. norvegica L. and P. artentea. L. Canadian Journal of Plant Science* 56:591-603.

Whitfield, P.H. and N.L. Wade. 1992. Monitoring transient water quality events electronically. American Water Resources Association. August 1992 . *Water Resources Bulletin* 28(4):703-711.

Exotic Submersed Aquatic Macrophytes in British Columbia

Peter R. Newroth
Water Quality Branch
Ministry of Environment,
Land and Parks
765 Broughton Street
Victoria, B. C.
V8V 1X5

Vascular aquatic plants can be broadly categorized as rooted floating, non-rooted floating, rooted emergent and rooted submergent. Rooted submergent aquatic plants grow in the littoral zone (along the shoreline) which is subject to disturbances through development, resource exploitation or natural events that create opportunities for invasion of this habitat by alien plants. This article briefly reviews the status of submersed aquatic macrophytes (the larger forms of attached algae and angiosperms) in the fresh waters of British Columbia, and then focuses on the impacts of alien or introduced plants on the aquatic ecosystem, using Eurasian watermilfoil (*Myriophyllum spicatum*) as a case study.

Submersed Aquatic Plants in British Columbia

Prior to 1972, most reports on submersed aquatic plants in British Columbia were limited to general floristic studies, without focus on ecosystems, exotic species or management of nuisance populations. Subsequent major efforts to document the submersed aquatic plant flora of British Columbia and to test, develop and apply management methods were made mainly a result of the introduction and subsequent nuisance impacts of Eurasian watermilfoil. Much of this work has been performed since 1972 by the Water Quality Branch of the B. C. Ministry of Environment, Lands and Parks. The Water Quality Branch has established an herbarium with about 10,000 aquatic plant specimens and an electronic database of aquatic plant distribution in about 2,000 water bodies.

Approximately 150 species of vascular freshwater aquatic plants (excluding algae, mosses and true emergents) have been recorded in British Columbia (Warrington, 1980). Of these 150 species, about 28 or nearly 20% are included in the rare vascular plant list for British Columbia (Straley, *et al.*, 1985). Eleven of these 28 species are classified in the most rare group, represented by a single or few known populations.

Perhaps this relatively high proportion of rare species reflects the small amount of research performed to date. Research has been limited by the large geographic area of British Columbia and by the difficulty of investigating underwater habitats before the advent of SCUBA. Also, to date there has been little public interest in aquatic plants, except when they have been collected for use in aquaria or ornamental ponds or when their growth has impaired water use. Generally, therefore, inventory and ecological studies of plants in freshwater ecosystems in British Columbia have been neglected, especially in comparison to studies of sport and commercial fish. Biodiversity of aquatic plants and longer term community studies remain promising areas for future research.

At least 7% of the recorded aquatic macrophytes are alien to British Columbia, and their presence represents a change in aquatic biodiversity. The introduced aquatic macrophytes (including several emergent species) recorded in British Columbia include:

o   Eurasian watermilfoil (*Myriophyllum spicatum*)

o   parrot-feather (*Myriophyllum aquaticum*)

180

- curlyleaf pondweed (*Potamogeton crispus*)
- fragrant water lily (*Nymphaea spp.*)
- Brazilian elodea (*Egeria densa*)
- pickerelweed (*Pontederia cordata*; a marginal emergent)
- water hyacinth (*Eichhornia crassipes*)
- pillwort (*Pilularia americana*)
- Oakes pondweed (*Potamogeton oakesianus*)
- water fern (*Azolla filiculoides*)
- floatingheart (*Nymphoides spp.*)

There is considerable debate about when and how most of these species were introduced. A number of them, such as curlyleaf pondweed and fragrant water lily, have been present in B.C. waters for decades and are local nuisances. Several species, including pickerelweed, pillwort and parrot-feather, are not killed by freezing conditions and have been found flourishing on local lake shores, probably after having escaped from cultivation. Water hyacinth and floatingheart are at present limited to enclosed water gardens in the mild climate of the coastal zone. Brazilian elodea has become well established in one small lake on Vancouver Island and has nuisance potential, like several other species on the above list.

The federal Plant Quarantine Act may be applied to ban transport into Canada of Eurasian watermilfoil and other watermilfoils, water chestnut (*Trapa natans*), hydrilla (*Hydrilla verticillata*) and Brazilian elodea, but there is no inspection of transboundary movement of boating equipment. There are no legislated regulations to deter aquatic plant shipments or transfers across Canada.

The mild, maritime climate in western British Columbia and the large number of water bodies along the coast provide conditions encouraging survival and dispersal of introduced aquatic plants. There is increasing concern about further introductions and the potential for some of the alien species already present in garden and lakefront settings to spread outside their cultivated areas. At the same time, there appears to be growing interest in the development of water landscape features in private gardens and aquatic displays as tourist attractions in public gardens. Several times in the past ten years, aquatic plants in over 40 public ornamental gardens and nurseries have been surveyed, and each time many exotic species have been identified (Warrington, 1989). Not only are exotics deliberately planted, they are also spread inadvertently through adherance to boating equipment and other items which are moved between freshwater bodies. For example, water lily rhizomes transferred from several private water gardens to other private ponds and public waters have been found to have been accompanied by viable fragments of Eurasian watermilfoil (Newroth, 1987).

Hydrilla (*Hydrilla verticillata*) is established in California and is subject to intensive management there. If introduced to British Columbia by boaters or aquarium shipments, it is likely to become a greater nuisance and be more difficult to control than Eurasian watermilfoil. Other noxious aquatic species which probably could survive well in British Columbia, but have not yet been found in-

181

clude: water chestnut (*Trapa natans*), fanwort (*Cabomba caroliniana*), Australian swamp stonecrop (*Crassula helmsii*) and African elodea (*Lagarosiphon major*).

White and Haber (1992) noted that southern parts of British Columbia and Ontario are experiencing the greatest impacts from invasive (both endemic and alien) plants in wetland habitats across Canada. Based on literature studies and a questionnaire returned by 33 botanists across Canada, they identified the following alien species as most significantly threatening natural wetlands: Eurasian watermilfoil, purple loosestrife (*Lythrum salicaria*), European frog-bit (*Hydrocharis morsus-ranae*), and three other grass, rush and shrub species. Frog-bit is not reported in British Columbian waters. Curlyleaf pondweed was listed among invasive species that were considered less threatening to natural areas.

In a number of cases, overabundance or invasiveness of endemic species such as American elodea (*Elodea canadensis*) also impairs our use of water and probably adversely affects the ecosystem (Spicer and Catling, 1988). Some research on the ecology of this plant in British Columbia has been sponsored by BC Environment and Environment Canada. Control of a number of native plants, in addition to exotic species, is now being implemented in a number of locations, including Harrison Lake, Hope River and Elk Lake, where water uses are impaired.

Because of the major environmental impacts and public concerns about Eurasian watermilfoil, the following sections review the characteristics of this plant and describe its expansion in British Columbia. Problems with Eurasian watermilfoil may be compounded if additional introductions occur. Since both hydrilla and African elodea have the physiological capacity to grow well in British Columbia waters, it is probable that these species would become additional nuisances if they were introduced.

Eurasian Watermilfoil- Impacts on Aquatic Ecosystems in British Columbia

The biological characteristics of Eurasian watermilfoil have been described in detail elsewhere and are only briefly discussed here (Aiken *et al.*,1979; Anon., 1981; Smith and Barko, 1990). Eurasian watermilfoil is a rooted perennial with long, flexible stems and branches bearing whorls of dissected leaves. Extensive, dense root systems develop to anchor the plant and are the main site for nutrient uptake from the substrate. This plant may grow in ponds, sloughs, rivers and lakes on a wide range of substrate types, from silts to coarse gravels. Water depths where growth is recorded range from about 1 to 10 metres; maximum root biomass has been recorded at depths of between 2 and 5 metres. Eurasian watermilfoil densities in many sites in the Okanagan Valley range up to over 130 individual plants per square metre, with over 100 to over 900 stems per square metre. While this plant produces seeds, its main method of reproduction is vegetative, by fragmentation. Fragments are released by the plant during the growing season by development of abscission, and extensive fragment release occurs through damage of surfacing populations by boaters and wave action.

Physiological characteristics of Eurasian watermilfoil include the capacity to concentrate biomass near the water surface in dense mats or canopies and to utilize both free carbon dioxide and bicarbonate ions. In the Okanagan Valley, Eurasian watermilfoil may overwinter with short stems well below the surface, but major stem growth commences in spring when water temperatures exceed 10°C. Summer rates of vertical growth have been recorded at about four centimetres per day. Because of its relatively plastic growth form, this species may outcompete many other submersed plants (Newroth, 1986; Madsen et al., 1991a).

Native to Europe, Asia and North Africa, this species was first officially recorded in British Columbian waters in 1972, probably following introduction in about 1970. Probably Eurasian watermilfoil came from the United States, where it was first documented in the 1940's. It is now considered an aquatic pest in most of the United States, British Columbia, Ontario and Quebec. It has not been reported in the three prairie provinces or the atlantic provinces of Canada. Nine watermilfoil species are recorded in British Columbia, including two new records for Canada and one for North America (Warrington, 1986).

Eurasian watermilfoil is now known in about 80 lakes, ponds, rivers and wetlands in southwestern and south-central British Columbia. Seven of these water bodies are in a small area of Vancouver Island, and two of the lakes are in the Kootenay region. While large areas of northern, western and southeastern British Columbia appear to be uninfested with Eurasian watermilfoil, it is difficult to be sure since surveys in northern B.C. have been limited. Since the Kootenay region adjoins the Province of Alberta, intensive management is particularly important because Eurasian watermilfoil is not known in the prairie provinces. As well, annual surveys check 30 to 50 lakes on the periphery of known infestations. Most southern British Columbian water bodies that are accessible by trailered boats and near Eurasian watermilfoil populations have been checked.

A large effort has been made to locate new Eurasian watermilfoil infestations, especially since 1977, when it became apparent that this species had the potential to thrive under a wide range of water quality conditions and altitudes. The rate of spread appears to have slowed during the past ten years, although this apparent trend may reflect limited surveys and the difficulty of finding new populations. Also, it probably reflects the rapid establishment and spread of this species into the most suitable habitats during the 1970's. Public information programs, including pamphlets, posters and Eurasian watermilfoil warning signs, were introduced about 1976. Combined with intensive control of Eurasian watermilfoil, which limits boater contact with this plant and its spread on boating equipment, such preventive actions are credited with reducing further spread. Eurasian watermilfoil was even eradicated in about six small ponds on Vancouver Island in 1986.

As reported by Kangasniemi (1983), routine ecological studies of Eurasian watermilfoil in six Okanagan Valley lakes detected declines in Eurasian watermilfoil populations, apparently associated with extensive insect damage. In many locations, Eurasian watermilfoil decline was apparent because surface canopies did not develop, thereby reducing the nuisance impact in these areas. Up to 80% of Eurasian watermilfoil meristems in these areas showed damage from larvae of the milfoil midge (*Cricotopus myriophylli*), weevil larvae (several species) and caddisflies (*Triaenodes tarda*). Unfortunately, the Water Quality Branch's research into the use of the milfoil midge as a biocontrol agent was discontinued in 1992 due to a lack of funds.

Interference with human uses

Newroth (1986) outlined the main water quality problems perceived by the public following the spread and establishment of Eurasian Watermilfoil. These problems were observed mostly in littoral areas in which native plants were not considered a nuisance. In the Okanagan Valley lakes, interference with public recreation, especially beach and boating activities, by dense populations of Eurasian watermilfoil led to demands for control as early as 1972, shortly after its initial introduction. This negative reaction was based mainly on observation of rapid Eurasian watermilfoil growth along previously unvegetated beach areas. The high plant density associated with Eurasian watermilfoil, development of mats on the surface, accumulation of fragments on beaches, and persistence from June to September, were new phenomena in many areas. Public safety also became a concern following several drownings in dense Eurasian watermilfoil populations.

Spread of Eurasian watermilfoil within southern British Columbia has led to ongoing control programs in about 16 lakes (Newroth, 1988) and several privately or municipally funded projects. Impacts on the tourism industry in the Okanagan Valley were assessed in 1991 by a socio-economic study (Anonymous, 1991). This study concluded that termination of annual harvesting and derooting treatments of Eurasian watermilfoil in about 150 hectares of littoral area would lead to a projected decline of $85 million in Okanagan tourist revenues.

Habitat Modification

Eurasian watermilfoil populations in flowing water grow sufficiently dense to restrict discharges and cause minor flooding, interfering with regulation and gauging of flows. Dense Eurasian watermilfoil growth also contributes to accumulation of sediment in shallow shoreline areas. As well, Eurasian watermilfoil tissues continue to slough off throughout the growing season. In some locations, Eurasian watermilfoil mats effectively form a breakwater along the shoreline, reducing normal current, storm and wave washing action and sediment sorting. The long term effect of this change probably is expressed in changes in the plants and benthic invertebrates in such areas. In some areas, beach quality has deteriorated because of this change in sediment texture.

Physico-chemical Impacts on the Aquatic Community

Dense populations of aquatic plants, especially those that form surface mats over large areas and in small water bodies, cause many physical and chemical effects, including light attenuation, reduction of oxygen during night-time respiration and saturation

184

during photosynthesis, and carbon dioxide limitation. Madsen et al., (1991b), demonstrated the shading effects of Eurasian watermilfoil in Lake George, New York, and concluded that native species were excluded by shading, reducing species diversity and irrevocably altering the littoral ecosystem. Studies of the water column and sediments underlying dense aquatic plant populations in Washington State have shown that dissolved oxygen, pH and phosphorus levels may be affected (Frodge *et al.*, 1990; 1991). Dissolved oxygen levels may fall below the minimum standard for several fish species and the pH changes also may be detrimental to aquatic life.

Eurasian watermilfoil produces a larger number of shoots and occupies the water column for a longer period of the year than some native species. Unusually high shoot turnover during much of the growing season also is characteristic of Eurasian watermilfoil (Smith and Adams, 1986). With its high productivity and high uptake of phosphorus through the roots from sediments, it acts as a significant pump for phosphorus. Kangasniemi and Hall (unpublished) calculated that the potential seasonal phosphorus release from Eurasian watermilfoil into Okanagan Lake exceeds that from point sources such as storm sewers, industrial sources, and fertilizers. Phosphorus loadings from Eurasian watermilfoil are less than, but in the same order of magnitude as, loadings from septic tanks, logging operations, dustfall or precipitation. As well, Eurasian watermilfoil tissue contains a number of phenolic compounds that are released during extracellular secretion and tissue decomposition. These compounds may be toxic or affect the growth of surrounding plants and animals, including phytoplankton (Planas *et al.*, 1981). Research on aquatic plants is beginning to address the implications of these and other effects.

Lake	Year of Observation		
	1975	1977	1979
Okanagan	233 hectares	258 hectares	403 hectares
Kalamalka	0.3 hectares	11 hectares	4 hectares
Skaha	4 hectares	64 hectares	71 hectares
Wood	traces	3 hectares	17 hectares
Vasseux	traces	73 hectares	101 hectares

Table 17-5: Encroachment of Eurasian Watermilfoil (*Myriophyllum spicatum*) in Five Okanagan Lakes, 1975-1979.
Source: Newroth, 1993.

Rapid Expansion and Displacement of Native Aquatic Plants

Expansion of Eurasian watermilfoil has been documented in many lakes in British Columbia. From 1975 to 1979, Eurasian watermilfoil and native plant populations in five mainstem Okanagan Valley lakes were documented and mapped in detail (Newroth, 1993). Despite control efforts in high public use areas, rapid encroachment by Eurasian watermilfoil was indicated by increases in the size of the affected areas, as illustrated in Table 17-5.

Observations at Eurasian watermilfoil study sites during ten years of biocontrol studies have shown remarkable consistency in the borders of colonies of this plant. While densities of Eurasian watermilfoil have been noted to vary from lake to lake and site to site over the years between 1979 and 1992, the total area affected in the Okanagan Valley lakes has expanded. In 1992, Eurasian watermilfoil populations in the mainstem Okanagan lakes now occupy most of the suitable habitat, which is estimated at about 750 hectares. Annual control using harvesting, derooting and bottom barrier techniques treats only about 150 hectares, or about 20% of the affected area, in eight Okanagan Valley lakes.

185

While dense Eurasian watermilfoil populations have grown in some areas where little native vegetation had been established, mapping also has shown considerable displacement of native plants, especially pondweeds (*Potamogeton* spp.), and successful competition with coontail (*Ceratophyllum demersum*). In many cases, there had been no human disturbance of the areas affected. This success is attributed mainly to shading of competitors, but dense Eurasian watermilfoil root development may also be implicated. In Lake George, New York, the total number of aquatic plant species found within Eurasian watermilfoil beds declined from twenty to nine during the 1987 to 1989 period (Madsen et al., 1991a). Also, this study noted that Eurasian watermilfoil colonized a stable and healthy plant community.

Impacts on Aquatic Invertebrates and Fish

Limited research has been performed on fish and invertebrate populations associated with Eurasian watermilfoil in British Columbia. Robinson (1981) found salmonid fishes absent from dense Eurasian watermilfoil colonies during summer months, and benthic invertebrates were less abundant in sites with Eurasian watermilfoil in the Okanagan Valley. Benthic invertebrate populations in dense Eurasian watermilfoil in Washington State were found to be limited in abundance, diversity and evenness of distribution (Gibbons *et al.*, 1987). In Ontario, Keast (1984) found three to four times more fish in *Potamogeton-Vallisneria* populations than in Eurasian watermilfoil populations; dense Eurasian watermilfoil apparently formed a barrier to fish movement. He also confirmed earlier work, which reported that Eurasian watermilfoil had the poorest invertebrate fauna of eight aquatic plant species, by showing that native plants supported twice as many invertebrates in late summer as Eurasian watermilfoil.

Eurasian watermilfoil has been reported to have adverse impacts on fish spawning in British Columbia (Truelson, 1986). Several species of salmon were affected by reduction of the available spawning area in Cultus Lake. Also, kokanee spawning and bass habitat have been adversely affected in Okanagan Valley lakes and rivers, requiring Eurasian watermilfoil control. There are growing concerns about the probable adverse impacts of Eurasian watermilfoil on salmonid rearing in the Shuswap Lake system and other areas of British Columbia. Juvenile salmonids may be subject to additional predation because the establishment of dense macrophyte populations creates spawning substrates and rearing habitats for fish that spawn in vegetation. Research by Fisheries and Oceans Canada in Cultus Lake indicates that populations of squawfish, which eat juvenile salmonids, may be encouraged by Eurasian watermilfoil colonies. Much work remains be undertaken by fisheries biologists to investigate the impacts of aquatic plants on spawning and rearing of salmonids.

Adverse Impacts of Eurasian Watermilfoil Control

Approximately 1500 hectares of littoral area are now affected by Eurasian watermilfoil; annual control of about 300 hectares in 16 lakes is performed by six local agencies with technical and funding assistance from the Water Quality Branch. The provincial and local costs of implementing Eurasian watermilfoil control were approximately $700,000 in 1992, which are expected to continue.

Areas treated are selected for their high public recreation value. Control operations may be classified as intensive, preventive or cosmetic. Intensive management minimizes disruption of the sites by primarily using bottom barrier systems and SCUBA divers to prevent Eurasian watermilfoil growth from establishing small colonies and to delay spread to new locations (Newroth, 1990). This approach is practical only where Eurasian watermilfoil is found soon after initial colonization and where reinfestation from upstream or adjacent areas is unlikely.

Where Eurasian watermilfoil is well established and containment is no longer practical, harvesters, rototillers and cultivators are used for mechanical control (Maxnuk 1986; Newroth 1986). Though more disruptive, these more cosmetic treatments usually prevent the development of dense surface mats, and encourage some repopulation of the sites by native plants. Harvesting removes large numbers of juvenile fish, which may or may not be advantageous, depending on the location. Several harvests may be required each summer for Eurasian watermilfoil management. In general, derooting provides longer-lasting control and may be used in spring, fall and winter.

Conclusions

Aquatic plant communities of British Columbia are poorly understood because of the difficulty associated with their study and the apparently low priority placed on their research. Many alien aquatic plant species have become established in British Columbia, especially in comparison with the prairie provinces, where none have been cited. Already, several of these species are affecting aquatic ecosystems and reducing biodiversity in some sites. The degree of impact remains unpredictable in many sites because of inadequate information on ecosystem effects. Without ongoing management, there is potential for additional impact, especially if other exotic plants invade the province.

Acknowledgements

I thank P. D. Warrington and B. J. Kangasniemi for some of the information presented and for commenting on the manuscript. V. Bartnik, Environment Canada, reviewed the manuscript.

References

Anonymous. 1981. *A Summary of Biological Research on Eurasian Water Milfoil in British Columbia*. Information Bulletin, Volume XI. B.C. Ministry of Environment, Aquatic Studies Branch, Victoria.

Anonymous. 1991. *Evaluation of the Socio-economic Benefits of the Okanagan Valley Eurasian Water Milfoil Control Program*. Ference Weicker & Company.

Aiken, S.G., P.R. Newroth and I. Wile. 1979. The Biology of Canadian Weeds. 34. *Myriophyllum spicatum L. Canadian Journal of Plant Science* 59:201-215.

Frodge, J.D., G.L. Thomas and G.B. Pauley. 1990. Effects of canopy formation by floating and submergent aquatic macrophytes on the water quality of two shallow Pacific northwest lakes. *Aquatic Botany* 38:231-248.

Frodge, J.D., G.L. Thomas and G.B. Pauley. 1991. Sediment phosphorus loading beneath dense canopies of aquatic macrophytes. *Lake and Reservoir Management* (7)1:61-71.

Gibbons, M.V., H.L. Gibbons, Jr. and R.E. Pine. 1987. *An Evaluation of a Floating Mechanical Rototiller for Eurasian Water Milfoil Control.* Department of Ecology, Water Quality Program, State of Washington.

Kangasniemi, B.J. 1983. Observations on herbivorous insects that feed on Myriophyllum spicatum in British Columbia. In: *Lake Restoration, Protection and Management.* Proceedings of the North American Lake Management Society Conference, October 27-29, 1982, Vancouver, British Columbia. US Environmental Protection Agency. 44015-83-001. pp. 214-218.

Keast, A. 1984. The introduced aquatic macrophyte, *Myriophyllum spicatum,* as habitat for fish and their invertebrate prey. *Canadian Journal of Zoology* 62:1289-1303.

Madsen, J.W., J.W. Sutherland, J.A. Bloomfield, L.W. Eichler and C.W. Boylen. 1991a. The Decline of Native Vegetation Under Dense Eurasian Watermilfoil Canopies. *Journal of Aquatic Plant Management* 29:94-99.

Madsen, J.W., C.F. Hartleb and C.W. Boylen. 1991b. Photosynthetic characteristics of *Myriophyllum spicatum* and six submersed aquatic macrophyte species native to Lake George, New York. *Freshwater Biology* 26: 233-240.

Maxnuk, M.D. 1986. Bottom tillage treatments for Eurasian water milfoil control. In: *Proceedings of the First International Symposium on Watermilfoil (Myriophyllum spicatum) and related Haloragaceae species.* July 23 and 24, 1985, Vancouver, B.C. Canadian Aquatic Plant Management Society, Inc. 1986. pp. 163-172.

Newroth, P.R. 1986. A review of Eurasian water milfoil impacts and management in British Columbia. In: *Proceedings of the First International Symposium on Watermilfoil (Myriophyllum spicatum) and related Haloragaceae species.* July 23 and 24, 1985, Vancouver, B.C. Aquatic Plant Management Society, Inc. 1986. pp. 139-153.

Newroth, P.R. 1987. Help protect lakes from exotic aquatic plants. *Lake Line* 7(4):p 44, 49.

Newroth, P.R. 1988. Review of current aquatic plant management activities in British Columbia In: *Proceedings; 22nd Annual Meeting, Aquatic Plant Control Research Program.* Miscellaneous Paper U.S.A.C.E. A-88-5, 66-71.

Newroth, P.R. 1990. Prevention of the Spread of Eurasian Water Milfoil. In: *Proceedings, National Conference on Enhancing the State's Lake and Wetland Management Programs.* Chicago, Illinois. USEPA, NALMS. pp. 93-100.

Newroth, P.R. 1993. *Application of Aquatic Vegetation Identification, Documentation and Mapping in Eurasian watermilfoil Control Projects.* Presented to Aquatic Vegetation Quantification Symposium, NALMS, 1990. (In Press).

Planas, D., F. Sarhan, L. Dube, H. Godmaire and C. Cadieux. 1981. Ecological significance of phenolic compounds of Myriophyllum spicatum. *Verhandlungen der Internationalen Vereinigung für Theoretische und Angerwandte Limnologie* 21:1492-1496.

Robinson, M.C. 1981. *Eurasian Water Milfoil Studies, Volume I: The Effect of Eurasian Water Milfoil (Myriophyllum spicatum L.) on Fish and Waterfowl in the Okanagan Valley, 1977.* APD Bulletin 15. British Columbia Ministry of Environment, Victoria, B.C.

Smith, C.S. and M.S. Adams. 1986. Phosphorus transfer from sediments by *Myriophyllum spicatum. Limnology and Oceanography* 31 (6):1312-1321.

Smith, C.S. and J.W. Barko. 1990. Ecology of Eurasian Watermilfoil. *Journal of Aquatic Plant Management* 28:55-64.

Spicer, K.W. and P.M. Catling. 1988. The biology of Canadian weeds. 88. Elodea canadensis Michx. *Canadian Journal of Plant Science:*1035-1051.

Straley, G.B., R.L. Taylor and G.W. Douglas. 1985. *The Rare Vascular Plants of British Columbia.* Syllogeus no. 59. World Wildlife Fund (Canada). UNESCO Program on Man and the Biosphere.

Truelson, R.L. 1986. Community and government cooperation in control of Eurasian water milfoil in Cultus Lake, B.C. In: *Proceedings of the First International Symposium on Watermilfoil (Myriophyllum spicatum) and Related Haloragaceae Species.* July 23 and 24, 1985, Vancouver, B.C. Canada. Aquatic Plant Management Society, Inc. pp. 154-162.

Warrington, P.D. 1980. *Studies on Aquatic Macrophytes. Part XXXIII.* Aquatic plants of British Columbia. Inventory and Engineering Branch Report No. 2916. British Columbia Ministry of Environment.

Warrington, P.D. 1986. Factors associated with the distribution of Myriophyllum in British Columbia. In: *Proceedings of the First International Symposium on Watermilfoil (Myriophyllum spicatum) and Related Haloragaceae Species.* July 23 and 24, 1985. Vancouver, B.C. Canada Aquatic Plant Management Society, Inc. 1986. pp. 79-94.

Warrington, P.D. 1989. *Aquatic Vegetation Survey of the Major Parks, Gardens and Other Tourist Attractions of Southern Vancouver Island and the Lower Fraser Valley, 1988.* Water Management Branch Report No. 3550. British Columbia Ministry of Environment.

White, D.J. and E. Haber. 1992. *Invasive Plants of Natural Habitats in Canada. Part 1: Wetland Species.* Report prepared for Canadian Wildlife Service.

Effects of Alien Insects and Microorganisms on the Biodiversity of British Columbia's Insect Fauna

Risa Smith
Ministry of Environment, Lands and Parks
810 Blanshard Street
Victoria, B.C.
V8V 1X4

Alien insects enter British Columbia via two major routes—either as inadvertent introductions associated with human activity or as purposeful introductions associated with biological control programs. There are also other, less common routes, such as the introduction of honeybees for honey production and crop fertilization. The public, scientists and the media have all expressed concern about the impact of exotic insects. However, this concern most often centres around the possibility of exotic insects becoming pests. Interest in the impact of exotic species on biological diversity has been very limited.

The major threat to biodiversity from alien insect species is the possibility of competitive displacement of native species by invaders. This displacement could lead to the extinction of native species and alterations in the structure of communities. However, major alterations in the biodiversity of insects in native habitats as a result of inadvertent introductions of alien insects are unlikely. There is some evidence that native habitats are protected from invasions by what Elton (1958) termed ecological resistance. In the unlikely event that a new immigrant is able to overcome obstacles to establishment, such as too few colonizing individuals to make a viable population, inappropriate climatic conditions, unsuitable habitats or unavailability of appropriate hosts (Simberloff, 1989), ecological resistance in native habitats is likely to overwhelm the invader. Generalist predators, parasites and diseases, as well as superior competitors, either kill invaders directly or quickly outcompete them. Initial data on this subject support theoretical expectations. In a unique set of experiments, Spence and Spence (1988) and Spence (1990) have demonstrated that, of 21 species of introduced carabid beetles in British Columbia, none have invaded native climax coniferous forests, although many dominate in the cultural steppe (habitats associated with human activity) and disturbed habitats. Because none of the introduced carabids were associated with the cultural steppe in Britain, where they originated (Spence, 1990) a convincing argument has been made that native habitats in British Columbia are impermeable to invasion. If alarm is to be raised over the loss of biodiversity of the British Columbian insect fauna, the solution is protection of native habitats.

Inadvertent Invasions

Most insects which have adventitiously been introduced into British Columbia have entered through ship ballast or on nursery stock. Before 1914, it was common practise to stabilize European ships en route to North America by filling the holds with ballast consisting of sand and rocks. For the return trip, the hold was emptied on shore and filled with North American raw materials. Many insect species must have invaded British Columbia through this route and it is testament to the resistance of native habitats that British Columbia has not been overrun with exotic species.

Until 1965, and the discovery of the golden nematode, *(Globodera rostockensis)*, in Newfoundland, quarantine restrictions on the entry of soil into Canada from outside North America were not in force (J. Bell, Plant Protection, Agriculture Canada, personal communication; Spence and Spence, 1988). The golden nematode is a serious pest of potatoes. Because it can persist in soil for up to 28 years, the only way to prevent its spread is to control the movement of infested soil. Through quarantine restrictions on both the crops that can be grown in infested soil and the movement of soil outside infested areas, the golden nematode's distribution in North America has been confined to New York State, Newfoundland and Vancouver Island, British Columbia (Baldwin and Mundo-Ocampo, 1991).

Nursery stock is the second major vehicle of entry for alien insects, in spite of restrictions imposed by the Plant Quarantine Act. For example, nursery stock was the likely route of entry for many recent introductions in British Columbia, including the European crane fly *(Tipula paludosa)*, the winter moth, *(Operophteron brumata)*, the apple ermine moth, *(Yponomeuta malinellus)*, and the cherry bark tortrix, *(Enarmonia formosana)*.

It is impossible to assess the total number of inadvertent introductions into British Columbia. It seems likely that most species arriving in the province never become established. Of those that do become established, an initial population explosion is usually followed by a reduction to a lower equilibrium density. The classic example of this process in British Columbia was the accidental introduction of the European crane fly. After an initial post-introduction population eruption, the European crane fly is now regulated by natural processes at low densities (Myers and Iyer, 1981). Although the cause of the decline in population density is not certain, protozoan parasites, which were inadvertently introduced along with the fly, probably play an important role.

Biological Control

In the past, the introduction of biological control agents has been deemed to be environmentally safe and risk-free. The data to support this supposition have not been systematically collected and some people claim there is little basis for the assertion of safety (Howarth, 1991). From the viewpoint of biological diversity, biological control is merely a subset of the problem of invasion of new areas by alien species.

Two approaches to biological control can involve the introduction of alien species—classical biological control and augmentative releases. Classical biological control involves the introduction of natural enemies, usually to control alien insects and weeds. Most often the introduced species originates from the same region as its introduced host. Introduced agents are expected to establish themselves in British Columbia and provide long-term population suppression of target species. Augmentative release, with alien species, involves the repeated introduction of species unable to establish themselves in British Columbia. Historically, augmentative release programs have been most popular in greenhouses. However, the technique is becoming more prevelant in non-greenhouse settings, such as orchards, berry crops and urban environments (Parrella et al., 1992).

In Canada, biological control agents for classical control are chosen with care. The general rule is to introduce predators and parasites with a narrow diet, thereby minimizing the impact on non-target species. Selection of agents to control weeds has been very stringent. To some extent these precautions have successfully protected the British Columbian insect fauna, as host transfer of biological control agents to native species has yet to be recorded in the province. However, attempts to assess the impact of biological control agents on native, non-target species are not usually made. In a recent review article, Howarth (1991) recorded only two examples, worldwide, of invertebrate biological control agents causing extinctions of native species. In one of those examples the species that became extinct, the Fijian coconut moth *(Levuana iridescens)*, was the target of the biological control program. In the other example the introduced parasite was known to be a generalist feeder before its introduction. The majority of extinctions caused by biological control programs involved the introduction of vertebrate predators.

In spite of attempts to be extremely cautious in the choice of biological control agents, unexpected changes in the diet of new species can occur. In California, several species of insects have been introduced to control the alien weed, St. John's wort, *(Hypericum perforatum)*. Although careful study prior to introductions demonstrated that the leaf beetle *(Chrysolina quadrigemina)*, did not feed on native species of St. John's wort, the leaf beetle is now feeding and reproducing on the California native, *H. concinnum* (Andres, 1985). The leaf beetle has also been introduced into British Columbia for biological control of St. John's wort. To date, there has been no documentation of the leaf beetle expanding its diet to include the two species of St. John's wort native to British Columbia.

The only example of a biological control agent displacing a native species in British Columbia has proven to be erroneous. The predatory Anthocorid, *Anthocoris nemoralis* (Family: Miridae) was introduced in the 1960s to control pear psylla, *(Psylla pyricola)*. Two species of native anthocorids, *A. melanocerus* and *A. antevolens*, seemed to disappear after the introduction of *A. nemoralis*, and were deemed to have been competitively displaced by the exotic species (McMullen, 1971; Debach and Rosen, 1991). Further experimentation, however revealed that *A. melanocerus* migrates out of pear orchards in late August, when pear psylla is becoming sparse, and onto nearby cottonwoods, where an alternative aphid host is abundant. This migration occurs whether or not the exotic species is present (Fields and Beirne, 1973).

The potential for the recently introduced Seven-spotted ladybird beetle, *(Coccinella septempunctata)*, to displace native North American ladybird beetles (family: Coccinellidae) has been raised (Howarth, 1991). In British Columbia there are 94 species of Ladybird beetles—89 are native species and 5 are exotic species (Bousquet, 1991). They prey on over 400 species of aphids (Forbes and Chan, 1989), and other soft-bodied insects. A complex of ladybird beetles is important in the regulation of some aphid populations

(Frazer, 1984). The seven-spotted ladybird beetle, with its very general feeding habits, could displace some native ladybird beetles, with their comparatively more restricted diets (B. D. Frazer, Agriculture Canada, personal communication). The effects of a reduction in native ladybird beetle diversity could potentially cascade into increases in some aphid populations and unforseen stress on native plants.

Permits are required to introduce agents into Canada for classical biological control. The permit system ensures strict adherence to some of the basic safety rules of biological control. Permits are not required for augmentative releases; hence these are not as well documented. The release of species, such as *Encarsia formosana* to control whiteflies (*Trialeurodes vaporariorum*) in greenhouses and the predatory mite (*Phytoseiulus persimilis*) to control two-spotted spider mites (*Tetranychus urticae*), are done augmentatively. Although *E. formosana* and *P. persimilis* have been known to escape from the greenhouse environment, they are unable to overwinter in British Columbia and hence do not, at present, constitute a long term threat to native species. Uncertain processes, such as global warming, may enable species which cannot currently survive a British Columbia winter to complete their life cycles in the province. The diversity of native predatory mites is particularly rich in British Columbia. Rather than risk a potential threat to native mite diversity, researchers in British Columbia are beginning to develop augmentative release programs using native species (D. Raworth, Agriculture Canada, personal communication).

Resistance does not usually occur in classical biological control using predators and parasites. However, there is one example worth noting. The ichneumonid parasite, *Mesoleius tenthredinis*, was introduced into Manitoba in 1927 to control the larch sawfly, *Pristiphora erichsonii*. Although the biological control program was initially successful, by 1940 this parasite no longer provided effective control of the larch sawfly. The larch sawfly had developed the ability to protect itself from parasitization by forming a capsule around parasite eggs. This process of encapsulation is a common method by which insects are able to fight parasitic infections. The encapsulation reaction spread quickly across North America. Presently, only larch sawfly populations in the mountainous areas of British Columbia and in Newfoundland remain susceptible to the parasite (Ives and Muldrew, 1984).

In this unusual case, the genetic make-up of the larch sawfly, an introduced pest, was altered as a result of the biological control program. Parasites were released by placing larch sawfly coccoons in the field. Most of the coccoons were parasitized, so that parasites, rather than the pest hosts, were introduced. However, some of the coccoons were not parasitized, resulting in the release of a new population of larch sawflies. Unfortunately, some of the adults from Europe probably had the encapsulation ability and their progeny are now the dominant strain in most areas across North America (Ives and Muldrew, 1984).

Effects of Introduced Microorganisms on Insect Biodiversity

A few species of microorganisms have been used as biopesticides. These organisms are not expected to persist in the environment and are often repeatedly sprayed over large areas. Several varieties of a bacterium, *Bacillus thuringiensis* (B.t.), have been used extensively in British Columbia to control defoliating caterpillars in forests, for eradication of the gypsy moth *(Lymantria dispar)* and for mosquito control. Because of its low mammalian toxicity, the use of *B.t.* for large scale aerial spray programs is becoming more popular. However, growing evidence suggests that *B.t.* can have a negative impact on the biodiversity of non-target species. For example, *Bacillus thuringiensis israelensis* (B.t.i.), is used extensively to control mosquitos, even though it is known to cause significant mortality of mayfly (Order: Ephemeroptera) and dragonfly (Order: Odonata) larvae (Howarth, 1991). Because the different life stages of mayflies and dragonflies can play an important role in both aquatic and terrestrial ecosystems, population reductions or extinctions would have an impact on many species, including fish, birds and flying insects.

Bacillus thuringiensis var *kurstaki* (B.t.k.) has been used extensively in North American and, in particular, in British Columbia, in programs to eradicate the gypsy moth. A detailed study conducted in Oregon, before and after *B.t.k.* spraying for the gypsy moth, demonstrated that, after only one season of spraying with *B.t.k.*, the diversity of butterflies and moths was significantly reduced for at least two subsequent years (Miller, 1990). Because the study was terminated after two years, longer term effects are unknown. Species emerging early in the season, with only one generation a year, were particularly vulnerable. As Miller's study was conducted in an area where the spray zone was fairly small (2,000 hectares) it was surprising that immigration from surrounding regions was not more rapid. Decisions to conduct widespread spray programs for alien species, such as gypsy moth, are highly controversial. Many factors are taken into consideration. However, it should be noted that concerns about the effects of spray programs on the biodiversity of non-target species are real. Spraying with *B.t.k.* in areas where rare or endangered moths or butterflies are located would, no doubt, place extreme stress on the populations of those species.

The use of nematodes as biological control agents is gaining in popularity. In particular, two families of nematodes (Steinernematidae and Heterorhabditis) occur naturally in soil and are associated with bacteria in the genus *Xenorhabdus* (Kaya, 1990). Although *Steinernema feltiae* (= *Neoplectana carpocapsae*) attacks at least 250 species of soil dwelling insects in several orders, it is considered environmentally safe (Howarth, 1991). In fact, because it is a native species, its use in British Columbia is not regulated by Agriculture Canada. Inspite of their wide host range, these nematodes are generally believed to have a low environmental impact because they remain at the point of placement. However, current research efforts are focussing on the development of strains with improved host-searching ability (Kaya, 1990). These more mobile strains will have the ability to disperse of their own volition, thereby having an impact beyond the point of introduction. Further concern

194

is raised because the survival of some species of non-target inverte-brate predators and parasites can be reduced by exposure to disease agents such as the nematode, *S. feltiae* (Vinson, 1990). The impact of introducing generalist disease agents, with enhanced dispersal abilities, into soil ecosystems is very poorly understood.

Microorganisms can be accidently introduced along with their hosts. In the case of the European crane fly, five species of proto-zoan parasites were apparently introduced along with their host. To my knowledge, investigations have not been conducted to determine whether these protozoans have extended their host range to include any of the native crane flies of British Columbia. Finally, baculoviruses, such as the nuclear polyhedral viruses of insects, have been of interest to biological control practitioners for many years. These viruses regularly cause natural declines in insect populations, are believed to be highly specific to their hosts, and are supposedly non-toxic to non-target species. Two viruses have been used as biological control agents in the province.

Current research is showing that perhaps the nuclear polyhe-dral viruses are not as host-specific as once believed. For example, the nuclear polyhedral virus of the Egyptian cotton leafworm *(Spodoptera littoralis)*, infects a number of related caterpillars as well as locusts (Bensimon et al., 1987). Because viruses of native insects are an integral part of natural population regulation, their use as biopesticides is unlikely to have a major impact on biodiversity of the insect fauna. However, caution is advised with the use of alien viruses against alien insect pests. Mechanisms of host specificity would have to be well understood before claims of safety could be justified.

Genetically Engineered Organisms

Regulatory agencies are in the process of determining how to assess the risk of introducing genetically engineered microrganisms into the environment. The discussion on the safety of genetically engineered microrganisms is broad and beyond the scope of this report. From the point of view of biodiversity, the release of geneti-cally engineered microrganisms is inherently different than the release of unaltered microorganisms. With genetic engineering techniques, genes from widely unrelated taxa can be combined. Particularly with insect diseases, genetic alterations are usually made in order to make a pathogen more virulent, to expand its host range or to increase its survivability (Fuxa, 1987). Limitations in these very characterisitcs are used as the basis for safety with unaltered microorganisms.

An intensive research effort is being conducted on the genetic manipulation of insect viruses. In particular, efforts to understand and manipulate the mechanisms of specificity of insect viruses are at the forefront of this research. Although manipulations rendering insect viruses more specific are attractive from an environmental protection viewpoint, manipulations reducing specificity are attrac-tive from an economic viewpoint. The effects of genetically engi-neered viruses on native insects will depend on the nature of the genetic manipulation performed on each virus. Regulating agencies will have to look at each agent with care if the protection of biodiversity in British Columbia's insect fauna is considered a priority. As with other biological control agents, the broader the host range of the agent, the greater the possibility that impacts on non-target species will occur.

Trophic Cascades

The term trophic cascade refers to the sequence of extinctions arising from interactions among organisms at different trophic levels in an ecosystem (Macdonald et al., 1989). A good example of a potential trophic cascade in British Columbia centres around the invasion of native grasslands by annual cheatgrass, *(Bromus tectorum)*. Caepitose (or tussock) grasses, such as the native perennial blue-bunch wheatgrass *(Agropyron spicatum)*, are particularly vulnerable to plant invasions by species such as annual cheatgrass (Mack, 1989). As native tussock grasses are displaced by annual cheatgrass, parasitic wasps, that depend on the low, dense tufts of tussock grasses for overwintering sites, tend to disappear (Wratten, 1992). One study in Britain has shown that as tussock grasses are replaced by invading annual species in hedgerows, the populations of parasitic wasps show a corresponding decline in numbers and many species become locally extinct (Wratten, 1992).

Lists of Alien Insects

Very little data exist on the direct effects of alien insects on biodiversity in British Columbia. The first step towards obtaining this data is to compile lists of alien species and compare them to lists of native species. This task is formidable. Although many checklists of insects have been compiled, most do not distinguish between native and alien species (for example, Forbes and Chan, 1989). Research on the origin of all classified species in all insect orders would have to be initiated. The task requires considerable knowledge of the distribution of each species and the ability to make inferences about origin from distributional data.

The four lists presented below (Tables 17-6 through 17-9) do not begin to present a full picture of alien insects and microorganisms in British Columbia. Rather, they are intended to illustrate the numbers of species introduced into the province. The list of beetles is complete (Table 17-6). It is based on information extracted from a recently compiled list of beetles in Canada (Bousquet, 1991). The list of lepidopterans (moths and butterflies) is only a partial list (Table 17-7). It includes the species of alien lepidopterous pests listed in: Beirne, 1971; Gillespie and Beirne, 1982; Gillespie and Gillespie, 1982; and Belton 1988. The lists of insects introduced as biological control agents for insect pests (Table 17-8) and weeds (Table 17-9) is fairly complete. It includes all liberations of biological control agents recorded in: The Canadian Insect Pest Review, 1937 to 1965; Summary of Parasite and Predator Liberations in Canada, 1966; Insect Liberations in Canada, 1967 to 1980; and Liberation News, 1981 to 1990 (Anonymous, 1938; Kelleher, 1960; MacNay, 1952; Maybee, 1958; Williamson, 1966; Williamson, 1984; Sarazin and Hamilton, 1988; Sarazin, 1991).

Introduced Species	Family	Common Names and Notes
Aderus populneus	Aderidae	antlike leaf beetles
Anobium punctatum	Anobiidae	furniture beetle
Microbregma emarginatum emarginatum	Anobiidae	
Stegobuim paniceum	Anobiidae	stored product pest
Omonadus floralis	Anthicidae	ant-like flower beetles
Omonadus formicarius	Anthicidae	ant-like flower beetles
Araecerus fasciculatus	Anthribidae	coffee bean weevil
Apion longirostre	Apionidae	pear-shaped weevils
Chaetophora spinosa	Byrrhoidea	moss feeder
Simplocaria semistriata	Byrrhoidea	
Acupalpus meridianus	Carabidae	ground beetles
Agonum muelleri	Carabidae	
Amara apricaria	Carabidae	
Amara familiaris	Carabidae	
Anisodactylus binotatus	Carabidae	
Bembidion lampros	Carabidae	
Bembidion tetracolum	Carabidae	
Bradycellus harpalinus	Carabidae	
Calathus fuscipes	Carabidae	
Carabus granulatus	Carabidae	
Carabus nemoralis	Carabidae	
Clivina collaris	Carabidae	
Clivina fossor	Carabidae	
Elaphorus parvulus	Carabidae	
Harpalus affinis	Carabidae	
Notiophilus biguttatus	Carabidae	
Pristonychus complanatus	Carabidae	
Pristonychus terricola	Carabidae	
Pterostichus melanarius	Carabidae	
Pterostichus strenuus	Carabidae	
Trechus obtusus	Carabidae	
Chrysolina hyperici	Chrysomelidae	biocontrol agent for weeds

Table 17-6: Alien Beetles in British Columbia. Source: Bousquet 1991.

Introduced Species	Family	Common Names and Notes
Chysolina quadrigemina	Chrysomelidae	biocontrol agent for weeds
Chrysolina varians	Chrysomelidae	
Crioceris asparagi	Chrysomelidae	
Crioceris duodecimpunctata	Chrysomelidae	
Longitarsus jacobaeae	Chrysomelidae	biocontrol agent for weeds
Phyllotreta striolata	Chrysomelidae	
Xanthogaleruca luteola	Ciidae	minute tree-fungus beetles
Ennearthron spenceri	Clambidae	minute beetles
Clambus gibbulus	Cleridae	stored-product pest?
Necrobia ruficollis	Cleridae	stored-product pest?
Necrobia rufipes	Cleridae	stored-product pest?
Necrobia violacea	Coccinellidae	ladybird beetles
Adalia bipunctata	Coccinellidae	ladybird beetles
Coccinella septempunctata	Coccinellidae	ladybird beetles
Stethorus impexus	Coccinellidae	ladybird beetles
Stethorus punctillum	Coccinellidae	ladybird beetles
Orthoperus atomus	Corylophinae	minute fungus beetles
Sericoderus lateralis	Corylophinae	
Atomaria pusilla	Cryptophagidae	silken fungus beetles
Cryptophagus cellaris	Cryptophagidae	
Cryptophagus distinguendus	Cryptophagidae	
Cryptophagus laticollis	Cryptophagidae	
Cryptophagus pilosus	Cryptophagidae	
Cryptophagus saginatus	Cryptophagidae	
Cryptophagus subfumatus	Cryptophagidae	
Ephistemus globulus	Cryptophagidae	
Pteryngium crenatum	Cryptophagidae	
Cryptolestes pusillus	Curcujidae	stored-product pest
Cryptolestes turcicus	Curcujidae	
Leptophloeus alternans	Curcujidae	
Pediacus depressus	Cucujidae	stored-product pest
Amalus scortillum	Curculionidae	weevils

Table 17-6: Alien Beetles in British Columbia (Continued).

Introduced Species	Family	Common Names and Notes
Barynotus obscurus	Curculionidae	
Barypeithes pellucidus	Curculionidae	
Ceutorhynchus assimilis	Curculionidae	
Ceutorhynchus erysimi	Curculionidae	
Ceutorhynchus punctiger	Curculionidae	
Ceutorhynchus rapae	Curculionidae	
Diocalanadra elongata	Curculionidae	
Gymnetron antirrhini	Curculionidae	
Gymnetron netum	Curculionidae	
Gymnetron pascuorum	Curculionidae	
Gymnetron tetrum	Curculionidae	
Hypera nigrirostris	Curculionidae	
Hypera punctata	Curculionidae	
Mecinus pyraster	Curculionidae	
Otiorhynchus ligustici	Curculionidae	
Otiorhynchus rugosostriatus	Curculionidae	
Otiorhynchus singularis	Curculionidae	
Otiorhynchus sulcatus	Curculionidae	
Philopedon plagiatum	Curculionidae	
Phyllobius oblongus	Curculionidae	
Rhinoncus castor	Curculionidae	
Rhinoncus pericarpius	Curculionidae	
Sciaphilus asperatus	Curculionidae	
Sciopithes obscurus	Curculionidae	
Sitona cylindricollis	Curculionidae	
Sitona hispidulus	Curculionidae	
Sitona lineatus	Curculionidae	
Sitona lineelus	Curculionidae	
Sitona tibialis	Curculionidae	
Sitophilus granarius	Curculionidae	
Sitophilus oryzae	Curculionidae	
Strophosoma melanogrammum	Curculionidae	

Table 17-6: Alien Beetles in British Columbia (Continued).

Introduced Species	Family	Common Names and Notes
Trachyphloeus bifoveolatus	Curculionidae	
Tychius picirostris	Curculionidae	
Tychius stephensi	Curculionidae	
Anthrenus pimpinellae	Dermestidae	dermestid beetle
Anthrenus scrophulariae	Dermestidae	
Anthrenus verbasci	Dermestidae	
Attagenus pellio	Dermestidae	
Attagenus unicolor japonicus	Dermestidae	
Dermestes carnivorus	Dermestidae	
Dermestes frischii	Dermestidae	
Dermestes maculatus	Dermestidae	
Dermests lardarius	Dermestidae	
Laricobius erichsonii	Derodontidae	biocontrol agent
Agriotes lineatus	Elateridae	click beetles
Mycetaea subterranea	Endomychidae	stored-product pest
Atholus bimaculatus	Histeridae	hister beetles
Carcinops pumilio	Histeridae	
Dendrophilus punctatus	Histeridae	
Dendrophilus xavieri	Histeridae	
Gnathoncus rotundatus	Histeridae	
Margarinotus merdarius	Histeridae	
Margarinotus obscurus	Histeridae	
Margarinotus purpurascens	Histeridae	
Anacaena limbata	Hydrophilidae	water scavenger beetles
Cercyon analis	Hydrophilidae	(subfamily: Sphaeridiinae)
Cercyon impressus	Hydrophilidae	terrestrial
Cercyon lateralis	Hydrophilidae	terrestrial
Cercyon pygmaeus	Hydrophilidae	terrestrial
Cercyon quisquilius	Hydrophilidae	terrestrial
Cercyon unipunctatus	Hydrophilidae	terrestrial
Sphaeridium bipustulatum	Hydrophilidae	terrestrial

Table 17-6: Alien Beetles in British Columbia (Continued).

Introduced Species	Family	Common Names and Notes
Sphaeridium lunatum	Hydrophilidae	terrestrial
Sphaeridium scarabaeoides	Hydrophilidae	terrestrial
Aridius nodifer	Lathridiidae	minute brown scavenger beetles
Cartodere constricta	Lathridiidae	
Corticaria pubescens	Lathridiidae	
Corticaria serrata	Lathridiidae	
Cortinicara gibbosa	Lathridiidae	
Dienerella filum	Lathridiidae	
Dienerella ruficollis	Lathridiidae	
Lathridius hirtus	Lathridiidae	
Lathridius minutus	Lathridiidae	
Thes bergrothi	Lathridiidae	
Lyctus brunneus	Lyctidae	powder-post beetle
Lyctus linearis	Lyctidae	powder-post beetle
Malachius aeneus	Melyridae	wheat pest
Brachypterus urticae	Nitidulidae	sap beetles
Carpophilus hemipterus	Nitidulidae	
Carpophilus mutilatus	Nitidulidae	
Epuraea aestiva	Nitidulidae	
Meligethes atratus	Nitidulidae	
Nitidula bipunctat	Nitidulidae	
Nitidula carnaria	Nitidulidae	
Nitidula rufipes	Nitidulidae	
Omosita colon	Nitidulidae	
Omosita discoidea	Nitidulidae	
Soronia grisea	Nitidulidae	
Acrotrichis cognata	Ptinidae	feather-winged beetles
Ptenidium pusillum	Ptinidae	
Mezium affine	Ptinidae	spider beetle
Niptus hololeucus	Ptinidae	
Pseudeurostus hilleri	Ptinidae	
Ptinus bicinctus	Ptinidae	

Table 17-6: Alien Beetles in British Columbia (Continued).

Introduced Species	Family	Common Names and Notes
Ptinus clavipes	Ptinidae	
Ptinus fur	Ptinidae	
Ptinus ocellus	Ptinidae	
Ptinus raptor	Ptinidae	
Ptinus villiger	Ptinidae	
Sphaericus gibboides	Ptinidae	
Trigonogenius globulus	Ptinidae	
Monotoma longicollis	Rhizophagidae	rhizophagid beetles
Montoma picipes	Rhizophagidae	
Aphodius distinctus	Scarabaeidae	scarab beetles
Aphodius fimetarius	Scarabaeidae	
Aphodius granarius	Scarabaeidae	
Aphodius prodromus	Scarabaeidae	
Ochodaeus luscinus	Scarabaeidae	
Onthophagus nuchicornis	Scarabaeidae	
Oxyomus silvestris	Scarabaeidae	
Pleurophorus caesus	Scarabaeidae	
Hylastinus obscurus	Scolytidae	bark beetles
Scolytus multistriatus	Scolytidae	
Scolytus rugulosus	Scolytidae	
Aleochara bilineata	Staphylinidae	rove beetles
Aleochara fumata	Staphylinidae	
Aleochara lanuginosa	Staphylinidae	
Anotylus nitidulus	Staphylinidae	
Anotylus rugosus	Staphylinidae	
Anotylus tetracarinatus	Staphylinidae	
Carpelimus bilineatus	Staphylinidae	
Cilea silphoides	Staphylinidae	
Creophilus maxillosus	Staphylinidae	
Elonium minutum	Staphylinidae	
Gabronthus nigritulus	Staphylinidae	
Gabronthus subnigritulus	Staphylinidae	

Table 17-6: Alien Beetles in British Columbia (Continued).

Introduced Species	Family	Common Names and Notes
Gyrohypnus angustatus	Staphylinidae	
Gyrohypnus fracticornis	Staphylinidae	
Gyrophaena affinis	Staphylinidae	
Habrocerus capillaricornis	Staphylinidae	
Hapalarea floralis	Staphylinidae	
Lathrobium fulvipenne	Staphylinidae	
Leptacinus batychrus	Staphylinidae	
Leptacinus intermedius	Staphylinidae	
Lithocharis obsoleta	Staphylinidae	
Lithocharis ochracea	Staphylinidae	
Lobrathium multipunctum	Staphylinidae	
Omalium rivulare	Staphylinidae	
Oxytelus laqueatus	Staphylinidae	
Oxytelus sculptus	Staphylinidae	
Philonthus agilis	Staphylinidae	
Philonthus carbonarius	Staphylinidae	
Philonthus cephalotes	Staphylinidae	
Philonthus cognatus	Staphylinidae	
Philonthus concinnus	Staphylinidae	
Philonthus cruentatus	Staphylinidae	
Philonthus debilis	Staphylinidae	
Philonthus politus	Staphylinidae	
Philonthus rectangulus	Staphylinidae	
Philonthus sanguinolentus	Staphylinidae	
Philonthus sordidus	Staphylinidae	
Philonthus umbratilis	Staphylinidae	
Philonthus varians	Staphylinidae	
Proteinus atomarius	Staphylinidae	
Quedius curtipennis	Staphylinidae	
Quedius fulgidus	Staphylinidae	
Quedius mesomelinus	Staphylinidae	

Table 17-6: Alien Beetles in British Columbia (Continued).

Introduced Species	Family	Common Names and Notes
Sepedophilus littoreus	Staphylinidae	
Staphylinus aeneocephalus	Staphylinidae	
Staphylinus ater	Staphylinidae	
Staphylinus winkleri	Staphylinidae	
Stenus fulvicornis	Staphylinidae	
Tachinus rufipes	Staphylinidae	
Tachinus subterraneus	Staphylinidae	
Tachyporus dispar	Staphylinidae	
Trichophya pilicornis	Staphylinidae	
Xantholinus linearis	Staphylinidae	
Xlodromus depressus	Staphylinidae	
Xylodromus concinnus	Staphylinidae	
Alphitobius diaperinus	Tenebrionidae	stored-product pest
Alphitobius laevigatus	Tenebrionidae	stored-product pest
Gnatocerus cornutus	Tenebrionidae	stored-product pest
Gonocephalum bilineatum	Tenebrionidae	
Palorus ratzeburgii	Tenebrionidae	stored-product pest
Tenebrio molitor	Tenebrionidae	stored-product pest
Tenebrio obscurus	Tenebrionidae	stored-product pest
Tribolium castaneum	Tenebrionidae	
Tribolium confusum	Tenebrionidae	stored-product pest
Tribolium destructor	Tenebrionidae	stored-product pest
Mycetophagus quadriguttatus	Tenebrionoidae	hairy fungus beetles
Typhaea stercorea	Tenebrionoidae	found in buildings
Tenebroides mauritanicus	Trogossitidae	stored-product pest

Table 17-6: Alien Beetles in British Columbia (Continued).

Introduced Species	Common Name	Family	Known B.C. Host Plants	Origin
*Tyria jacobaeae**	cinnabar moth	Arctiidae	tansy ragwort	Europe
Choreutis pariana	apple-and-thorn skeletonizer	Choreutidae	fruit trees	Europe
Coleophora fuscedinella	birch casebearer	Coleophoridae	birch, alder, apple, hawthorn	Europe
Coleophora laricella	larch casebearer	Coleophoridae	larch	Europe
Coleophora serratella	cigar casebearer	Coleophoridae	apple, cherry, quince, hawthorn, plum, pear	Europe
Anarsia lineatella	peach twig borer	Gelechiidae	peach, apricot, almond, plum	Asia
Dichomeris marginella	juniper webworm	Gelechiidae	junipers	Palaearctic
Phtorimaea operculella	potato tubeworm (not found in BC since 1966)	Gelechiidae	potatoes	Australia
Recurvaria nanella	lesser budmoth	Gelechiidae	Malus, Pyrus, Prunus	Europe
*Aplocera = (Anaitus) plagiata**		Geometridae	St. John's wort	Europe
Hemithea aestivaria	common emerald	Geometridae	woody trees and shrubs	Europe, E. Asia
Operophtera brumata	winter moth	Geometridae	forest, urban, orchard trees	N. Africa, Europe
Caloptilia syringella	lilac leafminer	Gracillariidae	lilac, privet, ash	Palaearctic
Thymelicus lineola	European skipper	Hesperiidae	grasses, especially Phleum pratense	Palaearctic
Lampronia rubiella	raspberry bud moth	Incurvariidae	raspberry, loganberry	Europe
Euproctis chrysorrhoea	browntail moth	Lymantriidae	fruit trees, cranberry, raspberry, loganberry, woodland trees	
Lymantria dispar	gypsy moth-target of annual eradication program	Lymantriidae	polyphagous	Europe, Russia
Stilpnotia salicis	satin moth	Lymantriidae	poplar, cottonwood (maybe poplars, aspen, crabapple, Saskatoon berry, willow, oak,)	Europe, Asia

Table 17-7: Accidental Introductions of Lepidopteran Pests. Inadvertently introduced moths and butterflies are included in this list, as are moths and butterflies that were introduced as biological control agents for weeds (marked with an asterisk). Sources: Beirne, 1971; Gillespie and Beirne, 1982; Gillespie and Gillespie, 1982; and Belton, 1988.

Introduced Species	Common Name	Family	Known B.C. Host Plants	Origin
Amphipyra tragopinus	the mouse	Noctuidae	apricot, plum, rose	Palaearctic
Caradrina morpheus		Noctuidae	lambsquarter, dandelion, knotweed	Europe
Hydraecia micacea	potato stem borer	Noctuidae	strawberry	
Melanchra picta	zebra caterpillar	Noctuidae	polyphagous	
Peridroma saucia	variegated cutworm	Noctuidae	polyphagous	unknown
Tricho plusiani	cabbage looper	Noctuidae	crucifers	Europe possible Palaearctic
Cheimophila salicella	blueberry leafroller	Oecophoridae	Salix, Spiraea, Alnus, Betula Acer, Prunus, Myrica, Berberis Cornus, Potentilla, Ledum, Kalmia, Rubus	Palaearctic
Depressaria pastinacella	parsnip webworm	Oecophoridae	umbelliferous plants	Europe
Papilio polyxenes asterius	black swallowtail	Papilionidae	polyphagous - wide host range	Europe
Aethes rutilana	pale juniper webworm	Phaloniidae	juniper (especially Juniperus communis)	Palaearctic
Artogeia (Pieris) rapae	imported cabbage worm	Pieridae	Crucifers	Europe
Etiella zinckenella	lima-beam pod borer	Pyralidae	peas (uncommon)	Europe?
Eurrhypara hortulata		Pyralidae	unknown - probably garden plants e.g. Stachys recta, Mentha sp., Calystegia sepium, Ribes sp.	Europe
Loxostega stricticalis	beet webworm	Pyralidae	polyphagous (broad leafed) plants	Asia
Syanthedon bibionipennis		Sesiidae	raspberry, loganberry, strawberry	
Deidamia inscripta	lettered sphinx	Sphingidae	grape	
Manduca quinquemanculat	tomato hornworm	Sphingidae	tomato	maybe immigrant?

Table 17-7: Accidental Introductions of Lepidopteran Pests (Continued).

Introduced Species	Common Name	Family	Known B.C. Host Plants	Origin
Acleris comariana	strawberry tortrix	Tortricidae	strawberry, blackberry, raspberry	Europe
Acleris variegana		Tortricidae	Rosa, Prunus, Malus, Crataegus, Vaccinium	Europe
Ancylis comptana	strawberry leafroller	Tortricidae	raspberry, loganberry, strawberry	Unknown
Archips podana		Tortricidae	Amelanchier sp., occasionally apple	Europe, Asia Minor
Archips rosanus	European leafroller	Tortricidae	apple, blueberry, raspberry	Europe
Argyrotaenia citrana	orange tortrix	Tortricidae	stonefruit, strawberry, raspberry, loganberry, blueberry	Europe
Badebecia urticana		Tortricidae	blueberry	
Cnephasia interjectana		Tortricidae	fruit trees, strawberry	Europe
Cnephasia longana	omnivorous leaf-tier	Tortricidae	strawberry, wheat, iris	Europe, Asia North Africa
Cnephasia longana		Tortricidae	clover, vetch, alfalfa, Douglas fir	
Cnephasia stephensiana		Tortricidae	raspberry, loganberry	
Croesia holmiana		Tortricidae	Malus, Pyrus, Carataegus, Rosa, Rubus, Cydonia	Europe
Cydia (Laspeyresia)pro	codling moth	Tortricidae	fruit trees	Europe
Ditula angustiorana	vine moth	Tortricidae	fruit, hops, apples, oak	Europe
Epinotia nanana	green spruce leafminer	Tortricidae	spruces	Western Europe
Epinotia solandriana		Tortricidae	red alder	Europe
Grapholitha molesta	oriental fruit moth (eradicated in B.C. 1956)	Tortricidae	peach, quince, pear, apples, plums, cherries	USA, Japan
Hedia nubiferana	green budworm	Tortricidae	Rosaceae	Europe

Table 17-7: Accidental Introductions of Lepidopteran Pests (Continued).

Introduced Species	Common Name	Family	Known B.C. Host Plants	Origin
Laspeyresia nigricana	pea moth	Tortricidae	peas	Europe
Pandemis cerasana		Tortricidae	blueberry, raspberry	
Pandemis heparana		Tortricidae	Quercus, Malus, Alnus, Acer spp	Palaearctic
Pardia cynosbatella		Tortricidae	Malus, Pyrus, Prunus, Crataegus, Lonicera	Europe
Rhopobota naevana		Tortricidae	Rubus, Vaccinium, Spirea	
Pardia cynasbatella		Tortricidae	Rosa spp.	Europe
Rhopobota naevana	black-headed fireworm	Tortricidae	cranberry	Unknown
Rhycarcionia buoliana	European pine shoot moth	Tortricidae	pines	Europe
Spilonota ocellana	eye-spotted budmoth	Tortricidae	fruit trees, hawthorn, larch, laurel	Europe
Argyreshia conjugella	apple fruit moth	Yponomeutiae	oak, mountain ash	Europe, Japan
Ocnerostoma piniariella	European needle miner	Yponomeutiae	Malus, prunus, huckleberry, mountain ash, serviceberry	Europe
Plutella xylostella	diamond-back moth	Yponomeutiae	pine	Europe
Swammerdamia caesiella		Yponomeutida	stonefruit	
Yponomeuta malinellus	apple ermine moth	Yponomeutida	Malus	Europe, Japan?

Table 17-7: Accidental Introductions of Lepidopteran Pests (Continued).

Introduced Species	Year(s)	Host Species	Common Name	Origin
A. Insects				
Aphidius ervi	Unknown	*Acyrthosiphon pisum*	pea aphid	Unknown
Aphidius smithii	Unknown	*A. pisum*	pea aphid	Unknown
Coccinella septempunctata	Unknown	*A. pisum*	pea aphid	Unknown
Coccinella undecimpunctata	Unknown	*A. pisum*	pea aphid	Unknown
Praon barbatum	Unknown	*A. pisum*	pea aphid	Unknown
Aphidecta obliterata	1960-63,65,68	*Adelges piceae*	balsam woolly adelgid	Czechoslovakia, Germany, Norway
Aphidoletes thompsoni	1962-63,65-66	*A. piceae*	balsam woolly adelgid	Germany
Cremifania nigrocellulata	1966,68	*A. piceae*	balsam woolly adelgid	Germany
Labricobius erichsonii	1960-61,63,65	*A. piceae*	balsam woolly adelgid	Germany
Leucopis(Leucopis) n.sp.nr.melanopus	1968	*A. piceae*	balsam woolly adelgid	Germany
Pullus impexus	1960,63	*A. piceae*	balsam woolly adelgid	Germany
Pullus impexus	1965-66,68	*A. piceae*	balsam woolly adelgid	Germany
Scymnus pumilio (flavifrons)	1960	*A. piceae*	balsam woolly adelgid	Australia
Tetraphleps abdulghani	1965	*A. piceae*	balsam woolly adelgid	India
Tetraphleps n.sp.nr. pilipes	1965	*A. piceae*	balsam woolly adelgid	Pakistan
Aphelinus sp.	1938	*Aphididae*	aphids	Ontario
Triaspis thoracicus	1942	*Bruchus pisorum*	pea weevil	Europe thru USA
Bracon sp.	1945-46	*Ceutorhynchus assimilis*	turnip seed weevil	BC
Habrocytus sp.	1949	*C. assimilis*	turnip seed weevil	Europe
Microbracon sp.	1944	*C. assimilis*	turnip seed weevil	BC
Trichomalus fasciatus	1949	*C. assimilis*	turnip seed weevil	Europe
Xenocrepis pura	1949	*C. assimilis*	turnip seed weevil	Europe

Table 17-8: Agents Introduced for Biological Control of Insect Pests. Source: The Canadian Insect Pest Review, 1937 to 1965; Summary of Parasite and Predator Liberations in Canada, 1966; Insect Liberations in Canada, 1967 to 1980; and Liberation News, 1981 to 1990 (Anonymous, 1938; Kelleher, 1960; MacNay, 1952; Maybee, 1958; Williamson, 1966; Williamson, 1984; Sarazin and Hamilton, 1988; Sarazin, 1991).

Introduced Species	Year(s)	Host Species	Common Name	Origin
Coccophagus scutellaris	1942	*Coccus hesperidium*	soft scale	California
Metaphagus stanleyi	1942	*C. hesperidium*	soft scale	California
Agathis pumila	1969,74-78,85	*Coleophora laricella*	larch casebearer	USA, Italy, Australia, Switzerland, France
Agrypon flaveolatum	1978-82	*C. laricella*	larch casebearer	Nova Scotia
Bassus cingulipes	1985-87	*C. laricella*	larch casebearer	Austria & Switzerland
Chilocorus kuwanae	1988-89	*C. laricella*	larch casebearer	USA
Chrysocharis laricinellae	1974,82-87	*C. laricella*	larch casebearer	Switzerland, France, Austria, Italy
Cybocephalus nipponicus	1988-	*C. laricella*	larch casebearer	USA
Diadegma laricinella	1974-76	*C. laricella*	larch casebearer	Italy, Austria
Dicladocerus japonicus	1974	*C. laricella*	larch casebearer	Japan
Dicladocerus westwoodii	1974	*C. laricella*	larch casebearer	France, Switzerland, Austria
Rhizophagus grandis	1984-87	*C. laricella*	larch casebearer	France, Belgium
Thanasimus formicarius	1988-89	*C. laricella*	larch casebearer	France
Ascogaster carpocapsae	1933-37	*Cydia pomonella*	codling moth	Unknown
Cryptus sexannulatus	1942,46-47	*C. pomonella*	codling moth	Europe thru Ontario
Ephialtes caudata	1941,46-47	*C. pomonella*	codling moth	Europe thru Ontario
Aphelinus mali	1929	*Erisoma lanigerum*	wooly apple aphid	Ontario
Bigonicheta setipennis	1954-55	*Forficula auricularia*	European earwig	BC
Bigonicheta spp.	1956	*F. auricularia*	European earwig	BC
Bigonocheta setipennis	1938-36,39	*F. auricularia*	European earwig	USA
Sturmia sp.	1946	*Gilpinia hercyniae*	European spruce sawfly	Europe
Calosoma sycophanta	1917-18	*Lambdina somniaria*	oak looper	USA
Angitia sp.	1937	*Laspeyresia nigricana*	pea moth	England, Europe
Ascogaster carpocapsae	1936-37,39	*L. nigricana*	pea moth	Unknown

Table 17-8: Agents Introduced for Biological Control of Insect Pests (Continued).

Introduced Species	Year(s)	Host Species	Common Name	Origin
Ascogaster quadridentatus	1937-39	*L. nigricana*	pea moth	England
Glypta haesitator	1937-39	*L. nigricana*	pea moth	England
Ichneumonid.sp.A. (near Pristomerus)	1938	*L. nigricana*	pea moth	England
Macrocentrus ancylivorus	1935-36	*L. nigricana*	pea moth	USA
Blastothrix confusa	1971	*Lecanium coryli*	Lecanium scale	Czechoslovakia
Blastothrix sericea	1928-29	*L. coryli*	Lecanium scale	England, Europe
Hemisarcoptes malus	1917	*Lepidosaphes ulmi*	oyster shell scale	New Brunswick
Hemisarcoptes malus	1953	*L. ulmi*	oyster shell scale	Ontario, BC
Mesoleius tenthredinis	1934-36	*Lygaeonematus erichsoni*	larch sawfly	England
Zenillia nox	1935	*L. erichsoni*	larch sawfly	Japan
Mantis religiosa	1937-38,54	*Melanopus spp., etc.*	grasshoppers	Europe, Ontario
Protodexis australia	1947	*Melanopus spp., etc.*	grasshoppers	Argentina
Spalangia endius	1978	*Musca domestica*	house fly	California
Microplectron fuscipennis	1941	*Neodiprion tsugae*	hemlock sawfly	Europe, Ontario
Agrypon flaveolatum	1979-82	*Operophtera brumata*	winter moth	
Cyzenis albicans	1978-82	*O. brumata*	winter moth	Nova Scotia, Germany
Allotropa utilis	1939-45	*Phenacoccus aceris*	orchard mealy bug	Nova Scotia
Chrysocharis gemma	1936-39	*Phytomyza ilicis*	holly leaf miner	England
Chrysocharis syma	1936-39	*P. ilicis*	holly leaf miner	England
Cyrtogaster vulgaris	1937-39	*P. ilicis*	holly leaf miner	England
Opius sp.	1938-39	*P. ilicis*	holly leaf miner	England
Sphegigaster flavicornis	1936-39	*P. ilicis*	holly leaf miner	England
Coeloides sordidator	1986-91	*Pissodes strobi*	white pine weevil	France

Table 17-8: Agents Introduced for Biological Control of Insect Pests (Continued).

Introduced Species	Year(s)	Host Species	Common Name	Origin
Cyrtogaster vulgaris	1937-39	*Phytomyza ilicis*	holly leaf miner	England
Opius sp.	1938-39	*P. ilicis*	holly leaf miner	England
Sphegigaster flavicornis	1936-39	*P. ilicis*	holly leaf miner	England
Coeloides sordidator	1986-91	*Pissodes strobi*	white pine weevil	France
Eubazus (=Allodorus) semirugosus	1987-89,91	*P. strobi*	white pine weevil	France
Scambus sudeticus	1986-89	*P. strobi*	white pine weevil	France
Bessa selecta	1942	*Pristiphora erichsonii*	European larch sawfly	New Brunswick, Ontario
Mesoleius tenthredinis	1941-42	*P. erichsonii*	European larch sawfly	Europe, New Brunswick, Ontario, USA
Leptomastix abnormis	1940	*Pseudococcus citri*	citrus mealybug	Cicily thru USA
Leptomastix dactylopii	1939-42	*P. citri*	citrus mealybug	South America thru USA
Zarhopalus corvinus	1939-40	*Pseudococcus martimus*	grape mealybug	Ontario
Dacnusa gracilis	1949-54	*Psila rosae*	carot rust fly	Europe
Loxotropa tritoma	1950-54	*P. rosae*	carot rust fly	Europe
Anthocoris nemoralis	1962-63	*Psylla pyricola*	pear psylla	Switzerland
Anthocoris nemorum	1963	*P. pyricola*	pear psylla	Switzerland
Anthocoris pilosus	1963	*P. pyricola*	pear psylla	Switzerland
Endopsylla agilis	1968-70,73	*P. pyricola*	pear psylla	Germany
Prionomitus mitratus	1963	*P. pyricola*	pear psylla	Switzerland
Apanteles californicus	1949	*Recurvaria milleri*	lodgepole needle miner	USA
Eriplatys ardeicollis	1949	*R. milleri*	lodgepole needle miner	Europe
Phaeogenes sp.	1949	*R. milleri*	lodgepole needle miner	USA
Pnigalio sp.	1949	*R. milleri*	lodgepole needle miner	Europe
Apanteles solitarius	1933	*Stilpnotia salicis*	satin moth	USA

Table 17-8: Agents Introduced for Biological Control of Insect Pests (Continued).

Introduced Species	Year(s)	Host Species	Common Name	Origin
Compsilura concinnata	1929-34	*S. Salicis*	satin moth	USA
Eupteromalus nidulans	1933-34	*S. Salicis*	satin moth	Europe, USA
Meteorus versicolor	1934	*S. Salicis*	satin moth	USA
Phytoseiulus persimilis	?	*Tetranychus urticae*	two-spotted spider mite	Unknown
Siphona geniculata	1967-68,70-73	*Tipula paludosa*	European cranefly	Switzerland, BC, Germany
Encarsia formosa	1951-55	*Trialeurodes vaporariorum*	greenhouse whitefly	Europe
Encarsia formosa	1934-37,39-50	*T. vaporariorum*	greenhouse whitefly	USA, Ontario
Ageniaspis fuscicollis	1988-89	*Yponomeuta malinellus*	apple ermine moth	Germany
Herpestomus brunicornis	1990	*Y. malinellus*	apple ermine moth	Japan & Europe
Herpestomus brunicornis	1991	*Y. malinellus*	apple ermine moth	Switzerland & Germany
B. Protozoans				
Actinocephaus tipula	inadvertent	*Tipula paludosa*	European cranefly	Europe
Diploycystis sp.	inadvertent	*T. paludosa*	European cranefly	Europe
Gregarina longa	inadvertent	*T. paludosa*	European cranefly	Europe
Hirmocystis ventricosa	inadvertent	*T. paludosa*	European cranefly	Europe
Nosema binucleatum	inadvertent	*T. paludosa*	European cranefly	Europe
C. Fungi				
Entomophthora elateridiphaga	1981-82	*Agriotes spp.*	wireworms	Switzerland
D. Viruses				
Douglas-fir tussock moth MNPV	1962,74-75	*Orgyia pseudotsugata*	Douglas-fir tussock moth	BC
Tipula iridescent virus	1975	*Tipula paludosa*	European cranefly	Switzerland

Table 17-8: Agents Introduced for Biological Control of Insect Pests (Continued).

Introduced Agent	Year(s)	Host Plant	Common Name of Host	Origin
Subanguina picridis-nematode	1985,91	Acroptilon repens	Russian knapweed	Saskatchewan
Puccinia acroptili-Sydow fungus	1985-86	A. repens	Russian knapweed	BC
Ceutorhynchidius horridus	1979	Carduus nutans	nodding thistle	Saskatchewan
Rhinocyllus conicus	1979,84-85, 87-88	C. nutans	nodding thistle	Saskatchewan, France
Trichosirocalus horridus	1986-88	C. nutans	nodding thistle	Saskatchewan
Urophora solstitialis	1991	C. nutans	nodding thistle	Austria
Urophora jaceana	1986	Centaurea debeauxii		Nova Scotia
Agapeta zoegana	1986-88	Centaurea diffusa	diffuse knapweed	Austria, Hungary
Cyphocleonus achates	1987	C. diffusa	diffuse knapweed	USSR, Austria
Larinus minutus	1991	C. diffusa	diffuse knapweed	Romania
Pelochrista medullana	1982,84-86	C. diffusa	diffuse knapweed	Austria, Hungary
Pterolonche inspersa	1986-87	C. diffusa	diffuse knapweed	Austria, Hungary
Sphenoptera jugoslavica	1976	C. diffusa	diffuse knapweed	Greece
Urophora affinis	1970	C. diffusa	diffuse knapweed	France
Agapeta zoegana	1982-88	Centaurea maculosa	spotted knapweed	Hungary, Austria
Chaetorellia acrolophi	1991	C. maculosa	spotted knapweed	Switzerland
Cyphocleonus achates	1987	C. maculosa	spotted knapweed	USSR, Austria
Metznaria metzneriella (=paucipunctella)	1973-74,78-80	C. maculosa	spotted knapweed	Switzerland, BC
Pelochrista medulla	1983,86	C. maculosa	spotted knapweed	Austria, Hungary
Pterolonche inspersa	1986,91	C. maculosa	spotted knapweed	Austria, Hungary
Terellia virens	1991	C. maculosa	spotted knapweed	Austria
Urophora affinis	1970-72,74	C. maculosa	spotted knapweed	France
Urophora quadrifasciata	1986	C. maculosa	spotted knapweed	BC
Subanguina picridis-nematode	1986-87	Acroptilon repens	Russian knapweed	Saskatchewan
Altica carduorum	1964-65,67-69	Cirsium arvense	Canada thistle	Switzerland
Ceutorhynchus litura	1975,87	C.arvense	Canada thistle	Germany, Saskatchewan
Urophora carduii	1974-76,87, 89,91	C.arvense	Canada thistle	Germany, France, Saskatchewan

Table 17-9: Agents Introduced for Biological Control of Weeds. Introduced insects, fungi and nematodes are included. The Canadian Insect Pest Review, 1937 to 1965; Summary of Parasite and Predator Liberations in Canada, 1966; Insect Liberations in Canada, 1967 to 1980; and Liberation News, 1981 to 1990 (Anonymous, 1938; Kelleher, 1960; MacNay, 1952; Maybee, 1958; Williamson, 1966; Williamson, 1984; Sarazin and Hamilton, 1988; Sarazin, 1991).

Introduced Agent	Year(s)	Host Plant	Common Name of Host	Origin
Rhinocyllus conicus	1985	*Cirsium vulgare*	bull thistle	Austria, Finland
Urophora stylata	1973,78	*C. vulgare*	bull thistle	France
Chelymorpha cassidea	1969	*Convolvulus sepium*	hedge bindweed	Europe, BC
Chirida guttata	1969-71	*C. sepium*	hedge bindweed	Canada
Metriona bicolor	1969-71	*C. sepium*	hedge bindweed	Ontario
Celerio euphorbiae	1966-67	*Euphorbia esula*	leafy spurge	Europe
Aphthona cyparissiae	1988,90-91	*Euphorbia esula-virgata complex*	spurge	Saskatchewan
Aphthona nigriscutis	1988,90-91	*Euphorbia esula-virgata complex*	spurge	Manitoba
Lobesia occidentis (=euphorbiana)	1990	*Euphorbia esula-virgata complex*	spurge	Saskatchewan
Agrilus hyperici	1955,67,77, 87-88	*Hypercium perforatum*	St. John's wort	USA
Aphis chloris	1979-80,90-91	*H. perforatum*	St. John's wort	Saskatchewan, Hungary, Austria
Aphthona flava	1990	*H. perforatum*	St. John's wort	Yugoslavia
Aplocera (=Anaitis) plagiata	1976	*H. perforatum*	St. John's wort	Europe
Chrysolina gemellata	1951-54	*H. perforatum*	St. John's wort	USA
Chrysolina hyperici	1951-52,71,73	*H. perforatum*	St. John's wort	USA, Ontario, Switzerland
Chrysolina quadrigemina	1981-82,85	*H. perforatum*	St. John's wort	BC, France, Ontario
Chrysolina varians	1957	*H. perforatum*	St. John's wort	Sweden
Zeuxidiplosis giardi	1955	*H. perforatum*	St. John's wort	USA
Calophasia lunula	1987	*Linaria lunula*	Dalmatian toadflax	Saskatchewan
Calophasia lunula	1963,65,68-69 71,85-86	*Linaria vulgaris*	yellow toadflax	Switzerland, Ontario, Saskatchewan
Mecinus janthinus	1991	*L. vulgaris*	yellow toadflax	France, Hungary
Calophasia lunula	1989	*Linaria dalmatica*	Dalmatian toadflax	Saskatchewan

Table 17-9: Agents Introduced for Biological Control of Weeds (Continued).

215

Introduced Agent	Year(s)	Host Plant	Common Name of Host	Origin
Eteobalea intermediella	1991	L. dalmatica	Dalmatian toadflax	Yugoslavia
Hylemya seneciella	1968,70	Senecio jacobaea	Tansy ragwort	California
Hypocrita jacobaeae	1962-64	S. jacobaea	Tansy ragwort	England, California, Switzerland, Oregon
Longitarsus jacobaeae	1970-71,73-74 76,78,86	S. jacobaea	Tansy ragwort	USA
Pegohylemia seneciella	1985	S. jacobaea	Tansy ragwort	Nova Scotia
Tyria jacobaeae	1976	S. jacobaea	Tansy ragwort	Sweden, Switzerland
Cystiphora sonchi	1984	Sonchus arvensus	Perennial sow-thistle	Austria
Microlarinus lareynii	1986	Tribulus terrestris	Puncture vine	Colorado

Table 17-9: Agents Introduced for Biological Control of Weeds (Continued).

Although lists of alien species of insects are not easily obtained, a new effort initiated by the United States Department of Agriculture should soon improve the situation. A computerized data base entitled the North American Immigrant Arthropod Data Base (NAIAD) is being compiled (Knutson et al., 1990). Agriculture Canada, Research Branch, is currently encouraging Canadian researchers to contribute to this database (Schmidt and Parker, 1992).

Alien Beetles

Most of the lists I have presented are self-explanatory. Because the list of alien beetles is the only list recording species which are not of economic interest, it is the most interesting list from a biodiversity viewpoint, and a few comments are in order. Thousands of species of beetles must have entered British Columbia via ship ballasts. Most, however, never became established. British Columbia harbours 3,626 species of beetles, in 100 families. Of these, only 248 species, in 36 families (less than 7%) are aliens. The aliens include 3 species of Chrysomelids introduced for biological control of weeds and 2 species of Coccinellids introduced for biological control of aphids. A study comparing introduced carabid beetles in disturbed and native habitats has already been mentioned (Spence and Spence, 1988; Spence, 1990).

Some habitats appear particularly impermeable to invaders. Bousquet (1991) lists 290 species of aquatic beetles in British Columbia in families or subfamilies of entirely aquatic beetles (Families: Dytiscidae, Gyrinidae, Haliplidae, Elmidae, Amphizoidae, Scirtidae and Hydrophilidae, all subfamilies except Sphaeridiinae). Of these, only 1 species, Anacaena limbata (Hydrophilidae), is alien. Some ecological niches also appear to resist invasions. For example, many native beetles are important contributors to the decomposition

216

process in British Columbia's forests. Of the 145 species of Cerambycids, which are known to play a key role in the decomposition of dead and dying wood, no species are alien. Other ecological niches are more vulnerable to invasions. For example, the predaceous beetles are significant regulators of herbivorous insect populations. Although there are many species of predatory beetles, species in seven families are almost exclusively predaceous (Coccinellidae, Carabidae, Staphylinidae, Histeridae, Pselaphidae, Cantharidae). Seven percent of the 1,303 species of predators in the predaceous families are alien.

Summary

Danger to the biodiversity of British Columbia's diverse insect fauna comes from the potential for competitive displacement of native species by alien species and from possible mortality of native species by alien microorganisms. The threat is identical whether the alien organism is an introduced biological control agent or an adventitious introduction. With purposeful introductions, however, the risk of extinctions of native insects can be reduced by using biological control agents with a very specialized diet.

Microorganisms have been introduced as both classical biological control agents and biopesticides. The focus of research on entomopathogens is sometimes at cross-purposes with the preservation of biodiversity. Whereas specialized agents with very restricted host ranges are preferable from a biodiversity viewpoint, more general and more virulent strains look promising from a commercial viewpoint. The introduction of genetically engineered microorganisms poses problems and risks similar to those imposed by the introduction of other microorganisms, with the added uncertainty of persistence and transmission of organisms containing genes from widely diverse taxa.

It is very difficult to assess the present extent of invasion of the insect fauna of British Columbia because checklists of insects are incomplete and seldom contain information on the origin of species. Although there is very little information on the effects of alien species on biodiversity, the initial data suggests that native habitats are less permeable to invasion than urban and disturbed habitats. Therefore, preservation of native habitats is probably the best strategy to ensure protection of the diverse insect fauna of British Columbia.

Acknowledgements

Dr. J. Myers, University of British Columbia reviewed the manuscript. Yves Bousquet and Mike Sarazin of Agriculture Canada and Elspeth Belton, Vancouver, provided essential references and engaged in invaluable discussions.

217

References

Andres, L.A. 1985. *Interaction of Chrysolina quadrigemina and Hypericum spp. in California.* Proceedings of the VI International Symposium on Biological Control of Weeds; Ottawa, Canada. pp. 235-239.

Anonymous. 1938. *The Canadian Insect Pest Review, 1937 to 1951.* Department of Agriculture, Canada.

Baldwin, J.G. and M. Mundo-Ocampo. 1991. Heteroderinae, cyst and non-cyst forming nematodes. In: *Manual of Agricultural Nematology.* W.R. Nickle. Marcel Dekker Inc., New York, New York.p p. 275-362.

Beirne, B.P. 1971. Pest insects of annual crop plants in Canada Part I, Lepidoptera; II, Diptera; III, Coleoptera. *Memoirs of the Entomological Society of Canada* 78:1-124.

Belton, E.M. 1988. *Lepidoptera on Fruit Crops in Canada.* Simon Fraser University, Pest Management Papers. No.30. pp. 1-104.

Bensimon, A., S. Zinger, E. Gerussi, A. Hauschner, I. Harpaz and I. Sela. 1987. "Darkcheeks", a lethal disease of locusts provoked by a *lepidopterous baculovirus. Journal of Invertebrate Pathology* 50:254-260.

Bousquet, Y. 1991. *Checklist of Beetles of Canada and Alaska.* Research Branch, Agriculture Canada. Publication 1861/D. pp. 1-430.

Debach, P. and D. Rosen. 1991. *Biological Control by Natural Enemies.* Cambridge University Press, Cambridge. pp. 1-386.

Elton, C.S. 1958. *The Ecology of Invasions by Animals and Plants.* Methuen, London.

Fields, G.J. and B.P. Beirne. 1973. Ecology of Anthocorid (*Hemipt:Anthocoridae*) predators of the pear psylla (*Homopt.: Psyllidae*) in the Okanagan Valley, British Columbia. *Journal of the Entomological Society of British Columbia* 70:8-9.

Forbes, A.R. and C.K. Chan. 1989. *Aphids of British Columbia.* Research Branch, Agriculture Canada. Technical Bulletin 1989-1E.

Frazer, B.D. 1984. *Acyrthosiphon pisum* (Harris), Pea aphid (Homoptera: Aphididae). In: *Biological Control Programmes Against Insects and Weeds in Canada 1969-1980.* J.S. Kelleher and M.A. Hulme. Commonwealth Agricultural Bureaux, Slough. p. 7.

Fuxa, J.R. 1987. Ecological considerations for the use of entomopathogens in IPM. *Annual Review of Entomology* 32:225-51

Gillespie, D.R. and B.P. Beirne. 1982. Leafrollers (*Lepidoptera*) on berry crops in the Lower Fraser Valley, B.C. *Journal of the Entomological Society of British Columbia* 79:31-37.

Gillespie, D.R. and B.I. Gillespie. 1982. A list of plant-feeding Lepidoptera introduced into British Columbia. *Journal of the Entomological Society of British Columbia* 79:37-54.

Howarth, F.G. 1991. Environmental impacts of classical biological control. *Annual Review of Entomology* 36:485-509.

Ives, W.G.H. and J.A. Muldrew. 1984. *Pristiphora erichsonii* (Hartig), Larch sawfly, (Hymenoptera: Tenthredinidae). In: *Biological Control Programmes Against Insects and Weeds in Canada 1969-1980.* J.S. Kelleher and M.A. Hulme. Commonwealth Agricultural Bureaux, Slough. p. 369-380.

Kaya, H.K. 1990. Entomopathogenic nematodes in biological control of insects. In: *New Directions in Biological Control: Alternatives for Suppressing Agricultural Pests and Diseases.* R.R. Baker and P.E. Dunn. Alan R. Liss Inc., New York. pp. 189-198.

Kelleher, J.S. 1960. *The Canadian Insect Pest Review, 1960.* Department of Agriculture, Canada.

Knutson, L., R.I. Sailer, W.L. Murphy, R.W. Carlson and J.R. Dogger. 1990. Computerized data base on immigrant arthropods. *Annual Entomological Society of America* 83(1):1-8.

Macdonald, I.A.W., L.L. Loope, M.B. Usher and O. Hamann. 1989. Wildlife conservation and the invasion of nature reserves by introduced species: a global perspective. In: *Biological Invasions: a Global Perspective.* J.A. Drake, H.A. Mooney,

F. di Castri, R.H. Groves, F.J. Kruger, M. Rejmanek and M. Williamson. John Wiley & Sons Ltd., Chichester. pp. 215-255.

Mack, R.N. 1989. Temperate grasslands vulnerable to plant invasions: Characteristics and consequences. In: *Biological Invasions: a Global Perspective.* Drake J.A., H.A. Mooney, F. di Castri, R.H. Groves, F.J. Kruger, M. Rejmanek and M Williamson. John Wiley & Sons, Chichester. pp. 155-179.

MacNay, C.G. 1952. *The Canadian Insect Pest Review, 1952 to 1957.* Department of Agriculture, Science Service.

Maybee, G.E. 1958. *The Canadian Insect Pest Review, 1958 to 1959.* Agriculture Canada, Research Branch.

McMullen, R.D. 1971. *Psylla pyricola* Forster, Pear psylla (*Hemiptera: Psyllidae*). In: *Commonwealth Agricultural Bureaux, Biological Control Programmes Against Insects and Weeds in Canada 1959-1968.* Commonwealth Agricultural Bureaux, Farnham Royal, England. pp. 33-38.

Miller, J.C. 1990. Field assessment of the effects of a microbial pest control agent on nontarget lepidoptera. *American Entomologist* 135-139.

Myers, J.H. and R. Iyer. 1981. Phenotypic and genetic characteristics of the European cranefly following its introdution and spread in Western North America. *Journal of Animal Ecology* 50:519-532.

Parella, M.P., K.M. Heinz, L. Nunney. 1992. Biological control through augmentative releases of natural enemeies: A strategy whose time has come. *American Entomologist* Fall:172-179.

Sarazin, M.J. 1991. *Insect Liberations in Canada, 1986 to 1991.* Agriculture Canada, Research Branch.

Sarazin, M.J. and N.A. Hamilton. 1988. *Insect Liberations in Canada, 1983 to 1985.* Agriculture Canada, Research Branch.

Schmidt, A.C. and D.J. Parker. 1992. *Memo to Canadians Involved with Biological Control.* (Unpublished) North American Biological Control Directory.

Simberloff, D. 1989. Which insect introductions succeed and which fail? In: *Biological Invasions: a Global Perspective.* J.A. Drake, H.A. Mooney, F. di Castri, R.H. Groves, F.J. Kruger, M. Remjanek and M. Williamson, John Wiley & Sons, Chichester. pp. 61-75.

Spence, J.R. 1990. Success of European carabid species in western Canada: preadaptation for synanthropy? In: *The Role of Ground Beetles in Ecological and Environmental Studies.* N.E. Stork. Intercept, Andover. pp. 129-141.

Spence, J.R. and D.H. Spence. 1988. Of ground-beetles and men: Introduced species and the synanthropic fauna of Western Canada. *Memoirs of the Entomological Society of Canada* 144:151-168.

Vinson, S.B. 1990. Potential impact of microbial insecticides on beneficial arthropods in the terrestrial environment. In: *Safety of Microbial Insecticides.* M. Laird, L.A. Lacey and E.W. Davidson. CRD Press, Boca Raton. pp. 43-64.

Williamson, G.D. 1966. *The Canadian Insect Pest Review, 1961 to 1966.* Agriculture Canada, Research Branch.

Williamson, G.D. 1984. *Insect Liberations in Canada, 1984 to 1985.* Agriculture Canada, Research Branch.

Wratten, S. 1992. Farmers weed out the cereal killers. *New Scientist* 1835:31-35.

Exotic Introductions into B.C. Marine Waters

Major Trends

Michael Waldichuk (deceased)
West Vancouver Laboratory
Department of Fisheries and
Oceans

Philip Lambert
Royal British Columbia Museum
675 Belleville Street
Victoria, B.C.
V8V 1X4

Brian Smiley
Institute of Ocean Sciences
Department of Fisheries and
Oceans
P.O. Box 6000
Sydney, B.C.
V8L 4B2

The entry of exotic species into Canada's Pacific marine ecosystem has been, as far as we know, related almost entirely to the introduction and maintenance of Japanese and Atlantic oysters for commercial production. The transplant of the hardy Japanese (now called Pacific) oyster (*Crassostrea gigas*) as early as 1914 led to this species becoming the mainstay of the oyster industry in British Columbia. However, the Pacific oyster cannot successfully sustain populations in coastal B.C. waters because water temperatures are usually too low for reproduction. Pacific oyster seed (spat) is, therefore, being imported regularly from Japan. It is mainly in association with the imported oyster spat that six species of bivalves, seven species of snails (gastropods), four polychaete worm species, and assorted other invertebrates have accidentally been introduced into B.C. coastal waters (Table 17-10). Most have remained fairly localized in their distribution, which is particularly fortunate in the case of the gastropod species that prey on oysters. These nuisance species include: the Japanese drill (*Ocenebra japonica*) introduced into Canada and U.S. Pacific waters as early as 1928; *Purpura (Mancinella) clavigera*, found in Ladysmith Harbour in 1951; and *Batillaria cumingi*, a Japanese species of gastropod associated with oyster beds planted with Japanese oyster seed in Boundary Bay, Ladysmith, Crofton, Fanny Bay and Comox. Gastropod predators introduced into Boundary Bay with the Atlantic oyster include the eastern oyster drill (*Urosalpinx cinerea*) and the eastern mud snail (*Nassarius obsoletus*)—both of which survived in Boundary Bay but apparently had not spread to other parts of the coast by 1964 (Quayle, 1964). This latter species now occurs from B.C. to central California (Ricketts et al, 1985).

Some species inadvertently introduced with foreign oysters have spread widely and contributed to commercial production in British Columbia. The soft-shell clam (*Mya arenaria*) generally is believed (not unanimously) to have been introduced with plantings of *C. virginica* in San Francisco Bay around 1807, and then migrated northward. The Manila (or Japanese) little neck clam (*Tapes philippinarum*) was first observed in Ladysmith Harbour in 1936, apparently introduced with Japanese oyster seed. It reproduced and spread rapidly and, by 1941, became included with the native Little Neck Clam (*Protothaca staminea*) in the commercial catch. The so-called Japweed seaweed (*Sargassum muticum*) is also believed to have been introduced with Japanese oyster seed. Now part of the B.C. coastal ecosystem, this seaweed is used by herring as a substrate for egg laying, but is a nuisance to oyster growers and boaters. Another immigrant from Japan, the parasitic copepod, *Mytilicola orientalis*, now occurs in the intestines of oysters and mussels from B.C. to California (Ricketts et al, 1985).

Attempts to introduce the Atlantic lobster (*Homarus americanus*) to Canada's Pacific coast date back to the late 19th century. Intensive experimentation on such a transplant was conducted during the 1960's in Fatty Basin on the west coast of Vancouver Island; all transplant attempts failed.

Algae
Gelidium vagum   (Renfrew et al. 1989)
Sargassum muticum   (DeWreede 1983)
Lomentaria hakodatensis   (South 1969)

Vascular plants
Zostera japonica   (Harrison et al. 1982)

Sponges
Microciona prolifera   (Kozloff 1987)
Halichondria bowerbanki   (Kozloff 1987)

Anemones
Haliplanella luciae   (Carl and Guiguet 1957)

Flat worms
Pseudostylochus ostreophagus   (Quayle 1964)

Polychaetes
Polydora ligni   (Hobson and Banse 1981)
Streblospio benedicti   (Hobson and Banse 1981)
Pionosyllis uraga   (Fournier and Levings 1982)
Tharyx tessalata   (Fournier and Levings 1982)

Bivalves
Mya arenaria   (Quayle 1964)
Musculus senhousia   (Bernard 1983)
Tapes philippinarum   (Quayle 1964; Bourne 1982)
Gemma gemma   (Bernard 1983)
Teredo navalis   (Quayle 1964)
Trapezium liratum   (Quayle 1964)
Crassostrea gigas   (Quayle 1964)
Crassostrea virginica   (Quayle 1964)

Gastropods
Ocenebra japonica   (Quayle 1964)
Nassarius obsoletus   (Quayle 1964)
Purpura clavigera   (Quayle 1964)
Ceratostoma inornatum   (Quayle 1988)
Urosalpinx cinerea   (Quayle 1964)
Batillaria cumingi   (Quayle 1964)
Crepidula fornicata   (Kozloff 1987)

Crustaceans
Homarus americanus   (Carl and Guiguet 1957)
Mytilicola orientalis   (Quayle 1964; Bernard 1969)
Limnoria tripunctata   (Quayle 1964)

Fish
Salmo salar   (Carl and Guiget 1957)

Table 17-10. Marine Plants and Animals Introduced into British Columbia. Published reference given in brackets.

An unsuccessful attempt was made in 1905 to introduce Atlantic salmon *(Salmo salar)* into B.C. coastal waters. However, some individuals of this species, presumably escapees from salmon farms, have recently been caught by commercial fishers.

To our knowledge, there has not yet been a comprehensive study of introductions associated with ships in Vancouver or Seattle harbours. Carlton (1985) reviewed the introduction of foreign organisms world-wide, via the ballast tanks of freighters. In San Francisco harbour, he found that almost 100 exotic marine invertebrates had been introduced from other parts of the world, presumably as fouling organisms on hulls of ships, in their ballast tanks, or with oyster introductions (Carlton 1979). We do not have comparable data for B.C. waters, but we do know that wood-boring animals introduced to these waters by vessels and other means include the Atlantic shipworm (*Teredo navalis*) and the crustacean woodborer (*Limnoria tripunctata*) (Carl and Guiget, 1957).

Management Responses

Introduction of exotic species is controlled by regulations under the <u>Fisheries Act</u>. The oyster seed inspection system was properly organized and working by 1940. The introduction of exotic species with oyster seed appears to have been well controlled since then, but most introductions had already occurred prior to 1940.

Carleton (1989) writes about the conventions governing the introduction of exotic marine species through shipping practises, especially the dumping of ballast and the use of antifouling agents on hulls. According to him, there is a series of international policies and conventions with clauses that pertain to the control of introductions. These policies and conventions include the 1973 Code of Practice of the International Council for the Exploration of the Sea, the 1973 position paper of the American Fisheries Society, the 1984 recommendation of the Council of Europe, and the Convention on the Law of the Sea. Carlton (1989) maintains that, although these clauses have brought international attention to the problem of alien introductions, most biologists have never heard of them and most of them are not really enforceable.

Conclusion

Introduced species have out-competed some local populations of native species, but no indigenous marine species seem to have become extinct as a consequence of introductions. However, in the southern end of San Francisco Bay, introduced species dominate benthic communities. We should expect similar introductions to occur in Vancouver because it is an international port. So far, introduced populations seem to be confined to bays and harbours and not to open coast, but they should be monitored for changes. In general, human-caused extinctions have been comparatively rare in the sea, except for large birds and mammals (Vermeij, 1986).

Acknolwedgements

Colin Levings, Fisheries and Oceans Canada reviewed the manuscript.

References

Bernard, F. R. 1969. Parasitic copepod Mytilicola orientalis in British Columbia Bivalves. *J. Fish. Res. Bd.* Canada 26: 190-191.

Bernard, F. R. 1983. *Catalogue of the Living Bivalvia of the Eastern Pacific Ocean: Bering Strait to Cape Horn.* Department of Fisheries and Oceans, Ottawa. 102 p. (Canadian Special Publication of Fisheries and Aquatic Sciences 61).

Bourne, N. 1982. Distribution, reproduction, and growth of Manila Clam, Tapes philippinarum (Adams and Reeves), in British Columbia. *J. Shellfish Res.* 2: 47-54.

Carl, G. C., and C. J. Guiget. 1957. *Alien Animals in British Columbia.* K.M. MacDonald, Queen's Printer, Victoria, B.C.

Carlton, J. T. 1979. Introduced invertebrates of San Francisco Bay. In T.J. Conomos, (ed.). *San Francisco Bay - The Urbanized Estuary.* Pacific Division AAAS, California Academy of Sciences, San Francisco.

Carlton, J.T. 1985. Transoceanic and interoceanic dispersal of coastal marine organisms: the biology of ballast water. *Oceanography and Marine Biology: an Annual Review* 23:313-371.

Carlton, J.T. 1989. Man's role in changing the face of the ocean: biological invasions and implications for conservation of near-shore environments. *Conservation Biology* 3:265-273.

De Wreede, R. E. 1983. Sargassum muticum (Fucales, Phaeophyta): regrowth and interaction with Rhodomela larix (Ceramiales, Rhodophyta). *Phycologia* 22: 153-160.

Fournier, J. A., and C. D. Levings. 1982. Polycheates recorded near two pulp mills on the coast of northern British Columbia: a preliminary taxonomic and ecological account. *Syllogeus* 40: 1.

Harrison, P. G., and R. E. Bigley. 1982. The recent introduction of the seagrass Zostera japonica Aschers. and Graebn. to the Pacific Coast of North America. *Can. J. Fish. Aquat. Sci.* 39: 1642-1648.

Hobson, K. D., and K. Banse. 1981. Sedentariate and archiannelid polychaetes of British Columbia and Washington. *Can. Bull. Fish. Aquatic Sci.* 209: 1-144.

Kozloff, E. N. 1987. *Marine Invertebrates of the Pacific Northwest.* University of Washington Press, Seattle and London.

Lindstrom, S. C. 1990. *Marine Plant Introductions in the Northeast Pacific: Antithamnionella spirographidis to Zostera japonica.* (Text of talk).

Quayle, D. B. 1964. Distribution of introduced marine mollusca in British Columbia waters. *J. Fish. Res. Bd.* Canada 21: 1155-1181.

Quayle, D. B. 1988. Pacific oyster culture in British Columbia. *Canadian Bulletin of Fisheries and Aquatic Sciences* No. 218. Department of Fisheries and Oceans, Ottawa.

Renfrew, D.E., P.W. Gabrielson and R.F. Scagel. 1989. The marine algae of British Columbia, northern Washington and Southeast Alaska: division of Rhodophyta (red algae), Class Rhodophyceae, order Gelidiales. *Can. J. Bot.* 67: 3295-3314.

Ricketts, E.F., J. Calvin and J.W. Hedgpeth. 1985. *Between Pacific Tides.* 5th Stanford University Press, Stanford, California.

South, G. R. 1969. Intertidal marine algae from Gabriola Island, British Columbia. *Syesis* 1: 177-186.

Vermeij, G.J. 1986. The biology of human-caused extinction. In: *The Preservation of Species.* B.G. Norton (ed.). Princeton University, Princeton, New Jersy.

Part III
ECOSYSTEM DIVERSITY

225

Chapter 18
Overview of Ecosystem Diversity

Lee E. Harding
Canadian Wildlife Service
P.O. Box 340
Delta, B.C.
V4K 3Y3

Emily McCullum
TerraMare
3257 West 2nd Avenue
Vancouver, B.C.
V6K 1K9

Odum (1969) defined ecological systems, or ecosystems, as "any area of nature that includes living organisms and non-living substances interacting to produce an exchange of materials between the living and non-living parts." We use the term ecosystem to denote these interactions on spatial scales ranging from microhabitats, such as the underside of a log, to broad physiographic regions, such as the coastal mountains or the dry interior plateau. Habitat is a closely related term which refers to the particular environment within which a creature lives.

Hence, a forest ecosystem contains many smaller ecosystems—meadows, closed-canopy forest, streams and lakes—and each of those ecosystems contains many habitats for different species.

Development of Ecosystem Diversity

The diversity of ecosystems is due to climate and physiography, time, and the degree to which their development has been disturbed.

The regional climate is determined primarily by latitude, proximity to large bodies of water, and physiography. British Columbia is beside the ocean, whose warm, moist winds moderate the weather, and it straddles the mountains, beyond which the weather, more influenced by Arctic air, is colder in winter and drier in summer. The province's mountainous character provides much more opportunity for ecosystem diversification—lee and windward sides and variations in slope and aspect—than do less physiographically diverse regions. The coastal rainforest ecosystem, for example, is completely different from the dry interior ecosystem found in the rain shadow of British Columbia's coastal mountains. On a smaller scale, physiographic diversity allows development of ecosystems that could not arise on uniform terrain. For example, some communities of plants and animals favour the sunny south-facing slope of a mountain, while others thrive on the shadier, colder north-facing slope.

The development of complex ecosystems requires time. Areas which were glaciated during the last Ice Age have, therefore, simpler ecosystems than those that were not. Recently glaciated areas tend to be populated by highly mobile species of invertebrates, fish, birds and mammals that move in quickly after deglaciation and have behaviours, such as annual migration, that can turn the severe climate to advantage. More complex ecosystems tend to have high numbers of species that use microhabitats, such as insects and birds that have evolved symbiotic and complex host-parasite relationships, and plants and animals with conservative reproductive and dispersal capabilities.

Finally, the frequency of disturbance affects the development of ecosystems. Natural perturbations, such as landslides, floods and fires, are relatively frequent in mountainous regions with strongly seasonal weather patterns that generate, for example, rapid spring runoff and summer lightning storms which often touch off forest fires. Such small scale disasters create high ecosystem diversity by providing habitats for early colonizers and for species that require a variety of habitat types within their environments.

Brief History of Ecosystem Diversity in B.C.

Throughout British Columbia's geological development, from the collision of crustal plates and building of mountains out of ancient seabed sediments to the more recent advance and retreat of continental glaciers, there have been massive fluctuations in biodiversity, which have parallelled worldwide trends. Worldwide mass extinctions occurred 505, 367, 208, and 65 million years ago (Gould, 1989). In each of these extinctions, a major proportion (possibly up to 96% in one case) of the earth's biodiversity was lost and the course of evolution was changed as surviving organisms rapidly diversified to fill vacated niches (Raup, 1988). Despite the vast passage of time, enough information has been gleaned from geological and archaeological records to suggest some causes for these extinctions—and they are sufficiently similar to events in our own time to give us food for thought.

About 570 million years ago in the Cambrian era, a proliferation of marine life forms known as the Cambrian explosion occurred. Gould (1989) has written eloquently of the incredible diversity of soft-bodied and jointed-legged life forms in a shallow sea near what is now Field, British Columbia. Many of these organisms had fantastic body designs, totally unlike anything that exists today. This sea's sediments have since been uplifted into the massive fault blocks of the Rocky Mountains, where fossil beds in a formation known as the Burgess Shale provide the world's best example of the Cambrian explosion. The few species that survived the worldwide mass extinction at the end of the Cambrian era (about 505 million years ago) later diversified, eventually giving rise to such modern taxonomic groups as insects and vertebrates. Still later, some 360 million years ago, after vertebrates had come to land as primitive amphibians, extensive swamps in the northeast and southeast of British Columbia laid down the decaying vegetation that would become our coal fields. Reptiles evolved, then dinosaurs. Coniferous forests developed, followed by a great diversification of flowering plants. After the demise of the dinosaurs some 65 million years ago, the mammals diversified, evolving into modern forms during the Pleistocene epoch.

The Pleistocene epoch lasted about two million years, during which time northern latitudes (including North America) experienced repeated glaciations. At least 15 cold periods are known from the Pleistocene, but it is unlikely that each was associated with extensive glaciation, and we currently do not know how many times glaciers advanced across northern North America. In British Columbia, there is good evidence for the last major glaciation, and sparse evidence for at least two earlier glaciations. During the Pleistocene, there were repeated episodes of climatic (and hence environmental) change, and there were world-wide changes in sea level as significant quantities of water were "locked up" in ice sheets and later released as melt water.

There is very little evidence of life in B.C. during the Pleistocene, mainly because the most recent glaciation either eroded fossils away or covered them with hundreds of metres of sediments. Most evidence which we do have relates to the last interglacial period, the last glacial period, and the subsequent Holocene period (10,0000 years ago). Prior to the last major glaciation, there was a long period (at least 60,000 to 25,000 years ago) when conditions were somewhat cooler than today. Temperatures fluctuated, and warmer and cooler intervals occurred, with a final colder period leading into the subsequent last glaciation. Vegetation included coniferous forest and steppe tundra. Animal fossils from this time period include mammoth, mastodon, ground sloth, wild ass, mule deer, caribou, moose, bison and muskox. These species would probably have been found in different habitats. For example, muskox and caribou suggest tundra, while mammoth, ass, and bison suggest a warmer steppe. The mastodon was probably a woodland animal.

The last major glaciation began about 25,000 years ago. Mountain glaciers probably expanded to form ice caps which eventually merged to form a single ice sheet that covered most of B.C. by 15,000 years ago. Although terrestrial life may have survived in a few refugia (such as southern Vancouver Island and the Queen Charlotte Islands), it was effectively eradicated in most places.

By about 13,000 years ago, temperatures had begun to rise and ice to melt rapidly, first in southern B.C. As the ice sheets melted, enormous volumes of water were released, forming vast lakes of meltwater in the many valleys whose drainage was still blocked by glaciers. The millions of tons of silts and sands deposited in these lakes by the melting glaciers formed layers hundreds of feet thick which still remain in many of B.C.'s valleys. As the climate continued to warm, the ice dams melted and the lakes drained away. Temperatures peaked at about 2-3°C warmer than now; terrestrial ecologists refer to this period as the xerothermic or dry-warm interval, while aquatic biologists call it the hypsithermal interval, meaning a time of greater warmth. During this period, much of B.C. became open for colonisation by plants and animals.

Humans were among the early colonizers (Horgan, 1992). The dominant human culture throughout North America, including the interior of British Columbia, was initially based on the Clovis point, a crude stone projectile tip suitable for killing the abundant woolly mammoth and other large, slow-moving herbivores (Carlson, 1989). For example, near Olympia, Washington, archaeologists found remains of a giant mastodon and other game, including caribou, that had been killed and butchered by people about 12,000 years ago (Gustafson, 1984). The early Holocene was, however, a time of change, of global warming which precipitated changes in the flora and fauna. At least 64 of the large North American mammal species (such as mammoths and mastodons, ground sloths, wild asses, camels, giant beaver, and a wide range of carnivores like the dire wolf, sabre-tooth cat and short-faced bear) that had survived the advance and retreat of Pleistocene glaciers became extinct during this period (Kurtén and Andersen, 1981). Although it has been suggested that humans were responsible for the extinction of some species (notably the mammoth), the scale of the extinction is too great to have been caused by humans, especially as there is little evidence that early North Americans hunted most of the species which were eradicated. Whatever the cause of the extinctions, by about 11,000 years before the present, the main resource of the interior people was nearing extinction, and the Clovis culture with it (Carlson, 1989).

People adapted by adopting a new culture based on the Folsom point, a more refined, fluted projectile tip designed for hunting bison and other swift, large mammals (Carlson, 1989). Though there is little evidence of which species recolonised B.C. following the extinctions, it is likely that the incoming species were mainly those found in the province today. Evidence from Charlie Lake Cave outside Fort St. John, the only area in B.C. with good evidence for early animal populations, shows that the first post-glacial environment in northeastern B.C. consisted mainly of a grassland which supported animals such as a large species of bison, jackrabbit, ground squirrels, and various water and upland birds (Driver, 1988). Archaeological findings show that, by 10,500 years ago, people at Charlie Lake, B.C.'s oldest site of human habitation, were already hunting bison (Driver, 1988). Coast people of this time turned to small pebble tools and marine resources, developing a new culture along with their new mode of survival. People of the far North developed a microblade, specialized for hunting small and medium-sized game, such as the abundant barren ground caribou, in the boreal forest regions and some of them moved into northeastern British Columbia and part of the coast (Carlson, 1989).

After the warm, dry period, arctic birch, willow, and black spruce extended their ranges northward, following the retreating glaciers, and coastal and interior mountains became more heavily forested. Previously extensive tundra, the habitat of the caribou and white-tailed ptarmigan, became isolated at the tops of mountains as forests advanced. Grasslands, extensive throughout the interior immediately after retreat of the glaciers, began to be overtaken by shrublands and, by 10,000 years ago, forests. As a consequence of these changes, all the animals (excepting the passenger pigeon) found at the Charlie Lake Cave site after 10,000 years ago can be found in the region today. About 8,000 to 7,000 before the present, Douglas-fir extended its range over much of the area previously occupied by spruce and jackpine (Hebda, 1982). About 6,000 to 8,000 before the present, garry oak colonized the south coast, and ponderosa pine extended its range into the interior from the south (Hebda, 1991).

Then, during the cool period of 5,000 to 2,000 years before the present, cedar-hemlock forests established themselves along the coast and human cultures diverged still further (Hebda and Mathews, 1984). The coastal peoples developed the rich woodworking-based cultures we know today, while the interior people founded their lifestyle on the migratory salmon, supplemented with small game and the remaining few large herbivores. There followed a long period of ecological and cultural stability (Kew and Griggs, 1991), until other human races, notably those from Europe and Asia, began immigrating about three centuries ago.

Climate change and species extinctions have left their imprint upon the landscape of British Columbia today. Deglaciation was sufficiently recent that complex ecosystems have not yet evolved, except insofar as existing ones were able to spread into the province as the climate warmed. British Columbia's large physiographic variation—25,700 kilometres of coastline, mostly mountainous land mass, some of the largest rivers in North America in both Arctic and Pacific drainages and climates ranging from mediterranean to subarctic—provides niches for an immense variety of life forms. The result is a high diversity of relatively simple ecosystems populated, for the most part, by very mobile species of invertebrates, fish, birds and mammals. Like fractal geometry, this diversity is maintained on provincial, regional and local (habitat) scales.

Ecosystem Classification

This diversity is also reflected in the different systems used to classify British Columbia's natural areas. Krajina (1965) and his students founded their system on an understanding of how B.C.'s climatic regions and physiographic units combine to create environments suitable for specific vegetation complexes. The Canadian Parks Service recognizes seven physiographically based natural regions: Pacific Coast Mountains, Strait of Georgia Lowlands, Interior Dry Plateau, Columbia Mountains, Rocky Mountains, Northern Coast Mountains, and Northern Interior Plateaux and Mountains (Finkelstein, 1990). Rowe (1977) categorized the forest regions of Canada, finding five different categories in B.C.: Boreal, Subalpine, Montane, Coast and Columbia and Tundra. Each of these systems is largely based on one of three features of B.C.'s diversity: climate, physiography or vegetation complexes. By contrast, the two ecosystem classification systems becoming most widely accepted in British Columbia, biogeoclimatic zones and ecoregions, are the ones which most successfully integrate these three features.

The biogeoclimatic zone system has been used for many years by the B.C. Ministry of Forests (Meidinger and Pojar, 1991), other resource and environmental management agencies and research institutions. This system is designed to aid in the improvement of forest management in British Columbia. According to Demarchi (1991), it is based on a hierarchy of plant associations which are used to describe nutrient and moisture regimes. The focus is on late successional stage vegetation, especially forest trees, in each unit. Soils, climate and wildlife characteristic of that vegetation complex are then described, as are local variables, such as slope, aspect, water tables and drainage. The information gathered at the site level is aggregated to provide a description of the zone as a whole. There are 14 biogeoclimatic zones in British Columbia (Figure 18-1), 13 of which have been subdivided into a total of 79 subzones. Variants of 41 of the subzones are also recognized, giving a total of 124 classified ecosystems in British Columbia. Technical descriptions given by Meidinger and Pojar (1991) include no less than 20 climatic variables, as well as soil moisture and nutrients, physiographic variables, tree species occurrence, zonal vegetation, and forest, grassland and wildlife resources. Similarly, Miedinger and Pojar (1991) list the following major habitat types for just one

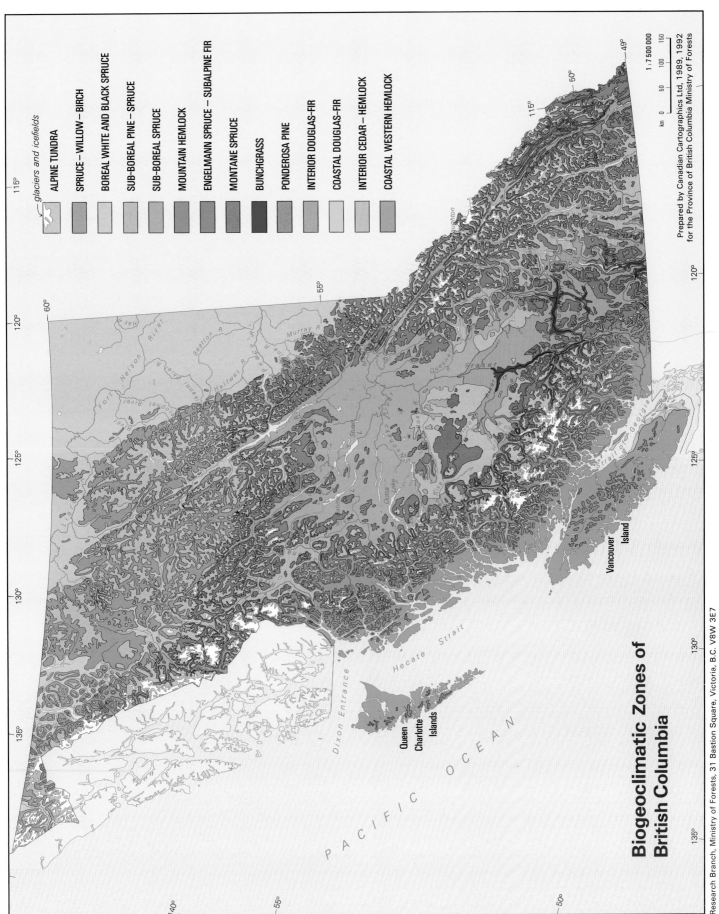

Biogeoclimatic Zones of British Columbia

glaciers and icefields

- ALPINE TUNDRA
- SPRUCE – WILLOW – BIRCH
- BOREAL WHITE AND BLACK SPRUCE
- SUB-BOREAL PINE – SPRUCE
- SUB-BOREAL SPRUCE
- MOUNTAIN HEMLOCK
- ENGELMANN SPRUCE – SUBALPINE FIR
- MONTANE SPRUCE
- BUNCHGRASS
- PONDEROSA PINE
- INTERIOR DOUGLAS-FIR
- COASTAL DOUGLAS-FIR
- INTERIOR CEDAR – HEMLOCK
- COASTAL WESTERN HEMLOCK

1 : 7 500 000

km   0   50   100   150

Prepared by Canadian Cartographics Ltd, 1989, 1992
for the Province of British Columbia Ministry of Forests

Research Branch, Ministry of Forests, 31 Bastion Square, Victoria, B.C. V8W 3E7

PACIFIC OCEAN

Dixon Entrance

Hecate Strait

Queen Charlotte Islands

Vancouver Island

Strait of Georgia

Hay River

Fort Nelson River

Beatton R.

Halfway R.

Murray R.

Fraser

Quesnel

Chilcotin R.

Stuart L.

Ootsa Lake

49°
50°
55°
60°

49°
50°
55°

115°
120°
125°
130°
135°
140°

biogeoclimatic zone, the Coastal Western Hemlock zone: old growth coniferous forest, young seral and managed second growth forests, mixed coniferous and deciduous forests, rocky cliffs, talus and sparsely vegetated rocks, avalanche tracks and seepage sites, upland grassy areas, agricultural areas, riparian areas, wetlands, meadows, floodplains, lakes and streams, offshore forested islands, offshore grassy and shrubby islands, marine cliffs and rocky islets, estuaries, shallow bays, intertidal and subtidal marine habitats. These habitat types are, to some extent, specific to each zone, subzone or variant. A wet meadow in the Finlay-Peace variant of the Sub-Boreal Spruce zone, for example, may not support the same species of plants and animals as a wet meadow in the Babine variant of the same zone, and will be very different from a wet meadow in the Coastal Douglas-fir zone.

The ecoregion system is used by the Ministry of Environment, Lands and Parks, other resource and environmental management agencies, and research institutions. This system is designed to "bring into focus the extent of critical habitats and their relationship with adjacent areas" (Demarchi et al., 1990). It is based upon the broad geographical relationships revealed by the interaction of macroclimatic processes and physiography. It, too, is hierarchical, stratifying landscape into ecosystems at a variety of spatial scales: ecodomain (global), ecoprovince (continental), ecoregion (regional), and ecosection (local). The 10 ecoprovinces in British Columbia are similar to Rowe's (1977) forest regions and Finkelstein's (1990) natural regions. Biogeoclimatic criteria are used to subdivide the 10 ecoprovinces into 30 terrestrial and marine ecoregions, 20 of which are further subdivided into 87 ecosections (Figure 18-2). This ecoregion system is roughly compatible with a more general ecological classification system for Canada which includes four major ecoclimatic regions and 25 subdivisions in British Columbia (CCELC, 1989). Like biogeoclimatic zones, this system can get quite complex, with detailed interpretations of climate, topography, soil and vegetation in the context of habitat and wildlife management (Demarchi, 1991).

The biogeoclimatic zone and ecoregion systems differ in three fundamental ways. First, the biogeoclimatic zone system is primarily provincially based, whereas the ecoregion system can be integrated with other North American or global ecoregion classifications, enabling us to view British Columbia's ecosystems in a continental as well as local context. Secondly, the biogeoclimatic zone system can be used to delineate the changes in ecosystem with altitude, while the ecoregion system cannot—its geographical units circumscribe all elevations (Demarchi, 1991). Finally, the biogeoclimatic zone system's emphasis on late successional vegetation as the distinguishing feature between ecological units makes it difficult to define a unit once a stand has been logged or burned. By contrast, the ecoregion system's use of relatively permanent landform features, such as parent materials, slope and aspect, as the basis for unit description means that the unit can be identified regardless of what happens to the vegetation (Demarchi, 1991).

Figure 18-1. (Opposite Page). Biogeoclimatic Zones of British Columbia.
Source:Meidinger and Pojar, 1991 (Reprinted with permission).

233

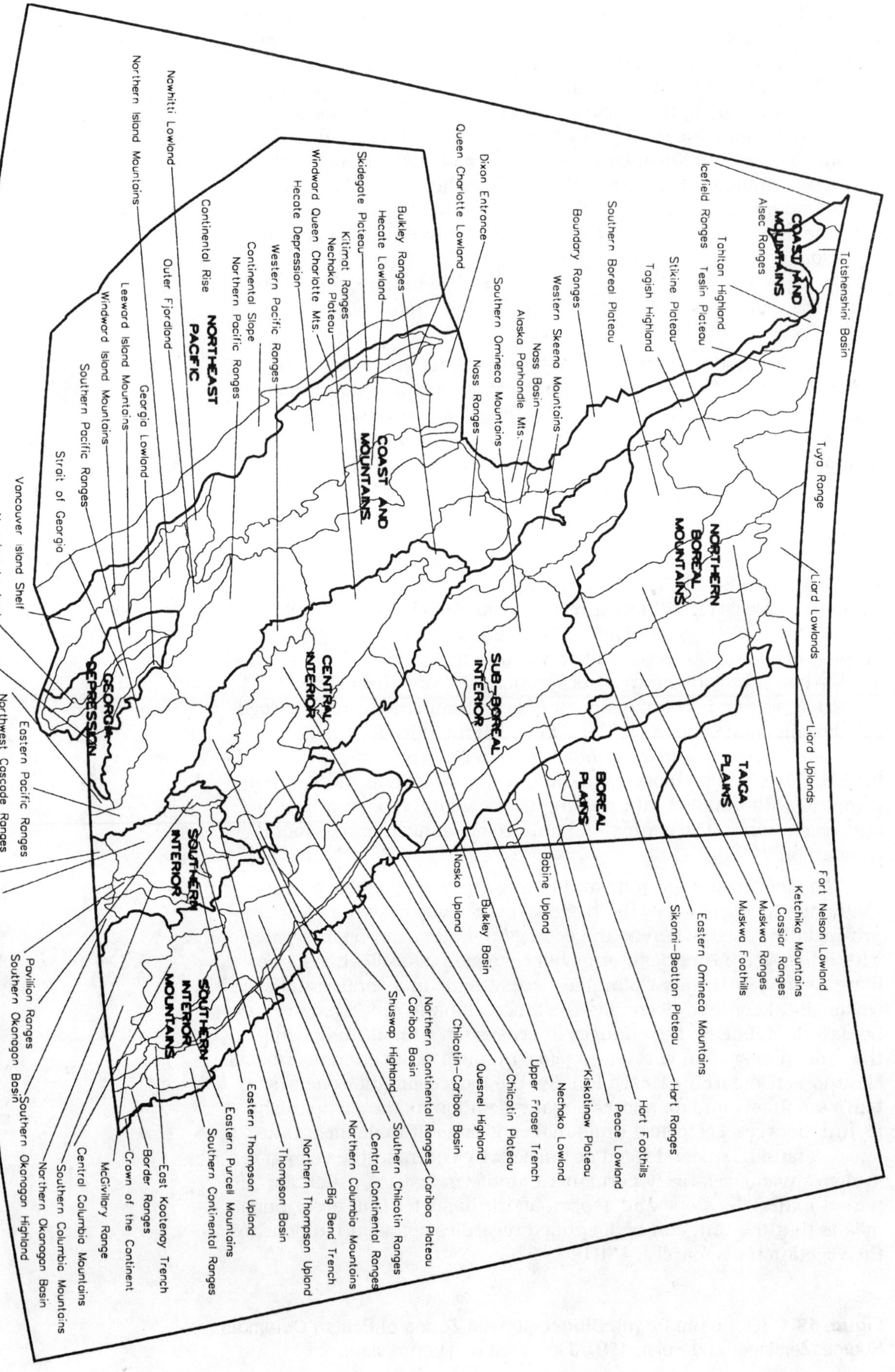

Figure 18-2. Ecoregions of British Columbia. Source: B.C. Ministry of Environment, Lands and Parks.

Despite the differences between these two systems, they serve compatible and complementary functions, and can be integrated, to a degree (Demarchi, 1991). The provincial Ministries of Forests and Environment, Lands and Parks have standardized their parameters for data collection, both use the biogeoclimatic sub-zone unit to represent zonal climates and ecosystems, and they are working on standardizing their mapping methodologies. Demarchi et al. (1990) show how the biogeoclimatic zones are integrated, mainly as elevational forest type zones, within ecoprovinces. It is now widely recognized that integration of these two systems is useful for provincial land use planning and resource allocation.

However the ecosystems of British Columbia are classified, maintenance of ecological diversity in the province would entail preservation of structure and function in each of them. For maximum ecosystem diversity, all successional stages as well as the climax community need to be represented.

Ecosystem Fragmentation

Ecosystems become fragmented when small, apparently insignificant increments are appropriated for other purposes. Roads are often the harbingers of fragmentation. They facilitate timber harvest, other resource extraction, hydroelectric dam construction, hunting, and settlement. As well, roads and vehicles are major vectors for the introduction of alien species of plants and animals, as discussed in Chapter 17. Brush along the right-of-way provides different bird, insect and small mammal habitat than in the forest interior, and these species, following a long, linear route as opposed to the more natural, sporadic forest openings created by landslides and forest fires, can intrude a great distance into an otherwise alien environment. Settlement, including farms and other rural land use, removes natural habitats, creates barriers to free movement of plant and animal species between remaining habitats, and creates more edge effect around the residual islands of habitat. Kimmins and Duffy (1991), stated that forest fragmentation by logging clear cuts increases the area of forest edge, which is good for many species of wildlife that use clear cuts, but can have a negative effect on species that have an absolute requirement for mature or old growth forest. The functioning of the original ecosystem may be quite altered, but look superficially similar. Predator-prey relationships may be disrupted, requiring active management; water may be diverted from streams and wastes introduced; and indigenous species edged out by those more adaptable to human disturbance. Biodiversity has changed and with it, though more subtly, ecosystem structure and function. These changes are progressive with increasing fragmentation.

The decline of caribou (*Rangifer tarandus*) populations in British Columbia provides a good example of the effect of fragmentation on just one species. The precipitous decline first publicized by Harding (1975) was later determined to be caused primarily by increased road access to hunters (Figure 18-3), increased wolf predation, and losses of habitat to logging in the Selkirks and fires in the northern ranges (Page, 1985). These impacts, which might have been supportable individually, were cumulatively devastating.

235

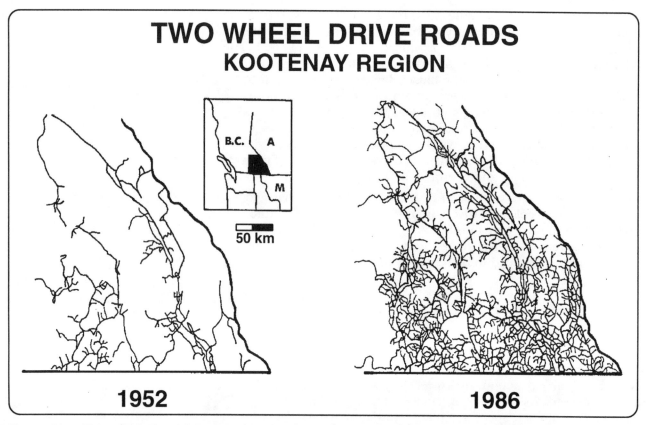

TWO WHEEL DRIVE ROADS
KOOTENAY REGION

1952

1986

Figure 18-3. Fragmentation of Caribou Habitat by Roads. Source: Hummel and Pettigrew, 1991.

For instance, wolves and caribou had always coexisted, but their balance was upset when large-scale habitat changes resulting from logging and forest fires improved the range for moose and deer, which can support higher predator populations than can the caribou (Bergerud, 1985). Caribou are dependent for their winter forage on lichens found only in old growth forests, so the habitat loss meant not only higher predation but less food. Though the Wildlife Branch now culls the wolf population, controls hunting, and works with other agencies to limit loss of mature forest winter range, caribou numbers still remain far below the management objective (Figure 18-4). Mountain caribou are now Blue-listed in British Columbia.

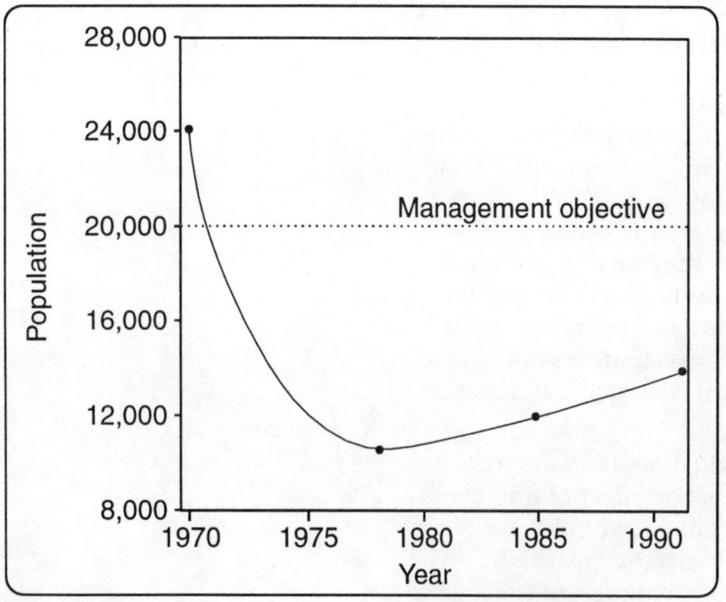

Unfortunately, some of the causes of caribou decline are irreversible. Loss of old growth forest to timber harvest is permanent because planned harvest rotation periods do not permit the growth of adequate lichen forage. Ritcey (1985) pointed out that mountain caribou habitat has declined by 23,310 square kilometres, or 45%, to only 29,000 square kilometres and is still

Figure 18-4. Trend in British Columbian Caribou Population. Sources: Halliday, 1985; Ministry of Environment, 1991.

shrinking. He also felt that the proliferation of roads throughout the province was a major threat to caribou populations. Fragmentation by roads is irreversible, because they bring increased use and development. Caribou populations are increasingly isolated from each other and from the remaining habitat, and several have been extinguished (Ritcey, 1985).

Harris (1984) provides an extensive analysis of island biogeography. As the size of an intact ecosystem dwindles, species diversity declines. When ecosystem structure and function are sufficiently altered by species decimation, the ecosystem becomes unstable, less resilient and less productive. Hence, a useful approximation of ecosystem fragmentation can be obtained by measuring the areas occupied by roads, resource extraction and settlements.

Vold (1992) chose to categorize wilderness by degree of road access. He defined primitive areas to be greater than 5,000 hectares in size and at least 8 kilometres from a four-wheel-drive road. Semi-primitive non-motorized areas were defined as greater than 1,000 hectares and more than 1 kilometre from a four-wheel drive road. Semi-primitive motorized areas were greater that 1,000 hectares and at least 1 kilometre from a two-wheel drive road (Figure 18-5). Based on Vold's analysis, 25% of British Columbia is primitive and 32% falls into the remaining two unroaded categories (Figure 18-6). These unroaded areas together with the 6% of the province's area devoted to provincial and national parks represent the proportion of B.C. with little or no habitat fragmentation. Much of this area is under permanent snow or glaciers, and has very low biodiversity. The remaining 37% of B.C. is subject to some degree of habitat fragmentation.

Some ecosystems are more fragmented than others. Figure 18-7 compares the amount fragmented with the amount protected in large parks and wilderness areas in each biogeoclimatic zone (from Vold, 1992). Four zones—Bunchgrass, Ponderosa Pine, Interior Douglas-fir and Coastal Douglas-fir—are more than 90% fragmented and have less than 1% of their area protected in large parks or wilderness. While important habitat remains in these zones, the conclusion that their ecosystems have been permanently altered is inescapable. In the rest of the British Columbia's biogeoclimatic zones where fragmentation by roads has not progressed too far, opportunities for protecting ecosystems remain.

When Vold's (1992) data are analyzed on a subzone/variant scale, 62 out of 124 ecosystems have less than 1% of their area protected, and 61 (not all of these are included in the above category) have less than 1% of their area in the primitive category—that is, they are almost completely fragmented. Thirty-nine of these 124 ecosystems have neither protected areas nor unfragmented habitat (primitive category) remaining (Table 18-1). In addition to the four badly fragmented biogeoclimatic zones noted above, fragmented subzones or variants with little or no large protected areas occur in the Boreal White and Black Spruce zone, the Coastal Western Hemlock zone, the Engelmann Spruce-Subalpine Fir zone, the Mountain Hemlock zone, the Sub-Boreal Pine-Spruce zone and the Sub-Boreal Spruce zone. These fragmented subzones or variants represent British Columbia's most endangered ecosystems.

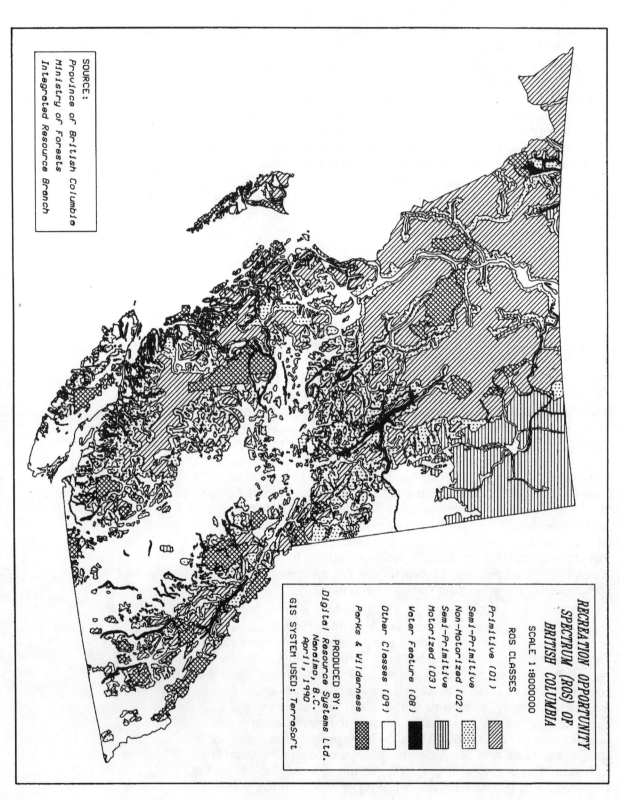

Figure 18-5. Wilderness Areas Categorized by Degree of Road Access. Source: Vold, 1992.

SOURCE:
Province of British Columbia
Ministry of Forests
Integrated Resource Branch

RECREATION OPPORTUNITY
SPECTRUM (ROS) OF
BRITISH COLUMBIA

SCALE 1:8000000

ROS CLASSES

Primitive (01)

Semi-Primitive
Non-Motorized (02)

Semi-Primitive
Motorized (03)

Water Feature (08)

Other Classes (09)

Parks & Wilderness

PRODUCED BY:
Digital Resource Systems Ltd.
Nanaimo, B.C.
April, 1990

GIS SYSTEM USED: TerraSoft

Zone	Subzone	Variant	% Roadless	% Protected
Bunchgrass (BG)	xh	Fraser	0.0	0.0
	xh	Okanagan	0.0	0.0
	xh	Thompson	0.0	0.0
Boreal White & Black Spruce (BWBS)	ms	Peace	0.1	0.2
	wk	Murray	0.6	0.0
Coastal Douglas Fir (CDF)	mt		0.0	0.0
Coastal Western Hemlock (CWH)	ds	Southern	0.9	0.8
	ws	Submontane	0.1	0.6
Engelmann Spruce-Subalpine Fir (ESSF)	dc	Okanagan	0.0	0.0
	dcp		0.0	0.0
	mv	Nechako	0.0	0.0
	wmp		0.0	0.0
	xv		0.0	0.0
Interior Cedar-Hemlock (ICH)	mc	Hazleton	0.0	0.1
	mk	Kootenay	0.5	0.5
	xw		0.0	0.0
Interior Douglas-Fir (IDF)	dk	Chilcotin	0.0	0.1
	dk	Fraser	0.0	0.2
	dk	Thompson	0.9	0.0
	dk1	Grassland Phase	0.0	0.0
	dm	Kettle	0.0	0.0
	dm	Kootenay	0.6	0.0
	mw	Okanagan	0.0	0.0
	xh	Thompson	0.0	0.0
	xh2	Grassland Phase	0.0	0.0
	xm		0.0	0.0
Montane Spruce(MS)	dm	Okanagan	0.0	0.0
Ponderosa Pine (PP)	dh	Kettle	0.0	0.0
	dh	Kootenay	0.0	0.0
	xh	Thompson	0.0	0.0
Sub-Boreal Pine-Spruce (SBPS)	dc		0.0	0.0
	mk		0.0	0.0
Sub-Boreal Spruce (SBS)	dk		0.4	0.5
	dw	Blackwater	0.0	0.0
	dw	Horsefly	0.0	0.0
	dw	Stuart	0.0	0.0
	mc	Moffat	0.0	0.0
	mh		0.0	0.0
	mw		0.0	0.0

Table 18-1: Percent of Protected Area and "Primitive Wilderness" in B.C.'s Most Endangered Ecosystems.

239

Managing Land Use

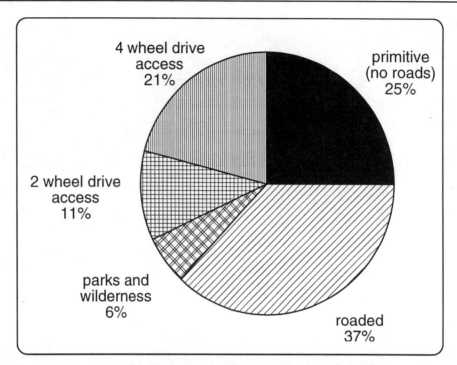

Figure 18-6. Proportion of British Columbia Subject to Habitat Fragmentation. Source: Vold, 1992.

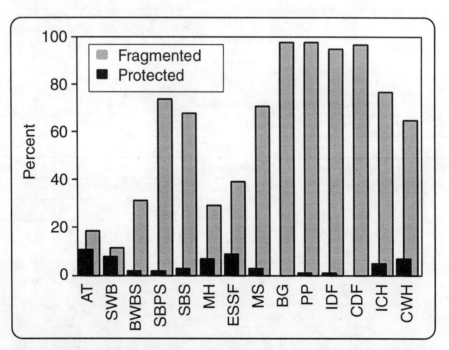

Figure 18-7. Protected and Fragmented Areas in each Biogeoclimatic Zone in B.C. Zones which are highly fragmented and have little protected area constitute our most endangered ecosystems. Source: Vold, 1992.

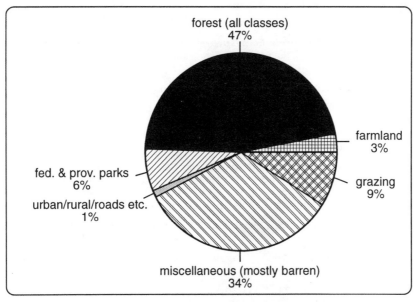

Figure 18-8. Land Use in British Columbia in Hectares (x 1000). Source: Ministry of Crown Lands, 1989.

About half (47%) of British Columbia is classified as forested (mature forest and forest land recovering from deforestation) and about a third (35%), mostly high elevation alpine land or glaciers, is classified as barren. Of the remaining 18%, 9% is allocated to grazing, 6% to federal and provincial parks, 3% to farms, and 1% to urban and rural developments (Ministry of Forests; Figure 18-8). Ecosystem diversity has been largely unaffected in the third of the province classified as barren (except perhaps in the few areas where livestock graze in alpine meadows) and has been most drastically altered in the urban and rural areas. It is, therefore, in the 56% of the province classified as forest and grazing land and in the waters of the coast that natural ecosystem diversity now hangs in the balance. Not only are these three areas important producers of economic and social benefits, but the opportunity to retain their natural ecosystem diversity has not been entirely foreclosed. In the next few chapters, we look closely at the challenges faced by these "endangered spaces" (Hummel, 1989). Recognizing that we cannot separate ourselves from our environment, we also consider urban biodiversity and some of the ways in which urban development can be modified to minimize its impact on natural ecosystems.

Acknowledgements

Michael Dunn of the Canadian Wildlife Service reviewed the manuscript. Richard Hebda, Royal British Columbia Museum, and J. Driver, Simon Fraser University, reviewed the portions of the manuscript relating to prehistory. Del Meidinger, British Columbia Ministry of Forests, and J. Driver provided additional information which has been incorporated into the text.

References

Bergerud, T. 1985. Caribou declines in central and southern British Columbia. In: *Proceedings: Caribou Research and Management in British Columbia.* B.C. Ministry of Forests, Victoria, B.C. pp. 201-225.

Carlson, C. 1989. The far west. In: Early Man in the Americas. R. Shutler (Ed.)

CCELC (Canada Committee on Ecological Land Classification). 1989. *Ecoclimatic Regions of Canada: First Approximation.* Environment Canada Ecological Land Class Serial No. 23. Ottawa, Ontario.

Driver, J.C., 1988. Late Pleistocene and Holocene vertebrates and palaeoenvironments from Charlie Lake, northeast British Columbia. *Can. J. Earth Sci.* 25:1545-1553.

Demarchi, D.A. 1991. Is it Coke or is it Pepsi—ecosystem classification in B.C. *Bioline* 10(1):2-5.

Demarchi, D.A., R.D. Marsh, A.P. Harcombe and E.C. Lea. 1990. The Environment. In: *The Birds of British Columbia Vol. 1. Non Passerines.* R. Campbell, N.K. Dawe, I. McTaggart Cowan, J.M. Cooper, G.W. Kaiser and M.C. McNall (eds.). Canada and Royal British Columbia Museum. pp. 55-144.

Finkelstein, M. 1990. *National Parks System Plan.* Environment Canada, Canadian Parks Service.

Gould, S.J., 1989. *Wonderful Life: the Burgess Shale and the Nature of History.* Norton Co., New York.

Gustafson, C.E. 1984. *The Manis Mastodon Site. An Adventure in Prehistory* (brochure).

Halliday, R. 1985. Caribou in British Columbia - management needs for the future. In: *Proc. Workshop on Caribou research and management in British Columbia.* R. Page, Ministry of Forests. pp. 251-263

Harding, L. 1975. Our mountain caribou...an endangered species? *B.C. Outdoors* 31(2):24-31.

Harrington, R.F. 1990. *To Heal the Earth: The Case for an Earth Ethic.* Hancock House, Surrey, B.C.

Harris, L.D. 1984. *The fragmented forest: Island biogeography theory and the preservation of biotic diversity.* University of Chicago press.

Hebda, R.J., 1982. Postglacial history of grasslands of southern British Columbia and adjacent regions. In: *Grassland Ecology and Classification, Symposium Proceedings.* June 1982. A.C. Nicholson, A. McLean and T.E. Baker (eds.). British Columbia Ministry of Forests, Victoria, B.C. pp. 157-191.

Hebda, R.J. 1983. Lateglacial and postglacial vegetation history at Bear Cove Bog, northeast Vancouver Island, B.C. *Canadian Journal of Botany* 61(2):3172-3192.

Hebda, R.J. 1991. Late quaternary paleoecology of Brooks Peninsula. In: *The Brooks Peninsula: an Ice-age Refuge.* R. Hebda and J.C. Haggarty (eds.). Royal British Columbia Museum, Victoria, B.C.

Hebda, R.J. and R.W. Mathews, 1984. Holocene history of cedar and native Indian cultures of the North American Pacific Coast. *Science* 225:711-713.

Horgan, J. 1992. Early arrivals: scientists argue over how old the New World is. *Scientific American* 266(2): 17-20.

Hummel, M. (Ed.). 1989. *Endangered Spaces: the Future for Canada's Wilderness.* Key Porter Books. Toronto, Ontario.

Hummel, M. and S. Pettigrew. 1991. *Wild Hunters: Predators in Peril.* Key Porter Books. Toronto, Ontario.

Kew, J.E.M. and J.R. Griggs, 1991. Native Indians of the Fraser Basin: towards a model of sustainable resource use. In: *Perspectives on Sustainable Development in Water Management: Towards Agreement in the Fraser River Basin.* Vol. 1. Dorcey (ed.). Westwater Research Centre, University of British Columbia. pp. 17-47.

Kimmins, J.P. and D.M. Duffy. 1991. Sustainable forestry in the Fraser River Basin. In: *Perspectives on Sustainable Development in Water Management: Towards Agreement in the Fraser River Basin, Vol. 1.* A.H.J. Dorcey (ed.). Westwater Research Centre University of British Columbia. pp. 189-215.

Krajina, V.J. 1965. Biogeoclimatic zones and biogeocoenoses of British Columbia. *Ecology of Western North America* 1:1-17.

Kurtén, B. and E. Anderson. 1981. *Pleistocene Mammals of North America.* Columbia University Press.

Hummel, M. (ed.). 1989. *Endangered Spaces: The Future of Canada's Wilderness.* Key Porter Books, Toronto, Ontario.

Meidinger, D. and J. Pojar. 1991. *Ecosystems of British Columbia.* B.C. Ministry of Forests, Victoria, B.C.

Ministry of Environment, 1991. Managing wildlife to 2001: a discussion paper. Wildlife Branch, B.C. Environment.

Odum, E.P. 1969. The strategy of ecosystem development. *Science* 164: 262-270.

Page, R. 1985. *Proceedings of a Workshop on Caribou Research and Management in British Columbia.* B.C. Ministry of Forests, Victoria, B.C.

Raup, D.M. 1988. Diversity crises in the geological past. In: *Biodiversity.* E.O. Wilson (ed.). National Academy Press, Washington, D.C.

Ritcey, R. 1985. Provincial approach by Ministry of Environment to caribou habitat management. In: *Proceedings of Caribou Research and Management in British Columbia.* Ministry of Forests. pp. 9-12.

Rowe, J.S. 1977. *The Forest Regions of Canada.* Minister of Supply and Services, Ottawa, Ontario.

Vold, T. 1992. *The Status of Wilderness in British Columbia: A Gap Analysis.* Appendix C: Representation by biogeoclimatic classification (Zone, Subzone, Variant). Ministry of Forests.

Chapter 19
Threats to Diversity of Forest Ecosystems in British Columbia

British Columbia's forest ecosystems are world famous for their diversity and beauty. They also provide the foundation for British Columbia's economy. Although First Nations people have used a great variety of forest products virtually as long as there have been forests in B.C. (since the last ice age) (Kew and Griggs, 1991), it is only since about 1870 that forests have been harvested commercially and, in some cases, replaced by development. The effects of human activities on British Columbia's forest ecosystems are reflected in the extent to which they have been fragmented by roads and urban development.

As outlined in Chapter 18, three of the 12 forested biogeoclimatic zones (Coastal Douglas-fir, Interior Douglas-fir and Ponderosa Pine) are more than 90% fragmented by roads—that is, large (greater than 5,000 hectares) "primitive" areas make up less that 10% of their area. Forest biogeoclimatic subzones or variants with little or no large unfragmented areas also occur in the Boreal White and Black Spruce zone, the Coastal Western Hemlock zone, the Engelmann Spruce-Subalpine Fir zone, the Mountain Hemlock zone, the Sub-Boreal Pine-Spruce zone and the Sub-Boreal Spruce zone.

Fragmentation by roads is a measurable index of forest ecosystem change, not only because of the ecosystem attributes it directly affects, such as the forest interior to edge ratio and the free movement of some organisms, but also because it admits other activities, such as renewable and non-renewable resource use, which have further consequences (Rosenburg and Raphael, 1986; Harris, 1984; Foreman and Gordon, 1986). Human activities that change fundamental ecological processes, such as primary production, nutrient cycling and losses, rates of decomposition (Woodley and Theberge, 1991) and carbon and hydrological cycles, may lead to changes in the physical and biological structural attributes of forest ecosystems, such as the size of trees and woody debris, canopy characteristics, species diversity, retrogression to an earlier successional stage, and habitat fragmentation (Woodley and Theberge, 1991). Management regimes, such as hunting regulations and timber harvest rules, that are meant to control environmental impacts can themselves cause changes in ecosystem structure, such as younger forests and altered species composition.

If changes in ecosystem structure are permanently maintained by management controls, new ecosystems are created and old ones lost. The effects of our use and management of forests—of forest loss, fragmentation and conversion from natural to managed forests—are a loss of natural forest ecosystem diversity. This diminished diversity is offset, to a degree, by the addition of new forest ecosystems with different characteristics. The challenge is to bal-

Lee E. Harding
Canadian Wildlife Service
P.O. Box 340
Delta, B.C.
V4K 3Y3

ance the loss of diversity with the benefits to society of managed forests. In this chapter, we examine the forest land base and the natural characteristics of unmanaged forests, then consider the effects of forest conversion and management, and conclude with some informed speculation about the future of British Columbia's forests.

Forest Land Base

More than half (51.8 million hectares) of the province's 94.8 million hectares of land is forested. Most of this land, or 42.7 million hectares (about 82%), is under provincial Ministry of Forests control; 3.3 million hectares (about 6%) is Crown land in Tree Farm Licences (TFL) under Ministry of Forests control but managed by private companies; another 3.3 million hectares is allocated to various uses such as Indian Reserves and local or federal administration; about two million hectares (4%) is in national and provincial parks, and the remaining one million hectares (2%) is in miscellaneous forest resource tenures, such as woodlots and shake permits (Ministry of Forests, 1991a; Figure 19-1).

British Columbia represents only 9.5% of Canada's land area, but contains 49.7% of Canada's softwood timber volume (Forestry Canada, 1990). Commercial timber is harvested from 11 of British Columbia's 14 biogeoclimatic zones (Figure 18-1), although the annual growth of new wood varies from less than 0.8 cubic metres per hectare per year in the Spruce-Willow-Birch zone to more than 6.4 cubic metres per hectare per year in the Coastal Western Hemlock, Coastal Douglas-fir and Interior Cedar-Hemlock zones (Meidinger and Pojar, 1991).

A great deal of the fertile valley bottom land which was originally forested (some would have been natural meadow or grassland) has been developed. For example, about 580,000 hectares of such land is now used for urban and rural settlement and transportation, 2,400,000 hectares for farm and pasture land, and another 330,000 hectares for hydroelectric development (flooded for reservoirs) (Ministry of Crown Lands, 1989). According to Meidinger and Pojar (1991), the productivity of such top quality forest land (in terms of timber biomass produced through the yearly growth of trees) is 6.3 cubic metres per hectare per year. So, if even half (about 1,650,000 hectares) of these high quality lands had remained forested rather than been developed, they would now produce about 10.4 million cubic metres of timber per year, or about 14% of the 1990 harvest of 74.3 million cubic metres (Ministry of Forests, 1992).

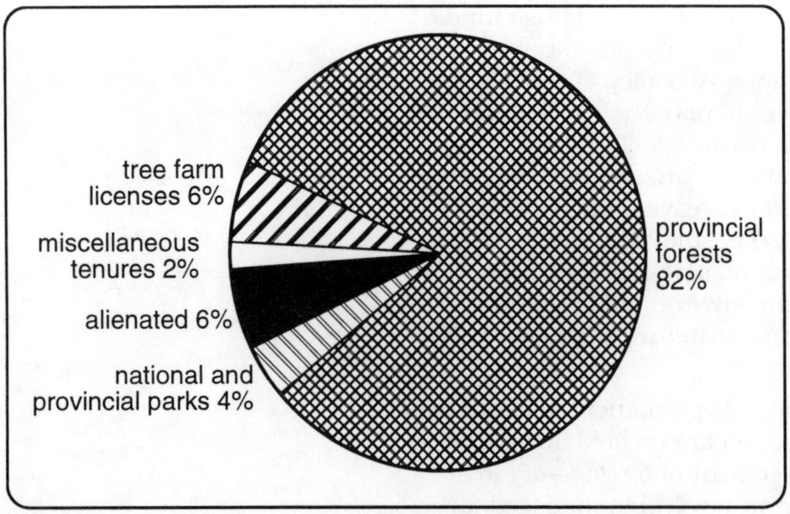

tree farm licenses 6%
miscellaneous tenures 2%
alienated 6%
national and provincial parks 4%
provincial forests 82%

Figure 19-1. Distribution of Forest Land Tenures in British Columbia. Source: Ministry of Forests unpubl. data, 1991.

Not only has the economic productivity of these forests been lost to development, but so also have the diversity of forest bottomland habitats and the plant and animal species associated with them.

Fires, Insects, and Pathogens

On a long time scale, disturbances such as fires and insect infestations are natural and necessary components of healthy forests, even though the organisms living in the forests and the trees themselves may be adversely affected in the short run. Such events provide the variety of habitat types needed for a diversity of organisms, some of which favour young forests, some old growth, and some both. Even irruptions of populations of insects that feed on trees—such as defoliators and bark beetles (*Dendroctonus spp.*)— foster ecosystem diversity by providing food for insectivorous birds and other forest wildlife, and by creating essential habitat elements such as snags and large woody debris. As well, disturbances facilitate adaptation of forest ecosystems to changing environments and, in the long term, the evolution of forest species.

Fire

Fire frequency is related to climate. In British Columbia, the moist coastal and interior wet zone forests have a much lower fire frequency than interior dry region forests (Pojar et al., 1990). Forest fire frequency and intensity are inversely related. In general, frequent (occur every few years or decades) surface fires are low in intensity because of limited fuel (woody debris and forest understory). Consequently, they leave most mature trees standing. Where fires are infrequent (occurring on a cycle of several centuries), accumulated woody debris can fuel an intensity that has a greater chance of reaching into the crowns to destroy mature trees. Weather, of course, is also a major factor: fires are more severe in hot, dry summers.

Although fires are natural components of forest ecosystems, human activities alter fire regimes. We accidentally set fires, which may increase the frequency above those naturally started by lightning, and we suppress those that do start with a sophisticated fire-fighting system. In recent years, forest fires have been increasing in number (Figure 19-2), though the total area burned has been decreasing (Figure 19-3), presumably because fire-fighting efforts have increased and the technology has improved. Large areas of the interior of the province are stocked with mature lodgepole pine, a fire-successional species that matures in about 80-100 years, suggesting a history of large-scale fires. Events that might have caused such fires include the 1856 gold rush, the 1880's construction of the railroad and the 1940's construction of the Alaska Highway. In the short term, our influence on natural fire regimes may be expected to result, overall, in increased fire frequency but decreased severity (in terms of the area burned). In the long term, however, the accumulation of fuel may increase the probability of larger, more intense, stand-replacing fires. The Ministry of Forests is currently undertaking a disturbance history analysis of forest regions.

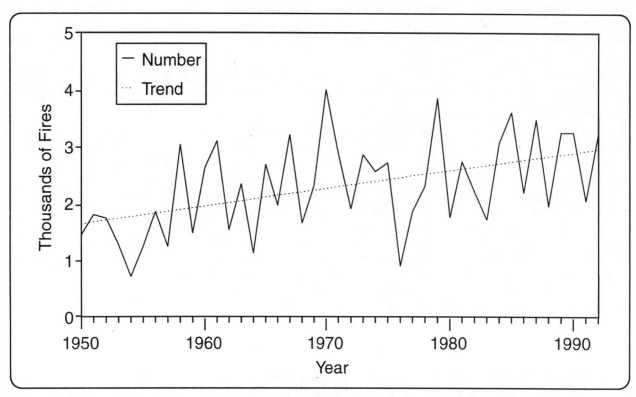

Figure 19-2. Frequency of Forest Fires (1950-1991). Source: Ministry of Forests, unpubl. data, 1992.

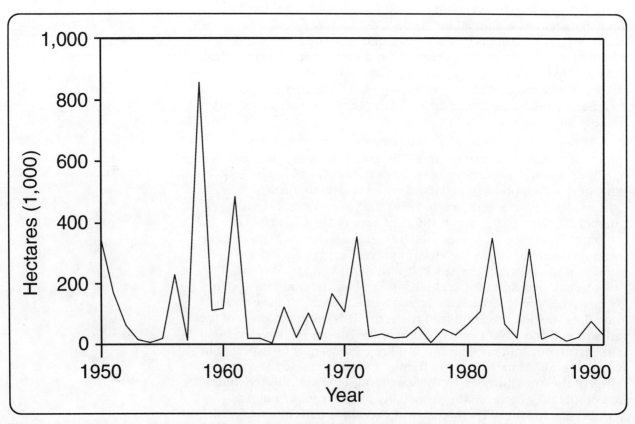

Figure 19-3. Total Area Burned by Forest Fires (1950-1991). Source: Ministry of Forests, unpubl. data, 1992.

Forest fire ecology has been a fertile field of research for many years, and prescribed fires are now used routinely as a wildlife habitat and range management technique (for example, see the review by Silva Management Services Ltd., 1992), as well as to remove logging slash. Certain western North American forests, like those dominated by lodgepole pine (*Pinus contorta*) or ponderosa pine (*Pinus ponderosa*), have always depended upon regular fires for regeneration (Habeck, 1988). Lodgepole pine forests are found throughout most of the interior and the north. Fires stimulate lodgepole pine cones to release their seed. The pines then grow quickly, maturing in less than a century. Since pine forests tend to be dry and ignite readily, they are replaced frequently, in some cases preventing natural succession to climax communities (which might be boreal white and black spruce, Engelmann spruce-subalpine fir, or another forest type, depending on local biogeoclimatic variables). Climax communities in these forests are, therefore, relatively rare and, hence, more valuable in terms of ecosystem diversity.

Unlike lodgepole pine, which depends on fire for frequent regeneration, ponderosa pine needs fire to attain old growth characteristics. Ponderosa pine forests are found in the dry southern and southeastern parts of the province. Old growth ponderosa pine stands are maintained by frequent low-intensity surface fires which burn brush and small trees but leave the mature pine, preventing the maturation of the more shade-tolerant Douglas-fir (*Pseudotsuga menziesii*), which otherwise would often succeed the pine (Habeck, 1990). There is some evidence that effective fire suppression, in combination with selective harvesting of the old growth ponderosa pine, has contributed to very dense regeneration of almost pure Douglas-fir in many formerly ponderosa pine stands (MacLauchlan, 1992). Also, Pitt (Chapter 20, this volume) reports an estimate that forest encroachment due to fire prevention may have absorbed about 30% of grassland in the Cariboo-Chilcotin region. Hence, the ponderosa pine and grassland ecosystems, already severely fragmented by roads and other developments (Chapter 18, this volume), are further threatened by forest management practices.

Insect Infestations and Pathogens

Like fires, insect infestations and pathogens (disease agents) contribute to the structural and biological diversity of the ecosystem. They are necessary ecological processes in healthy forest ecosystems. However, because they damage commercial timber, much of our data on populations and distribution, as well as our management responses, relate to their potential for economic losses.

Forest damage by insect and disease infestation is highly variable in space and time. Major infestations have been recorded since before the turn of the century. Since 1973 alone, British Columbian forests have been infested by at least 48 species of insects, 25 of which caused major forest damage, according to the most recent review by Forestry Canada (Van Sickle, in press). These insects can be broadly categorized as defoliators, cone and seed insects, and bark beetles. Defoliators tend to affect young trees more than mature ones and may only kill a small proportion of the host species of trees in the forest area infected. Cone and seed insects are particularly troublesome in tree nurseries and seed orchards. Bark beetles mainly attack mature or recently fallen (or felled) trees. Some insects work in concert with pathogens. For example, the effects of the mountain pine beetle (*Dendroctonus ponderosae*) and the blue stain fungi (*Ceratocystis montia* and *Europhium clavigerum*) it carries combine to make them the worst forest pests in British Columbia. Mortality is often nearly 100% in trees infected, and infestations spread rapidly in the pine beetle's preferred habitat, mature pine forests. Forest diseases, such as root rot, stem decay, dwarf mistletoe and rust, are mainly of concern in plantations and in some high-value seed tree stands.

Infestations have been increasing in recent years, particularly those caused by the mountain pine beetle, which reached a peak in 1984 before declining, and the western spruce budworm (*Choristoneura occidentalis*), which peaked in 1987. Figure 19-4 illustrates recent trends in forest damage caused by forest insects. In some regions, bark beetle infestations have influenced logging schedules because timber harvest has been redirected from planned harvest areas to infested areas (personal communication in 1992 with G. Allan Van Sickle, Pacific Forestry Centre). On a volume basis, approximately one-third as much timber is destroyed annually by insects and pathogens as is logged (personal communication in 1992 with G. Allan Van Sickle, Pacific Forestry Centre).

There are many possible reasons for insect and disease outbreaks, which in the long term must be balanced by regenerative capacity for forest ecosystems to be sustained. In natural forests of mixed tree species and varied physical structures, insect and disease infestations usually persist only a few years, until their host tree supply is exhausted or the insects are killed by weather or by other pathogens, such as parasites or a virus (Van Sickle, in press). However, human activities, such as fire management and forest harvest practices, have modified forest ecosystems in several ways that may have affected forest insect population dynamics. As well,

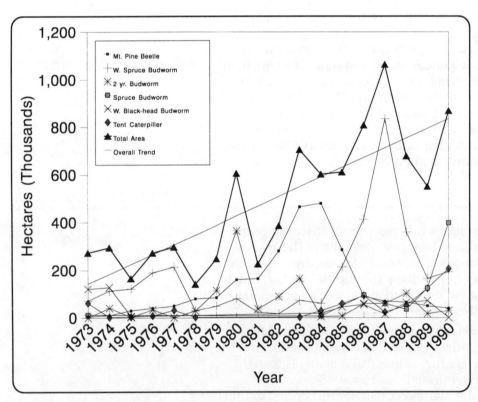

Figure 19-4. Trends in Infestations by Six Major Forest Pests in British Columbia. The solid line represents the total area of damage by all species, and the trend line is a linear correlation of the total area damaged over time. Source: data from Van Sickle (in press).

global changes such as climate change and tropical deforestation may be involved. These possible factors in insect pest outbreaks are discussed below, as hypotheses for further research.

Large areas of the central interior are covered with near monocultures of maturing lodgepole pine forest which grew following the extensive forest fires that accompanied European and Asian settlement. This extensive host tree supply is doubtless a factor in recent mountain pine beetle infestations (personal communication in 1992 with G. Allan Van Sickle, Pacific Forestry Centre). There is some evidence that these infestations are part of a natural cycle: the mountain pine beetle kills the mature pine, providing fuel for stand-replacing fires, which are followed by regeneration of the pine (personal communication in 1992 with A. McKinnon, Ministry of Forests).

MacLauchlan (1992) felt that past forestry practices in interior Douglas-fir ecosystems may have exacerbated the impact of the western spruce budworm. He noted that selective harvesting of ponderosa pine in the drier portions of these ecosystems and effective fire suppression have left predominantly Douglas-fir forests. In a more natural scenario, frequent ground fires would prevent much of the Douglas-fir regeneration, promoting a higher proportion of ponderosa pine on these sites. The combination of reduced species mix and fire-exclusion "has created a high-hazard environment in terms of the budworm" (MacLauchlan, 1992).

Similar harvest-related factors may also be involved in insect infestations in other forest types. The extensively logged and reforested areas in interior and northern forests are more susceptible than natural forests to infestations of defoliators, such as the two-year-cycle budworm (*Choristoneura biennis*) and the western balsam bark beetle (*Dryocetes confusus*) of northern and high elevation spruce and fir forests (Ebata, 1992a). These insects attack young as well as mature trees. Van Sickle (in press) notes that, as more young stands develop following harvest of the higher-elevation spruce-fir types, the potential need for control of the two-year-cycle budworm is increasing. This statement may also apply to the spruce budworm (*C. fumiferana*) of the northern boreal forests. Similarly, forests regenerating following timber harvest on the coast are susceptible to defoliators, such as the western black-headed budworm (*Acleris gloverana*) and green striped forest looper (*Melanolophia imitata*).

A second harvest-related factor is that clear-cut logging removes nest trees for cavity-nesting birds, many of which, such as woodpeckers, nuthatches and creepers, feed on insects in or under bark. Two cavity-nesting birds, Lewis' Woodpecker and the White-headed Woodpecker (*Picoides albolarvatus*), are currently endangered in British Columbia (Chapter 2, this volume), and another, the Northern Flicker (*Colaptes auratus*) is declining (Erskine et al., 1990). The standing snags which house cavity-nesting birds are not replaced during planned rotation periods.

A third harvest-related factor is the loss of predacious and parasitoid insects, which are a feature of old growth forests. Winchester (in press) and his colleagues found that the canopies of old growth coastal forests in British Columbia support a surprisingly

251

rich invertebrate community. Moss and litter accumulations on the large branches provide a habitat which is both previously unrecognized and unique to old growth forests. This habitat harbours a much larger proportion of predator and parasitoid insects than that of second growth forests, whose arthropod fauna is mainly herbivorous. Predator and parasitoid insects prey on or parasitize herbivorous insect species, including those that have caused so much damage to B.C. forests in recent years.

Changing weather patterns also affect insect populations. Severe winters can kill overwintering broods, while cold or wet spring weather can kill eggs, reduce hatching success and retard growth of larvae (Van Sickle, in press). Milder winter temperatures may have permitted northward range extensions of some insect pests, such as the spruce bud moth (*C. fumiferana*), which was recorded for the first time north of Mackenzie in 1989 (Van Sickle, in press). Hot, dry summers encourage outbreaks of Douglas-fir tussock moth (*Orgyia pseudotsugata*) in the Ponderosa Pine-Bunchgrass and Interior Douglas-fir biogeoclimatic zones (Shepherd et al., 1988). Likewise, infestations of white pine weevil (*Pissodes strobi*) are so temperature-related that infestation hazard ratings based on weather records (degree-days) have been developed for the Prince George and Prince Rupert Forest Regions (Ebata, 1992b). The discussion of climate change in Chapter 24 shows that recent (late 1970s, 1980s) trends to warming in the interior occurred more or less in step with increasing insect pest outbreaks.

Insect populations may also have benefited from the decline in neotropical migrant birds, which are important predators of forest insect pests, though much more information is needed to support this hypothesis (Chapter 23, this volume). The reverse is also possible: high insect populations may benefit insectivorous birds, possibly offsetting the adverse effects of fragmentation discussed in Chapter 23.

Determining exactly which of the above impacts had the most, if any, influence on insect outbreaks in British Columbia's forests remains a subject for future research. There is no question, however, that the cumulative effects of local land use and global change have provided the conditions for sudden irruptions of insect populations. Already, such irruptions have influenced the rate and location of forest conversion by timber harvest. If these recent phenomena turn into ongoing trends, they could have serious implications for forest ecosystem diversity.

Old Growth Forests

The definition of old growth is currently a matter of some debate. Because reforestation has been underway for less than 60 years in British Columbia, almost all the stands currently being cut are the result of natural regeneration, and almost all can be termed old growth when considering only tree age in relation to the longevity of the tree species composing these stands (Klinka et al., 1990a). In general, old growth forests have matured beyond the stage at which they first become economically harvestable to the point where growth has slowed and some trees have died and fallen (though the remaining trees are still very valuable). The death of trees creates standing snags, fallen logs, and other large woody debris, and leaves openings in the forest canopy through which sunlight can stimulate the growth of understory plants and young trees of various species.

Pojar et al. (1990) provided definitions of old growth based on economic, mensurational (maturation of trees), conceptual/ecological and spiritual/aesthetic attributes. They note that definitions may have to vary between ecological zones. For example, the wetter coastal zones in B.C. can often be described in terms of climax characteristics, while some dry interior zone forests (such as lodgepole pine forests) with high fire frequencies may never attain climax conditions. They give minimum ages of climax forests as 150 years for coastal old growth, 200 years for Ponderosa Pine-Douglas-fir zones and 300 years for the Interior Cedar-Hemlock zone.

The definition of old growth currently accepted in British Columbia is provided by the Forest Land Use Liaison Committee (Ministry of Forests, 1992):

> Old growth is a forest that contains live and dead trees of various sizes, species composition and age class structure that are part of a slowly changing but dynamic ecosystem. Old growth forests include climax forests, but do not exclude sub-climax or even mid-seral forests. The age structure of old growth varies significantly by forest type and from one biogeoclimatic zone to another.

> The age at which old growth develops the specific structural attributes that characterize old growth will vary widely according to forest type, climate, site characteristics and disturbance regime. However, old growth is typically distinguished from younger stands by several of the following attributes:

> o   large trees for species and site;

> o   wide variation in tree sizes and spacing;

> o   accumulations of large size dead standing and fallen trees;

> o   multiple canopy layers;

> o   canopy gaps and understory patchiness; and

> o   decadence in the form of broken tops or boles and root decay.

As noted in this definition, old growth is not synonymous with climax forest. Studies by Krajina and his students (1978) and by the British Columbia Forest Service have found that tree species composition depends on several factors. Comparison of ecologically equivalent sites showed that the same site may support old growth stands of different tree species composition, implying that some old growth stands can be replaced by different, more tolerant, tree species (Klinka et al., 1990a). To achieve a complete representation of forest ecosystem diversity in a region, then, a conservation strategy would have to include not only old growth, but also the latter stages of climax development.

The British Columbia Forest Service inventories forests in 9 age categories and classifies them as immature, mature or overmature. Mature forests are 121 to 140 years old, except for those comprised primarily of lodgepole pine, other pine and deciduous trees, which are mature at 81 years. Overmature forests, which are used as a surrogate for old growth until definitions are clarified (a task force has been established to do this), are those in age classes 8 (141 to 250 years old) and 9 (over 250 years old).

It should be emphasized that age alone is not a good criterion for old growth. For example, less than 50% of the age class 8 forests in the southern interior meet the structural criteria for old growth (in literature received on October 8, 1992, from A. McKinnon, Ministry of Forests). Old growth forests in British Columbia are currently being inventoried and mapped through a program funded under the Forest Resource Development Agreement.

Since the definition of old growth is not well standardized and the Ministry of Forests' forest data inventory does not yet categorize forests by old growth character, it is difficult to estimate how much old growth is left in British Columbia. What we do know is that British Columbia has a vast amount of forest, 58.3% of which is mature (calculated from Table 4 in Ministry of Forests, 1992) and all of which harbours wildlife and other non-timber values. The proportion of mature versus immature forest and deforested areas varies from 38.3% in the Nelson forest region to 68% in the Prince Rupert region (Ministry of Forests, 1992; Figure 19-5). Mature forest (age classes 8 and 9, except for lodgepole pine which includes age class 7) includes all of our old growth, as well as younger forests that have matured to harvest stage but have not yet developed the characteristics of old growth.

Species Diversity within Forest Ecosystems

The question of how much old growth remains in British Columbia's forest ecosystems is highly relevant to the diversity of species within these ecosystems. Many of the threatened or endan-

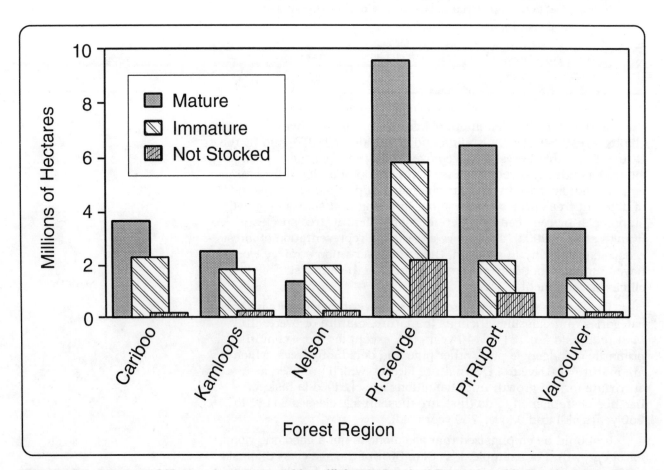

Figure 19-5. Amount of Mature, Immature and Insufficiently Stocked Forest Land in Provincial Forests (FSA's and TFL's). Source: Ministry of Forests, 1991a.

gered (Red List) and vulnerable (Blue List) wildlife species discussed in Chapters 4-16 at times require some aspects of old growth forest habitat. For example, red-listed species relying on old growth forest include: long-eared Keene's myotis bat (*Myotis evotis*), Marbled Murrelet (*Brachyramphus marmoratus*), sharp-tailed snake (*Contia tenuis*), Spotted Owl (*Strix occidentalis*), and Ancient Murrelet (*Synthliboramphus antiquus*). Blue-listed species include: Bald Eagle (*Haliaeetus leucocephalus*), caribou (*Rangifer tarandus*), Cassin's Auklet (*Ptychoramphus aleuticus*), fisher (*Martes pennanti*), Flammulated Owl (*Otus flammeolus*), Fork-tailed Storm Petrel (*Oceanodroma furcata*), Great Blue Heron (*Ardea herodias*), Rhinoceros Auklet (*Cerorhinca monocerata*), White-headed Woodpecker (*Picoides albolarvatus*), and Williamson's Sapsucker (*Sphyrapicus thyroideus*) (B.C. Environment, 1991).

Bunnell and Kremsater (1990) have shown that the pattern of diversity in birds and mammals in British Columbia's forests is striking. About 69% of the approximately 420 native Canadian bird species and 73% of the 140 native terrestrial mammal species breed in British Columbia, and most are forest dwellers. By comparison, Ontario (which has 66% of the Canadian bird species and 48% of the Canadian mammal species) and Quebec (63% of the birds and 47% of the mammal species) are larger than British Columbia but have less vertebrate species diversity. Morgan and Wetmore (1986) showed that bird species richness (number of species in a genus) and diversity (total number of taxa in the class of birds) increased with the number of tree species, the size of the site and the presence of snags and deciduous trees and shrubs. Chapter 16 provides more detail on the bird species most dependent on old growth forest habitat.

The loss or dramatic reduction of such forest dwelling mammal species as cougar (*Felis concolor*), wolverine (*Gulo gulo*) and fisher (*Martes pennanti*) in the east illustrates the role of forests in providing refuge from hunting, trapping and other land-use activities. Regional differences in diversity are accentuated for species requiring special forest characteristics or large tracts of forest. For example, B.C. has about 87% of Canada's cavity-nesting bird species and 85% of the mammal species whose average body weight is greater than one kilogram. Ontario, on the other hand, has 63% of both the cavity-nesting bird and the large mammal species, and Quebec has 62% and 58%, respectively. Within British Columbia, Bunnell and Kremsater (1990) found, "repeatable patterns in the relative abundance of wildlife species as natural succession progresses." That is, each stage of forest succession has a different and characteristic assemblage of wildlife species.

Species diversity though, goes far beyond the high-profile vertebrates. Most (84.9% in one study; Figure 19-6) of the species found in the forest are arthropods (invertebrates like insects and spiders). A growing body of research is beginning to demonstrate the importance of invertebrate diversity in Pacific northwest forest ecosystems (Schowalter, 1990; Asquith et al., 1990; Lattin and Moldenke, 1990; Moldenke and Lattin, 1990a, and 1990b, Winchester, in press). Franklin (1988) refers to research in Pacific coast temperate forests, where more than 1,000 species of invertebrates have been identified within a single old growth stand, the upper bole

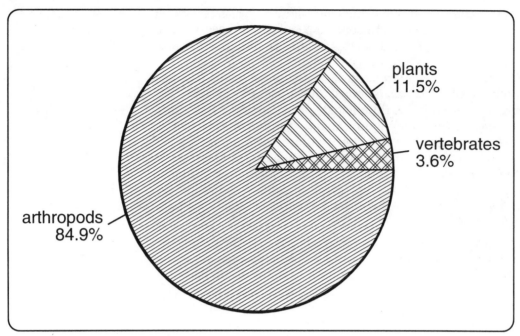

Figure 19-6. Arthropod (jointed-legged invertebrates, such as insects and spiders), Plant and Vertebrate Diversity in Northwest Pacific Forests. Source: Asquith et al., 1990.

and crown providing particularly rich habitat. Similarly, recent research into insect populations in the canopy of old growth coastal forests of British Columbia has demonstrated a startling diversity, including species new to science, and new functions of old forest ecosystems. Winchester (in press) studying biodiversity in the moss and litter layer on the upper side of large branches high in the crown of coastal sitka spruce (*Picea sitchensis*), notes that, "we are dealing with a largely undescribed, complex system that in several aspects rivals many of the tropical forest studies..." Similarly, Marshall (in press) notes that the high diversity of British Columbian soil fauna, though not classified and adequately inventoried, is essential to forest ecosystem function.

In addition to animal species diversity, other kingdoms represented in British Columbia forests include viruses, Monera ((bacteria and blue-green algae), Algae, Fungi, and vascular and non-vascular plants. Fungi and vascular plants were discussed in chapters 9 and 10. Except for vascular plants, all of these taxa are too poorly known to analyze their conservation status in terms of specific forest habitats. Hence, maintenance of species diversity must depend on maintaining ecosystem diversity at its various scales: a mix of early, mid-, and late-successional stages in each type of ecosystem.

Conversion of Natural to Managed Forests

Timber harvest in some respects mimics natural disturbances, but with important differences. Natural disturbances generally have a patchy distribution, as with insects that attack only one species of tree or fires that burn better in some areas than in others, leaving much of the natural forest ecosystem intact to regenerate the disturbed areas. In lands managed for timber production, clear-cut logging and reforestation convert large tracts of mature or old growth forests to managed forests, which do not support the same type of ecosystem as naturally disturbed forest. In effect, the natural forest ecosystem in such areas is permanently lost, along with such ecosystem attributes as progression to climax and species diversity. Managed forests may be healthy and productive for timber and for many species of plants and animals, but the species mix is quite different from that found in a natural forest.

The aspects of this conversion which most obviously affect biodiversity are: the rate of conversion, the method of harvest, reforestation practises, and the use of pesticides. After discussing each of these subjects, we consider the consequences of and some alternatives to current conversion practices.

Rate of Conversion

The total operable volume (the amount of provincially managed forest that is available and suitable for growing commercial timber) of timber is 4,030 million cubic metres (Min. of Forests, unpubl. 1992 data). Harvest rates increased exponentially between 1912 and 1989 (Min. of Forests, 1991a; Figure 19-7), when 86.9 million cubic metres were harvested from all tenures. In 1990, 74.3 million cubic meters were harvested from 181,530 hectares (Ministry of Forests, 1992).

A major review of the timber supply analysis process (Ministry of Forests, 1991b) identified a number of problems with the inventory process and planning assumptions, which generally tended to overestimate the amount of timber available and the rates of forest replacement. The Ministry of Forest's determination of the allowable annual cut is based on its estimation of the long run sustained yield of timber in each administrative region. In 1992, the Ministry's preliminary estimate of the long run sustained yield level for all regions was about half of the 1989 harvest level (Ministry of Forests, unpublished 1992 data; Figure 19-7). Current calculations for the long run sustained yield include areas that cannot be reforested, such as logging roads, reserves to be set aside for the protection of biodiversity (for which guidelines are being prepared by the Ministry) and possible future park and wilderness areas, so the actual long run sustained yield level will likely be somewhat lower than these calculations suggest.

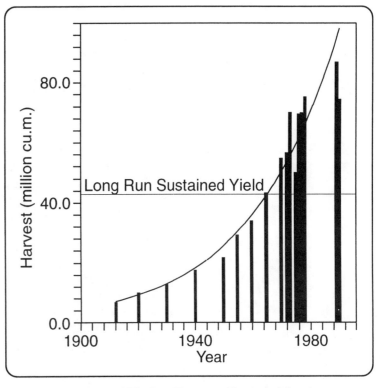

Figure 19-7. Annual Timber Harvest. The trend line is a statistical best fit for the data. Source: data for 1912-1978: Ministry of Forests, 1984; data for 1989 and 1990: Minstry of Forests, 1991a and Ministry of Forests, 1992; Long Run Sustained Yield level was a 1992 Ministry of Forests' preliminary estimate, which was still under review at that time.

257

Recognizing that annual growth of new timber will produce less volume than was being harvested in the province's original forests up until 1989, the Ministry of Forests has begun a gradual reduction of the annual harvest towards sustainable levels. The allowable annual cuts were further reduced in 1991, 1992 and 1993. Reductions are, however, proving difficult to implement for three reasons. First, companies may appeal their allocations, as was done successfully by one company in 1992. Secondly, the allowable cuts in some interior districts have been temporarily increased to deal with insect infestations—that is, infested stands are quickly logged to salvage the timber and remove the pests' habitat. Some very large, well publicized clear-cuts near Bowron Lakes and the Skeena area were logged for this reason. Thirdly, the Forest Act requires that the Ministry consider mill capacity in setting allowable annual cuts. Since mill capacity (and the rate of cut) has increased dramatically in the past two decades, especially during the 1980s (in part, to handle the volume of timber salvaged from insect infestations), there is considerable pressure to maintain high allowable annual cuts.

Future rates of timber harvest, and consequently the time frame for conversion of the province's original forests to managed forests, are difficult to predict for several reasons. First, the amount of natural forest left varies by forest tenure holder. For example, one company may hold large tenures which enable it to continue harvesting coastal old growth for several decades, while other companies with smaller holdings will run out much sooner. Hence, the overall cutting rate could decline, effectively extending the length of time it will take to cut down the remaining mature forest (personal communication in 1992 with J. Crover, Ministry of Forests). Secondly, the amount left is actually the amount that can be economically harvested, and economics change. As forest products become scarcer, the value of timber goes up and the inventory of forests that can be economically harvested increases. Thirdly, the definition of harvestable timber changes as new uses are found for wood products (personal communication in 1992 with J. Crover, Ministry of Forests). Through the 1950s, for example, western hemlock (*Tsuga heterophylla*) was considered a weed species, and was not harvested; now it is a major commercial product. Similarly, after the 1960s, markets were found for the extensive tracts of lodgepole pine (*Pinus contorta*) in the interior, greatly increasing the harvestable timber inventory. In the 1980s, aspen (*Populus tremuloides*) became marketable for use in manufactured products. These factors lend uncertainty to the planning assumptions. In particular, in the current climate of changes in timber harvest due to business, political and resource management decisions, it is not possible, based on past trends, to accurately predict the time remaining before all harvestable old growth outside of protected areas is gone.

Though updated timber supply analyses have not been completed for all areas, there is one for the Kalum North area of the Nass River drainage north of Prince Rupert (Integrated Resources Branch, 1993). Figure 19-8 shows the expected time frame for conversion of the natural forests in the Kalum North area to managed forests under current management practices and planning assumptions. In managed forests, ecological processes that control the distribution of age classes of forests are replaced by planned

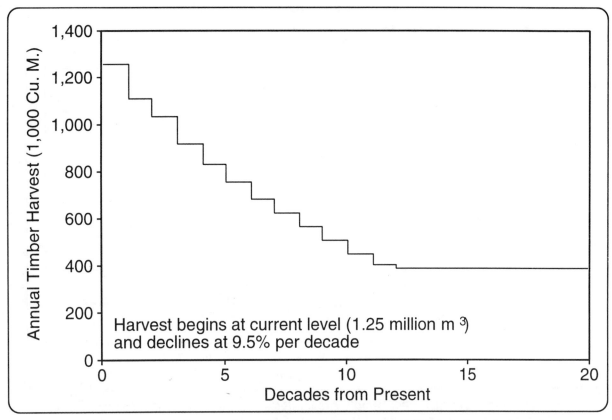

Figure 19-8. Time Frame for Conversion of Forests in Kalum North Timber Supply Area to Managed Forests (Based on Current Management Practices). Source: Integrated Resources Branch, Ministry of Forests, 1993.

cutting cycles designed to keep most of the forest young and growing rapidly, and a proportion maturing to harvest age in any given year. This change in age distribution alters the physical structure of the forest ecosystem from a mosaic in which most (more for moist coastal forests, less for interior forests where fires are more frequent) of the forest is old to a mosaic in which age classes within the available, operable portion of the forest are evenly distributed. Figure 19-9 illustrates, for the Kalum North Timber Supply Area, how the physical structure is altered by changing the age distribution. This figure shows two management scenarios (Integrated Resources Branch, 1993): (a) current practices, in which trees are harvested after 60 to 190 years of growth, depending on the species and area, and (b) a hypothetical regime of reduced harvest age (allowing harvest of younger trees). Note the large reduction in trees over 250 years old in both scenarios, and the virtual absence of trees between 121 and 250 years old after decade 10 in the second scenario. The conversion to managed forests would be more abrupt in areas like the Vancouver and Nelson regions, where a greater proportion of the natural forests have already been harvested.

Sustaining the yield of timber is not the same as sustaining the biological productivity of a forest ecosystem. Current plans call for timber harvest rotation periods of 80-140 years, whereas it may take 150-300 years for a forest to attain old growth characteristics (Pojar et al., 1990). Many forest scientists have recommended managing forests on the basis of ecological rotations, taking into account the

259

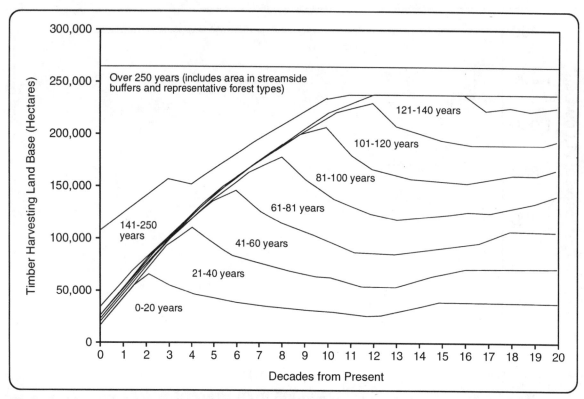

Figure 19-9a. Kalum North Timber Supply Area. Effects of current harvesting practices on the physical structure of the forest. Under current practices, trees are harvested every 60 to 190 years. Source: Integrated Resources Branch, Ministry of Forests, 1993.

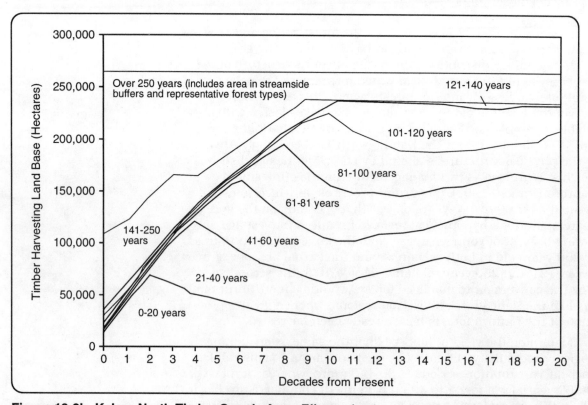

Figure 19-9b. Kalum North Timber Supply Area. Effects of reduced (by 20 years) harvest age on the physical structure of the forest. Source: Integrated Resources Branch, Ministry of Forests, 1993.

temporal and spatial scales of forest ecosystem development (Kimmins and Duffy, 1991). Extending the time scale of harvest rotation in this way would further reduce the long range sustained yield.

On a provincial scale, several decades of old growth cutting remain, but on a regional or local scale the old growth will be depleted sooner in many areas. Kimmins and Duffy (1991) felt that, if the sustainable rate of cutting is calculated over too large a region, and if the calculated cut is not distributed into several local cutting cycles (a cutting cycle is a complete set of stand age classes), local areas may experience boom and bust logging cycles, even though over the entire region the cut may be occurring at a sustainable rate. Social and economic sustainability at a smaller scale (a small local community), therefore, may be affected by forest management which ensures a sustained flow of logs at a larger geographic scale (a large timber supply area). The same can be said for ecological sustainability.

Method of Harvest

There are two basic ways of logging: partial-cutting and clear-cutting. Partial-cutting and its various refinements (see Klinka et al., 1990b for a description), along with the associated silvicultural treatments (such as site preparation and planting), alter the structural and compositional attributes of forests, though much less so than clear-cutting. Modern methods of partial-cutting can be selected and designed to maintain the structural attributes of old growth forest, wildlife habitat, or other non-timber values. However, although partial-cutting methods are used in some interior regions, about 90% of the harvested forests in British Columbia, including virtually all the harvested coastal forests, are clear-cut (Figure 19-10).

Clear-cut logging disrupts forest ecosystems more than partial-cutting, though it, too, varies widely in its impact. In winter logging, snow cover or frozen ground prevents or greatly reduces damage to soil, soil microorganisms, plants and even small mammals and other wildlife. Likewise, logging by helicopter or by cables attached to a high pole (high lead logging) limits disturbance to the land. By contrast, after summer clear-cutting using wheeled or tracked vehicles, followed by scarification (tilling to expose mineral soil) of

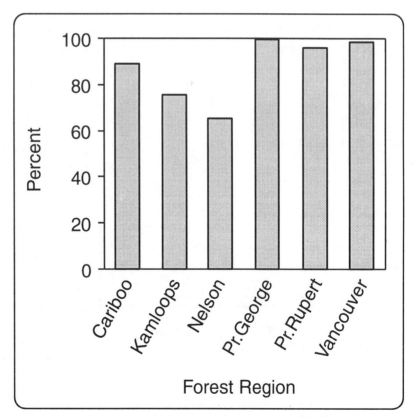

Figure 19-10. Percentage of Forest Harvested by Clear-cut in each Forest Region. Source: Ministry of Forests, 1991a

the land or burning, succession may be set back to the very early stage of bare ground available for colonization by opportunistic species of plants and animals. In 1990-91, 50% of the logged areas were treated with broadcast burning or mechanical site preparation (Ministry of Forests, 1992). However, many plants and soil microorganisms do survive burning or scarification to persist in the newly developing forest (even after slash burning, many plant species resprout from below-ground tissues).

Some aspects of old growth forest ecosystems, though, may not survive clear-cutting, whatever mitigation measures are taken. Franklin (1988) notes that some differences between old growth and earlier successional stages are qualitative, rather than quantitative. He points out that old growth forests in the coastal Douglas-fir region are extremely low (compared to earlier successional stages) in nutrient losses; they provide several important sites for nitrogen fixation (for example, epiphytic lichens and rotting wood); they are particularly effective at regulating water flows; they increase precipitation (the large crown surfaces and high volume of wood and organic material extract water from clouds and fog); and they influence the amount and spatial distribution of snowfall, thereby minimizing the potential for the damaging rain-on-snow floods that characterize the Pacific Northwest. These comments may not necessarily apply to our north coastal and interior old growth forest types.

Nevertheless, the overall productivity and plant species diversity in clear-cut areas can be very high, as the increase in sunlight reaching the soil stimulates the growth of browse that benefits some ungulates, such as deer, elk and moose, and berries that benefit species from grouse to grizzly bears. The early successional stages that follow clear-cutting often have as much or more diversity of wildlife (Sadoway, 1986) and plants (Schoonmaker and McKee, 1988) as old growth stages. Indeed, many species require disturbed habitats, and thrive on regenerating clear-cuts. But some species of plants, animals and fungi cannot survive without old growth forest habitat. Bunnell and Kremsater's (1990) work on Vancouver Island showed that 65% of bird species and sizeable proportions of mammals, amphibians and reptiles require some structural elements of old growth forests (Figure 19-11). They also noted that wildlife species associated with old growth forests often depend on one or more of three special features of such forests: variability or heterogeneity, big pieces, and age. Older stands have been exposed longer to conditions creating gaps and, for any tree species, older stands generally contain bigger pieces, whether these pieces be boles or crowns. Specific silviculture practices, however, can create some features in younger stands by making openings or by thinning trees so they grow larger more quickly. Other

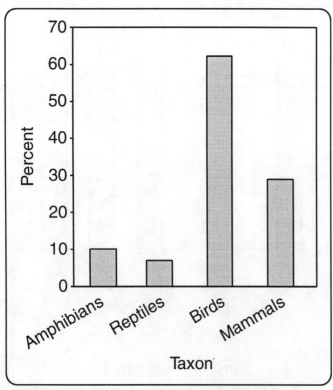

Figure 19-11. Proportions of Wildlife Species Requiring Habitat with Snags or Logs. Source: Bunnell and Kremsater, 1990.

features, however, can not be created by management practices. For example, the slow process of decay helps create suitable places for species breeding in logs or snags. Bunnell and Kremsater (1990) also noted that some arboreal lichens, an important food for wildlife species such as black-tailed deer and caribou, grow well only on older, slower-growing trees that do not shed bark rapidly.

McIver et al. (1990) studied spiders as bio-indicators of recovery after clear-cutting. They found that, of 93 identified species, the 36 most common ones exhibited a clear pattern of succession from clear-cut to old growth. Certain species rely on litter-inhabiting insects and mites for food and require relatively constant environmental conditions to survive. When a stand is clear-cut, these forest litter spiders disappear, due to the loss of available prey and to the increase in microenvironmental variability characteristic of habitats with little or no canopy cover. These spiders are replaced after clear-cutting by an entirely different community composed primarily of several species of diurnal pursuit spiders, which disperse readily and prefer sunny, open habitats. The gradual return of leaf litter 10-15 years after clear-cutting allows colonization by shrub-associated spiders, funnel-web spiders and crab spiders. As succession proceeds further, spiders characteristic of young forests become more common. Recovery of a typical forest spider community requires at least 30 years for the fastest growing forests and much longer for the slower growing types.

As noted in the section on forest insect pests, Winchester (in press) and his colleagues found that the canopies of old growth coastal forests in British Columbia support an invertebrate community particularly rich in predator and parasitoid insects, including many previously unidentified species. He suggested that continued removal of this habitat, which is not replicated even by the most intense silvicultural practices, may be causing species extinctions and a decrease in the genetic diversity contained in these largely undescribed arboreal communities, as well as removing natural controls on forest insect pests.

Timber has only been harvested for about 120 years in British Columbia, and reforested for half that. Consequently, issues important to forest replacement, such as soil organic matter content, water retention, nutrient cycling and maintenance of the soil organisms necessary for tree growth, may remain unresolved until several cycles of cutting and replanting have been completed. For example, the slash burning and herbicide use that sometimes accompany clear-cut logging may reduce or destroy the soil mycorrhizae needed by tree roots for nutrient cycling. Mycorrhizae are beneficial soil fungi that live in a symbiotic relationship with vascular plant roots, providing nourishment by breaking down organic debris and protecting the roots from harmful fungi and disease. Amaranthus and Perry (1987) and Perry et al. (1987) showed that some ericaceous (belonging to the heather family) shrubs and conifers host the same mycorrhizal fungi species, and that conifer seedlings are disproportionately associated with these shrubs, suggesting that the rich concentration of mycorrhizae enhances survival. In Europe, clear-cutting followed by the planting of monocultural forests significantly lowered tree productivity and the diversity of mycorrhizal and other fungi (see review by Mosquin et al., 1992). There is some evidence, but little quantitative data, that clear-cutting and site preparation through systematic burning of logging debris or herbicide use may cause permanent loss of forest productivity in British Columbian forests as well (Fyles and Feller, 1991). However, Roth and Berch (1992) and Berch and Roth (1993) showed that a variety of micorrhizae persist in clear-cuts and colonize planted tree seedlings, although comparative studies of clear-cut and non-clear-cut areas are not available for British Columbia.

Similarly, Marshal (in press) discussed the importance of, and current lack of knowledge about, soil animals, including protozoans (single-celled organisms), nematodes (round worms), gastropod molluscs (snails), arthropods (invertebrates like insects and spiders) and annelids (segmented worms). He feels that human activity increasingly threatens soil faunal diversity, which can be reduced by many agricultural and forestry practices, including the use of pesticides, fertilizers and fire. The fauna is also adversely affected by deletion of organic matter, acid deposition, forest harvesting and soil compaction.

Invertebrate studies (Schowalter, 1990; Asquith et al., 1990; Lattin and Moldenke, 1990; Moldenke and Lattin, 1990a and 1990b; Franklin, 1988) show that "reduced diversity in younger, managed forests promotes pests and deprives us of valuable biological resources" (Schowalter, 1990). Sadoway (1986) also felt that some reptiles and all amphibians may be adversely affected by intensive forestry.

Associated, unresolved issues are the susceptibility of re-planted, managed forests to disease, insect infestations and fires. For example, Hopkins and Stewart (1992) noted that many Douglas-fir plantations in the Interior Cedar-Hemlock Biogeoclimatic Zone will not reach free-growing status in part because of armillaria (*Armillaria ostoyae*) root disease, which also poses serious problems for reforestation on the coast. Continued research is, therefore, necessary before we can say with confidence that forests can be sustained over the long term under a clear-cutting and reforestation regime.

The loss in biodiversity that results from eliminating the old growth and other late successional phases of forest succession by clear-cutting can potentially have economic and social, as well as ecological repercussions. For example, the Pacific yew (*Taxus brevifolius*) grows in the interior and in the temperate rainforest of British Columbia and other parts of the Pacific Northwest (Hartjell, 1991). When the taxol produced by its bark was found to suppress certain types of cancers, these trees, which were once considered waste, suddenly became commercial commodities of considerable value (Hartjell, 1991; personal communication in 1992 with A. Mitchell, Pacific Forest Centre). The Pacific yew grows too slowly to be included in reforestation programs, as it does not mature or attain harvestable size (even for bark) during planned rotation periods.

By law, public lands that have been logged must be reforested. The British Columbia Ministry of Forests manages reforestation by: specifying procedures and standards for the development of silviculture plans; managing the seed collection and tree nursery infrastructure; approving pre-harvest silvicultural prescriptions (see Lavender et al., 1990 for a comprehensive review); and requiring acceptable reforestation of Crown lands. The Ministry's reforestation program covers a variety of silvicultural treatments, including site surveys, site preparation, seeding, planting, mechanical and chemical brushing (to reduce competition from broadleaf, or deciduous, trees), fertilizing and spacing. The Federal Department of Forestry supports this program via the Canada-British Columbia Forest Resource Development Agreements of 1985 and 1991 (FRDA I and FRDA II). The overall objective of reforestation is to initiate development of a managed forest after timber harvest.

Forests have not been replanted as rapidly as they have been disturbed. Figure 19-12 shows the cumulative impacts of logging and fires during 1980-88, and the amount of forest land successfully replanted. This figure does not show the amount regenerated by natural seeding, estimates of which vary widely from about 35% to 50% of the harvested area. By 1982, 738,000 hectares of land with medium to high timber production capability were still "not satisfactorily restocked," a legacy of past logging, wildfires, insect

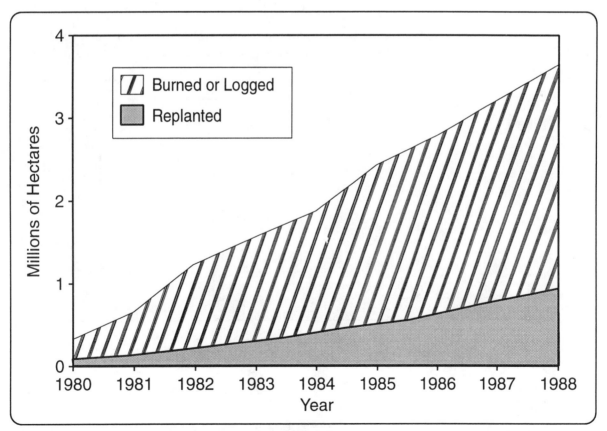

Figure 19-12. Forest Loss Due to Logging and Fires Compared to Replanted Area in British Columbian Forests. Source: Forestry Canada, 1990.

infestations and other forest damage (Mitchell et al., 1990). Despite reforestation efforts, these lands grew further, to 1,666,728 hectares by 1988-89. In 1989-90, reforestation reduced the total medium to high productivity not satisfactorily restocked lands to 1,555,560 hectares (Ministry of Forests, 1991a). An additional 310,293 hectares of low productivity lands remained not satisfactorily restocked, bringing the total of amount of these lands up to 1,862,853 hectares. According to the Ministry of Forests, replanting of denuded areas peaked in about 1991 and will decline to a steady state by about the year 2000, by which time approximately 65% of all not satisfactorily restocked areas will have been replanted (Mitchell et al., 1990). However, the 1991 extension of federal assistance (under the Forest Resource Development Agreement) for reforestation of the not satisfactorily restocked backlog will advance this schedule somewhat.

Reforestation involves natural regeneration (often with silvicultural assistance in the form of site preparation, fertilization, and brush reduction), seeding, and replanting with seedlings. Species replanted are selected according to the forest production capability of each site. They include cypress or yellow cedar (*Chamaecyparis nootkatensis*), Douglas-fir, sub-alpine fir (*Abies lasiocarpa*), silver fir (*Abies amabilis*), grand fir (*Abies grandis*), mountain hemlock (*Tsuga mertensiana*), western hemlock, western larch (*Larix occidentalis*), lodgepole pine, western white pine (*Pinus monticola*), yellow (ponderosa) pine, western red cedar (*Thuja plicata*), Engelmann spruce (*Picea engelmannii*), white spruce (*Picea glauca*) and sitka spruce (*Picea sitchensis*). Lodgepole pine, Engelmann spruce, white spruce and Douglas-fir accounted for 91.7% of all trees planted in 1989 (Ministry of Forests, 1991a; Figure 19-13). Although Douglas-fir may have been planted in

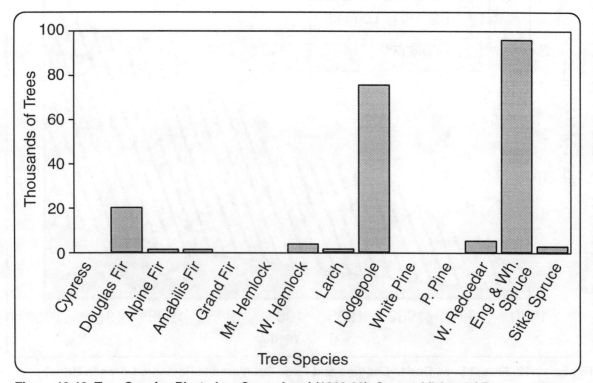

Figure 19-13. Tree Species Planted on Crown Land (1989-90). Source: Ministry of Forests, 1991a.

ecologically inappropriate locations in the past, species now are selected according to ecological and management factors and site-specific selection criteria for maximum growth and development of commercial coniferous species (Klinka et al, 1990b). Klinka et al (1990b) notes that "even-aged mixed species stands of no more than two or three species are most desirable." Of the 21 conifer species prescribed for various ecological conditions, three (sub-alpine fir, Engleman spruce and white pine) are prescribed for mixed stands only, and four (tamarack [*Larix laricina*], western larch, lodgepole pine and ponderosa pine) are prescribed for single-species stands; the others are considered suitable for either mixed or single-species plantations (Klinka et al., 1990b). The number of species of trees prescribed for mixed species reforestation is becoming less limited as the economic and ecological benefits of forest diversity are increasingly recognized. New species selection guidelines are currently being developed.

Seeds for both seedling nurseries and for direct seeding (very little direct seeding is done in B.C.) are obtained from seed orchards and from collections in natural forests. Although there has been some selective breeding of trees for reforestation, there is sufficient natural genetic variation built into the seed collection and production system to allay concerns about restricting genetic diversity in British Columbia's forests (Lester et al., 1990); indeed, the Ministry is working to further broaden the genetic base of managed forests (personal communication in 1992 with A. Yanchuk, Min. of Forests).

Use of Pesticides

Pesticides (herbicides and insecticides) are used to prepare land for planting, encourage the growth of young conifers by removing competing vegetation, treat insect infestations, and for a variety of other purposes. They affect biodiversity if: they alter the ecological structure of forests (which is the purpose of herbicides); their toxicity affects soil microorganisms, with consequent effects on plant growth; or they concentrate in forest food webs. Modern pesticides are selected and approved for use on the basis of their low persistence in the environment, low potential for bioaccumulation and low toxicity to non-target organisms.

Figure 19-14 shows the amount of land treated with herbicides and insecticides from 1980 to 1990. Only about 60,000 hectares of land (including 6,500

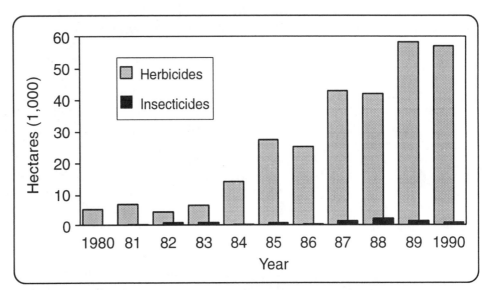

Figure 19-14. Total Area Treated with Herbicides and Insecticides for Forestry Purposes (1980-1990). Source: Humphrys, 1991.

267

hectares of rangelands treated for weed control) are sprayed annu-ally with herbicides, and much less land is sprayed with insecticides (Humphreys, 1991). This amount is small in comparison with the about 270,000 hectares logged annually (out of a total forest land base of 51,820,049 hectares). About 85% of the 60,000 hectares was treated with glyphosate herbicide; the remaining 15% was sprayed with chemicals such as 2,4-D, hexazinone, picloram and the biologi-cal control agent B.t.k. (*Bacillus thuringiensis kurstaki*)(Humphreys, 1991).

Insecticides are used much less in British Columbia than in some other forest jurisdictions (such as New Brunswick), partly because British Columbia's mountainous terrain makes spraying difficult and because sprays are ineffective for bark beetles, British Columbia's major pest. The effect of insecticides on non-target organisms has not been quantified in British Columbia (Risa Smith briefly discusses this issue in Chapter 17). Other alternatives to insecticide application used in British Columbia and elsewhere include: improved tree utilization standards (using smaller trees, so that fewer are left behind as waste); sanitation harvesting (logging and slash-clearing in such as way as to avoid the accumulation of tall stumps and woody debris in which beetles breed); mass trapping with trap logs or aggregating pheromones (sexually attractive scents); modification of dryland sorting procedures; release of bio-logical control agents; and quarantine of imported products to prevent the introduction of alien pests (Van Sickle, in press.).

Herbicides alter forest ecosystems by shortening early succes-sional stages. For example, Hamilton et al. (1991) studied the effects of glyphosate application on grizzly bear (*Ursus arctos*) forage in the Coastal Western Hemlock zone. They found that fewer of the berry-producing plants used by bears in the summer and fall were pro-duced in the short term, and that canopy closure reduced grizzly bear habitat in the long term. They made a number of recommenda-tions for maintaining grizzly habitat while practising forestry, in-cluding relaxing stocking standards to maintain an open canopy throughout the rotation and using silvicultural alternatives to herbicide. Lloyd (1990) reported that glyphosate damages willow (*Salix spp.*) and other moose browse after application and results in "lasting reductions in browse" when the canopy grows over. She also made recommendations for moose browse management which would result in reduction of the allowable annual cut.

Herbicides may also have unintended toxic effects on insects and soil micro-organisms. In detailed studies of herbicide applica-tions along a coastal stream, Feng and Thompson (1989) found that only about 1% of post-application residues persisted on leaves and leaf litter after 29 days, but 6-18% of initial levels persisted in soils after one year. Kimmins et al. (1989) found that herbicide applica-tion reduced nitrogen accumulation in the soil by removing nitro-gen-fixing plant species (primarily red alder, *Alnus rubra*). Soil microorganisms were shown to be affected by herbicide application in a short term (one month), intensive study, but not in longer term (six month) and laboratory studies (Preston and Trofymow, 1989). In the short-term study, all microflora populations were temporarily depressed 3 days after application; fungal populations in litter were reduced 10-25 days after spraying, but those in humus were unaffected; and actinomycete (bacteria which are structurally similar to fungi) and nitrogen-fixing bacteria populations were significantly reduced for the first three days. In the longer term study, the effects on soil fauna (such as mites and nematode worms) were small and transitory.

These studies suggest that, at the relatively limited scale of current use, herbicides may exacerbate the ecological effects of timber harvest by shortening the period and reducing the benefits of broadleaf colonization, while insecticides have little overall effect on forest ecosystems.

The Future of British Columbia's Forest Ecosystems
Avoiding the Loss of Biodiversity

Forest conversion does not have to mean loss of biodiversity. Many forest ecologists (such as Bunnell and Kremsater, 1990; Klinka et al., 1990a) believe that forest conversion can be managed to maintain biodiversity by:

o   retaining throughout British Columbia old growth forests in sufficient spatial distribution (including migration corridors) and density and to support wildlife species that require old growth characteristics (Bunnell and Kremsater, 1990); and

o   planning reforestation, forest management and harvest cycles on an ecological basis to re-create the structural conditions of old growth forest (Klinka et al, 1990a; Kimmins and Duffy, 1991).

Bunnell and Kremsater (1990) further suggest that wildlife diversity in selectively harvested forests can be maintained by creating the structural conditions needed by various wildlife in various forest types. They gave considerable attention to varying the spatial pattern of silvicultural treatments to optimize biodiversity on a watershed scale. "The resource manager's challenge," they note, "is to implement management practices through space and time such that a viable population of each species always occurs some-where in the area." They analyzed this challenge in terms of develop-ing processes to determine each species' needs: its minimum viable population size and habitat requirements in terms of size, number, and spatial and temporal distribution. Kimmins and Duffy (1991) further note that, where the scale of the mosaic of forest patches of various ages in a region is small enough, animals are able to move from patch to patch as conditions in any one patch change, thereby maintaining themselves in the appropriate habitat conditions. Where the patches are too large (as in a large valley continuously clear-cut over a few years), some species may not be able to move to an appropriate new area as habitat conditions change or as weather conditions require a change in habitat. This inability to adapt fast enough can result in large variations in their abundance in the area over time and suggests the need for small, geographically dispersed clear-cuts (many cutting cycles). However, Kimmins and Duffy (1991) felt that the few species in British Columbia that are known to depend exclusively on mature second growth or old growth for habitat may not be able to survive in the small patches (20-40

269

hectares on the coast and 5-20 hectares in some parts of the interior) that would result from having the current annual cut distributed over many cutting cycles. Maintenance of maximum regional wildlife diversity, they feel, will require a compromise between these two management approaches and will probably require a diversity of sizes of harvested areas.

Fortunately, some recent government initiatives will help to improve our use of the forests. Forestry Canada, under a new $200 million, four year, cost-shared agreement with the Ministry of Forests (FRDA II), is sponsoring research into many aspects of forest and range biodiversity, including: "gap analysis" (determining the gaps between the amount of land in threatened ecosystems that is unprotected and the amount that is protected); old growth mapping; forest floor and canopy arthropod populations; rare lichen and macrofungal diversity; diversity of soil fauna in old and second growth; mycorrhizal diversity and microbial activity; comparison of natural and managed disturbance regimes; effects of logging on wildlife diversity; impacts of vegetation treatment on wildlife diversity; habitat requirements of birds, small mammals and amphibians in old versus second growth; bird communities in old growth; and the genetic diversity of forest trees. Information gleaned from this research is published in newsletters, as well as in the usual scientific literature, so that researchers and forest managers can share information on this rapidly growing field (Anonymous, 1992).

The Ministry of Forests has begun to integrate protection of non-timber values into its forest management practices by, for example, introducing a new forest practices code, a protected area strategy, biodiversity guidelines, wildlife tree (tree that is important to wildlife) management guidelines, riparian management guidelines, and other procedural guidelines designed to reduce environmental damage and encourage biodiversity. The Ministry itself has established an Integrated Resources Branch, where the protection of wildlife habitat and other forest resources is planned in an ecological context. In early 1992, a land use commission was established to further the objective of managing forests for all uses. As well, the Ministry has established various public forums to ascertain community visions of forest use, and to deal with specific topics, such as old growth forests and wilderness areas. These initiatives amount to a virtual revolution in forest management in British Columbia. The reductions in the allowable annual cuts alone are a clear indication of the Ministry of Forests' progress towards sustainable forestry.

However, much progress in these areas is still needed. Forests that are disturbed through logging or fragmentation and development can be quickly lost if the disturbance is so widespread or frequent that the original ecosystem does not have an opportunity to recover. Loss of the original diversity of natural ecosystems means loss of their genetic and species diversity. Further measures to protect ecosystem integrity would include:

o   shifting our focus from managing trees and wildlife species that are currently perceived to have economic value to managing our use of the forest ecosystem as a whole;

o   ensuring that timber production and other forest-based resource development activities are managed to preserve a mix of seral stages, including old growth and climax forest communities of adequate size to support species diversity within the forest ecosystems of each biogeoclimatic zone, sub-zone and variant, with connecting forest pathways between them;

O selecting harvest methods to maintain forest ecosystem integrity; and

O monitoring and managing fragmentation by roads and development to minimize their impact on the forest ecosystems of B.C.

Recognizing the Potential Impacts of Global Change

While alternative management practices provide some flexibility in conserving biodiversity of forests in British Columbia, global events (such as global warming, ozone depletion and tropical deforestation) are less subject to intervention, at least by local and regional governments. With global warming, further increases in forest fire frequency are predicted (Perry and Borchers, 1990). Flannigan and Van Wagner (1991) noted the relationships between burned area and the Seasonal Severity Rating, a fire hazard index computed from analysis of climatic variables and available fuel (woody debris and underbrush). They suggest that a possible future 40% increase in national burned area may result from global warming. They show that the average national burned area has already increased more than 50% since 1970, and assert that the potential economic and ecological effects of climate change on the Canadian forest fire regime deserves continuing attention.

Forest ecologists such as Pollard (1991b), Bergvinson (1988) and Franklin et al. (1991) predict that, with global warming, forest pests may increase. Borden (1991) has given some examples of how the cumulative impacts of global warming and ozone depletion could encourage specific forests pests in British Columbia. Climate change will cause increased physiological stress because of moisture deficits and ultraviolet radiation on trees, especially those at higher latitudes. At the same time, climate change would relieve some of the stress on pest organisms (both pathogens and insects). This combination of effects could have a severe impact on indigenous tree species, causing their natural ranges to move northward and upward in elevation. Trees left in the wake of shifting ranges "will be increasingly ravaged by pests, creating forest pest management and silvicultural problems of unprecedented magnitude" (Borden, 1991). Page (1981) gave a concrete example of how weather has affected forest pest infestations in forests similar to British Columbia's. During the very dry years of 1975 and 1977 in northern California, several species of bark beetle (*Dendroctonus spp.*), dwarf mistletoe (*Arceuthobium spp.*) and root rot (*Fomes annosus*) invaded stands of ponderosa pine (*Pinus ponderosa*), destroying 45 million cubic metres of wood in 2.5 million hectares of national forest land.

These effects may be synergistic, creating positive feedback loops that compound the magnitude of change. Not only may climate change increase both fire frequency and insect infestations, but the two may exacerbate each other. Fires kill trees, creating habitat and food for bark beetles, which also kill trees, creating the dead wood that fuels fires. Defoliating insects also kill trees, contributing to increases in both bark beetles and fires. Tropical deforestation may contribute to defoliation by eliminating the winter habitat of avian predators of defoliating insects.

Government management programs will have to respond to these changes, which may occur on a time scale of decades: greenhouse gases are expected to double within the next century, perhaps by as early as 2050, giving rise to seasonal temperature increases in British Columbia of several degrees (Chapter 24, this volume); and, at current rates of tropical deforestation, the last hectare of unprotected tropical forest will fall between 2030 and 2045 (Chapter 23, this volume). This time scale is roughly the same as that of the planned conversion from natural to managed forests in British Columbia. Silvicultural prescriptions—the technical and scientific expertise contributed by professional foresters to the planning and management of commercial forests—have been based on assumptions of stability in climate regimes. A Douglas-fir zone has always been thought capable of growing new Douglas-fir trees after harvest; likewise western red cedar, sitka spruce, and so on. However, we already exist in a period of ecological instability, as evidenced by changes in insect damage and fire frequency, when several standard planning assumptions may turn out to be false.

First, the long run sustained yield levels currently established by the Ministry of Forests may be affected by climate change. The long run sustained yield is calculated from the annual increment in tree growth minus the expected unsalvageable losses. The rate of tree growth is taken from empirically determined standard growth curves. With increased summer moisture stress in the drier areas of the province and increasing ultraviolet radiation, particularly in the north, tree growth may be slower than at present. On a provincial scale, this growth reduction may be compensated by faster growth in the coastal and northern areas, which will have more moisture and warmer winters. At the same time, the unsalvageable losses due to insect and fire damage may be greater than expected, combining with slower growth to lower the long run sustained yield. Consequently, the options for conserving biodiversity could be foreclosed more rapidly than expected.

Second, reforestation is tailored to site characteristics based in part on biogeoclimatic zones, which are themselves delineated by their characteristic late successional stage vegetation. This vegetation may shift northward and upward in elevation and may also change qualitatively as the climate changes (Chapter 25, this volume), significantly altering the site characteristics of existing zones. Unless the zones are redefined in accordance with such changes, silvicultural prescriptions may no longer be appropriate to the sites for which they were intended. Results of the extensive provenance trials that determine how different genetic strains of each tree species perform in various environments may also become invalid. Tree species selected for particular sites today may be out of their element in the climate of tomorrow. As Pollard (1991b) noted, the 300 million seedlings planted annually in British Columbia will experience climates of the year 2050 and beyond.

Third, the purpose and use of protected areas may change. At present, a common objective is to preserve representative examples of ecosystems for study or recreation. We may find that, no matter how well protected such areas are, the large-scale effects of land use (such as elimination of the habitat of species with low mobility outside of protected areas) and of global change (warming, ozone depletion, and tropical deforestation) will alter their diversity. Pollard (1991a) stated that a northward movement of several hundred kilometres by a climatic region would leave most natural areas stranded in a climate to which they are not adapted. He suggested that, in the future, protected areas may best used as genetic reservoirs for the dispersal of species to damaged areas beyond their boundaries. To serve this function, protected areas would have to be differently designed and located. A few large protected areas would be needed for wide-ranging species and many small protected areas would have to be interspersed throughout the unprotected parts of the province. These protected areas would have to be connected by corridors of healthy habitats between valley bottoms and mountain tops, along riparian zones, and along north-south migration routes.

To face these challenges, forest, grazing, water and wildlife managers will need new ways of collecting and interpreting information on ecosystem changes as a result of cumulative impacts, and they will have to develop systems for cooperatively using this new information. However, new knowledge and new technology may not be sufficient to conserve the biodiversity of forests in British Columbia. Short term economic and social objectives may have to be recast in the light of long term sustainability if biodiversity of forest ecosystems is to be conserved.

Acknowledgements

Dag Kristensen and Darrel Ericco, British Columbia Ministry of Forests, were extremely helpful in providing information on current timber inventory. Silvia Pang, British Columbia Ministry of Forests, provided information on forest fire frequency. G. Allan Van Sickle, Pacific Forestry Centre, provided information on forest insect pest infestations. Information was also provided by Lynn Husted, Pat Humphreys, Miles Manna, Del Meidinger, Warren Mitchel, J.B. Nyberg, Barbara von Sacken, Dave Spittlehouse, Terje Vold and Denise Young of the British Columbia Ministry of Forests; James Harrington of Forestry Canada (Ottawa); Shannon Berch of the University of British Columbia; and Neville Winchester of the University of Victoria. The following forestry specialists reviewed the manuscript: Don Blumenauer and B. Wikeem of the British Columbia Ministry of Agriculture, Fisheries and Food (Kamloops); Ralph Archibald, Rob Bowden, Phil Comeau, Jim Crover, Evelyn Hamilton, Andy McKinnon, Brian Nyberg, John Parmenter, Mike Pedersen, Pasi Puttonen, and Alvin Yanchuk of the British Columbia Ministry of Forests (Victoria); Harry Hirvonen of Environment Canada; Douglas Pollard and G. Allan Van Sickle of the Pacific Forestry Centre; A.C. Allaye-chan of the University of British Columbia; and Neville Winchester of the University of Victoria.

References

Amaranthus, M.P. and D.A. Perry. 1987. Effect of soil transfer on ectomycorrhizal formation and the survival and growth of conifer seedlings on old, non-reforested clear-cuts. *Canadian Journal of Forest Resources* 17:944.

Anonymous. 1992. *Forest Research News*. Quarterly publication by the British Columbia Ministry of Forests and Forestry Canada, Victoria, B.C.

Asquith, A., J.D. Lattin and A.R. Moldenke. 1990. Arthropods: The invisible diversity. *Northwest Environmental Journal* 6(2):404-405.

B.C. Environment. 1991. Planning for the future. *Managing wildlife to 2001: A Discussion Paper.* Victoria, B.C.

Berch, S.M. and A.L. Roth. 1993. Ectomycorrhizae and growth of Douglas-fir seedlings preinoculated with *Rhizopogon vinicolor* and outplanted on eastern Vancouver Island. *Canadian Journal of Forest Research.* 23: (in press.).

Bergvinson, D. 1988. *The Green Spruce Aphid, Elatobium abietinum, (Homoptera: Aphididae): A Review of its Biology, Control and Status in British Columbia.* Unpublished M.Sc. Thesis, Simon Fraser University, Burnaby, B.C.

Borden, J.H. 1991. *Ozone Depletion, Global Warming and the Potential Impact of Forest Pests.* President's Lecture Series, Simon Fraser University, Burnaby, B.C.

Bunnell, F.L and L.L. Kremsater. 1990. Sustaining wildlife in managed forests. *Northwest Environmental Journal* 6:243-169.

Ebata, T. 1992a. *Western Balsam Bark Beetle - Tree Killer or Natural Recycler?* Ministry of Forests, Forest Health Progress XI(2): 7-9.

Ebata, T. 1992b. *White Pine Weevil Hazard Ratings for the Prince Rupert Forest Region.* Ministry of Forests, Forest Health Progress XI(2): 11-12.

Erskine, A.J., B.T. Collins and J.W. Chardine. 1990. *The Cooperative Breeding Bird Survey in Canada, 1988.* Progress Notes No. 188. Canadian Wildlife Service, Ottawa.

Feng, J.C. and D.G. Thompson. 1989. Persistence and dissipation of glyphosate in foliage and soils of a Canadian coastal forest watershed. In: *Proceedings of Carnation Creek Herbicide Workshop, 7-10 December 1987.* P.F. Reynolds (ed.). Forest Pest Management Institute, Forestry Canada, Sault Ste. Marie, Ontario.

Flannigan, M.D. and C.E. Van Wagner. 1991. Climate change and wildfire in Canada. *Canadian Journal of Forestry Research* 21(1):66-72.

Foreman, R.T. and M.G. Gordon. 1986. *Landscape Ecology.* John Wiley and Sons, New York, New York.

Forestry Canada. 1990. *The State of Forestry in Canada.* 1990. Report to Parliament. Forestry Canada, Ottawa, Ontario.

Franklin, J.F. 1988. Structural and functional diversity in temperate forests. In: *Biodiversity.* E.O. Wilson (ed.). National Academy Press, Washington, D.C.

Franklin, J.F., F.J. Swanson, M.E. Harmnon, D.E. Perry, T.A. Spies, V.H. Dale, A. McKee, W.K. Ferrell, J.E. Means, S.V. Gregory, J.D. Lattin, T.D. Schowalter and D. Larsen. 1991. Effects of global climate change on forests in northwestern North America. *Northwest Environmental Journal* 7: 233-254.

Fyles, J.W. and M.C. Feller. 1991. Forest floor characteristics and soil nitrogen availability on slash-burned sites in coastal British Columbia. *Canadian Journal of Forest Research* 21:1516-1522.

Habeck, J.R. 1988. Old growth forests in the northern Rocky Mountains. *Natural Areas Journal* 8:202-211.

Habeck, J.R. 1990. Old growth ponderosa pine-western larch forests in western Montana: ecology and management. *Northwest Environmental Journal* 6:271-292.

Hamilton, A.N., C.A. Bryden and C.H. Clement. 1991. *Impacts of Glyphosate Application on Grizzly Bear Forage Production in the Coastal Western Hemlock Zone.* Canada-British Columbia Forest Resource Development Agreement Report No. 165.

Harris, L.D. 1984. *The Fragmented Forest: Island Biogeography Theory and the Preservation of Biotic Diversity.* University of Chicago Press.

Hartjell, H. Jr. 1991. *The Yew Tree: A Thousand Whispers.* Biography of a Species. Hulogosi, Eugene, Oregon. p.319.

Hopkins, K. and R. Stewart, 1992. *Root Disease Impact Analysis in Fir Plantations of the Arrow and Kootenay Lake Timber Supply Areas.* Ministry of Forests internal report, Nelson, B.C.

Humphreys, P. 1991. *Pesticide Use in British Columbia Forestry, 1990.* Unpublished Ministry of Forests report.

Integrated Resources Branch. 1993. Kalum North timber supply analysis. Ministry of Forests, Victoria, B.C.

Kew, J.E.M. and J.R. Griggs. 1991. Native Indians of the Fraser Basin: towards a model of sustainable resource use. In: *Perspectives on Sustainable Development in Water Management: Towards Agreement in the Fraser River Basin, Vol. 1.* A.H.J. Dorcey (ed.). Westwater Research Centre, University of British Columbia. pp. 17-47.

Kimmins, J.P. and D.M. Duffy. 1991. Sustainable forestry in the Fraser River Basin. In: *Perspectives on Sustainable Development in Water Management: Towards Agreement in the Fraser River Basin, Vol. 1.* A.H.J Dorcey (ed.). Westwater Research Centre, University of British Columbia.

Kimmins, J.P., K.M. Tsze and R.E. Bigley. 1989. Development of non-commercial vegetation following clear-cutting and clear-cutting plus slash burning and soil biochemical effects of herbicidal control of the vegetation. In: *Proceedings of Carnation Creek Herbicide Workshop, 7-10 December 1987.* Forest Pest Management Institute, Forestry Canada, Saul Ste. Marie, Ontario.

Klinka, K., R.E. Carter and M.C. Feller. 1990a. Cutting old growth forests in British Columbia: ecological considerations for forest regeneration. *Northwest Environmental Journal* 6(2):221-242.

Klinka, K., M.C. Feller, R.N. Green, D.V. Meidinger, J. Pojar and J. Worral. 1990b. Ecological principles: applications. In: *Regenerating British Columbia's Forests.* D.P. Lavender, R. Parish, C.M. Johnson, G. Montgomery, A. Vyse, R.A. Willis and D. Winston (eds.). University of British Columbia Press, Vancouver, B.C. pp.55-72.

Krajina, V.J., J.B. Foster, J. Pojar and T. Carson. 1978. *Ecological Reserves in British Columbia.* Ministry of Environment, Victoria, B.C.

Lattin, J.D. and A.R. Moldenke. 1990. Moss lacebugs in northwest conifer forests: adaptation to long-term stability. *Northwest Environmental Journal* 6(2):406-407.

Lavender, D.P., R. Parish, C.M. Johnson, G. Montgomery, A. Vyse, R.A. Willis and D. Winston (eds.). 1990. *Regenerating British Columbia's Forests.* University of British Columbia Press.

Lester, D.T., C.C. Ying and J.D. Konishi. 1990. Genetic control and improvement of planting stock. In: *Regenerating British Columbia's Forests.* D.P. Lavender, R. Parish, C.M. Johnson, G. Montgomery, A. Vyse, R.A. Willis and D. Winston (eds.). University of British Columbia Press, Vancouver, B.C. pp. 180-192.

Lloyd, R. 1990. *Herbicide Effects on Moose Browse in Northern British Columbia.* Forest Resource Development Agreement Memo No. 161.

Marshall, V.G. Sustainable forestry and soil fauna diversity. In: *Proceedings of a Symposium on Biological Diversity in British Columbia.* Victoria, B.C., February 28 to March 2, 1991. (In press).

Maclauchlan, L. 1992. Kamloops Forest Region western spruce budworm spray program. Ministry of Forests, Forest Health Progress XI(2): 21-27.

McIver, J.D., A.R. Moldenke and G.L. Parsons. 1990. Litter spiders as bio-indicators of recovery after clear-cutting in a western coniferous forest. *Northwest Environmental Journal* 6(2):410-412.

Meidinger, D. and J. Pojar. 1991. *Ecosystems of British Columbia.* B.C. Ministry of Forests, Victoria, B.C.

Ministry of Crown Lands 1989. British Columbia land statistics. Province of British Columbia.

Ministry of Forests, 1984. *Forest and Range Resource Analysis 1984.* Ministry of Forests.

Ministry of Forests. 1991a. *Annual Report, 1989-90.* Victoria, B.C.

Ministry of Forests. 1991b. *Review of the timber supply analysis process for B.C. Timber Supply Areas.* Final Report (Vol. I).

Ministry of Forests. 1992. *Annual Report, 1990-91.* Victoria, B.C.

Mitchell, W.K., G. Dunsworth, D.G. Simpson and A. Vyse. 1990. Planting and seeding. In: *Regenerating British Columbia's Forests.* D.P. Lavender, R. Parish, C.M. Johnson, G. Montgomery, A. Vyse, R.A. Willis and D. Winston (eds.). University of British Columbia Press, Vancouver, B.C. pp. 180-192.

Moldenke, A.R. and J.D. Lattin. 1990a. Density and diversity of soil arthropods as "biological probes" of complex soil phenomena. *Northwest Environmental Journal* 6(2):409-410.

Moldenke, A.R. and J.D. Lattin. 1990b. Dispersal characteristics of old growth soil arthropods: the potential for loss of diversity and biological function. *Northwest Environmental Journal* 6(2):408-409.

Morgan, K.H. and S.P. Wetmore. 1986. *A Study of Riparian Bird Communities from the Dry Interior of British Columbia.* CWS Technical Report Series. No.11.

Mosquin, T. and P.G. Whiting. 1992. *Canada Country Study of Biodiversity: Taxonomic and Ecological Census and Economic Benefits, Conservation Costs and Unmet Needs (Draft Version 1.1).* Prepared for Delegations to the International Convention on Biodiversity. Rio de Janeiro, Brazil. June 1992. The Canadian Centre for Biodiversity, Canadian Museum of Nature, Ottawa, Ontario.

Page, J.M. 1981. Drought Accelerated Parasitism of Conifers in the Mountain Ranges of Northern California. *Environmental Conservation* 8:217-226.

Perry, D.A. and J.G. Borchers. 1990. Climate change and ecosystem response. *Northwest Environmental Journal* 6:293-313.

Perry, D.A., R. Molina and M.P. Amaranthus. 1987. Mycorrhizae, mycorrizospheres, and reforestation: current knowledge and research needs. *Canadian Journal Forest Resources* 17:929.

Pojar, J., E. Hamilton, D. Meidinger and A. Nicholson. 1990. Old growth forests and biological diversity in British Columbia. In: *Symposium on Landscape Approaches to Wildlife and Ecosystem Management.* Vancouver, B.C.

Pollard, D.F.W. 1991a. The Role of Natural Areas in a Changing Climate. *USA-Canada Symposium on the Implications of Climate Change for Pacific Northwest Forest Management.* Seattle, October 23-25, 1991.

Pollard, D.F.W. 1991b. Climate change as a current issue for the Canadian Forest Sector. *The Environmental Professional* 13:37-42.

Preston, C.M. and J.A. Trofymow. 1989. Effects of glyphosate (roundup) on biological activity of two forest soils. In: *Proceedings of Carnation Creek Herbicide Workshop, 7-10 December 1987.* Forest Pest Management Institute, Forestry Canada, Saul Ste. Marie, Ontario.

Rosenburg, K.V. and M.G. Raphael 1986. Modelling habitat relationships of terrestrial vertebrates. In: *Wildlife 2000.* J. Verner, L. Morrison and C.J. Ralph (eds.). University of Wisconsin Press. pp. 327-329.

Roth, A.L. and S.M. Berch, 1992. Ectomycorrhizae of Douglas-fir and western hemlock seedlings outplanted on eastern Vancouver Island. Can. J. For. Res. 22: 1646-1655.

Sadoway, K.L. 1986. *Effects of Intensive Forest Management on Amphibians and Reptiles of Vancouver Island: Problem Analysis.* Research, B.C. Ministries of Environment and Forests. IWIFR-23. Victoria, B.C.

Schoonmaker, P. and A. McKee. 1988. Species composition and diversity during secondary succession of coniferous forests in the western Cascade Mountains of Oregon. *Forest Science* 34(4):960-979.

Schowalter, T.D. 1990. Invertebrate diversity in old growth versus regenerating forest canopies. *Northwest Environmental Journal* 6(2):403-404.

Shepherd, R.F., G.A. Van Sickle and D.H.L. Clarke. 1988. Spatial relationships of Douglas-fir tussock moth defoliation within habitat and climatic zones. In: *Proceedings of Lymantriidae: A Comparison of Features of New and Old World Tussock Moths.* June 26-July 1, 1988. New Haven, Connecticut.

Silva Management Services Ltd. 1992. *Prescribed Burning for Wildlife Habitat Enhancement.* Report to Ministry of Environment, Lands and Parks, Victoria, B.C.

Van Sickle, G.A. The status of important forest insects in the Pacific and Yukon region. In: *Control of Forest Insects in Canada.* G. Armstrong and W. Ives (eds.). Forestry Canada, Ottawa, Ontario. (In press).

Winchester, N. (in press, 1993). Coastal sitka spruce canopies: conservation of biodiversity. Accepted by BioLine.

Woodley, S. and J. Theberge. 1991. Monitoring for Ecosystem Integrity in Canadian National Parks. In: *Science and the Management of Protected Areas.* J.H.M. Willison, S. Bondrup-Nielsen, C. Drysdale, T.B. Herman, N.W.P. Munro and T.L. Pollock (eds.). Elsevier, New York, New York.

Chapter 20
Threats to Biodiversity of Grasslands in British Columbia

Distribution of Grassland Ecosystems in British Columbia

Eighteen thousand years ago, much of British Columbia lay beneath ice several kilometres thick. The recolonization of the land by plants and animals which followed deglaciation (about 15,000 years ago) has been dynamic, as climatic fluctuations have been common and often pronounced (Pielou, 1991). Maximum temperatures occurred approximately 6,000 - 10,000 years ago (Hypsithermal period) and were appreciably warmer (1.5 - 2.0° C) than today. Grasslands in British Columbia likely reached their maximum extent during this Hypsithermal period, and may have merged with alpine grasslands in very arid areas (Hebda, 1982). Since then, the grassland/forest edge has retreated to lower elevations, as the climate has generally become cooler and wetter (Pielou, 1991).

Grasslands currently occur throughout the interior of British Columbia in generally dry regions, particularly in the Bunchgrass and Ponderosa Pine biogeoclimatic zones (Figure 18-1). As well, forested ecosystems, including those in the Interior Douglas-fir, Montane Spruce, Engelmann Spruce Subalpine Fir and Boreal White and Black Spruce biogeoclimatic zones, support grassy understories or clearings due to a combination of soil and topographic conditions and fire history.

This chapter focuses on those grassland communities dominated at climax by bunchgrass vegetation, particularly in the Bunchgrass, Ponderosa Pine and Interior Douglas-fir biogeoclimatic zones. These grasslands are the most well-known in British Columbia, and currently face the greatest threats to their biodiversity.

The climate of the Bunchgrass and Ponderosa Pine zones is typified by hot, dry summers and cold winters with relatively little snowfall. Plant growth depends on winter precipitation because spring months are dry and summer precipitation evaporates before it replenishes soil moisture (Nicholson et al., 1991). Low precipitation, warm temperatures, high winds, and heat-retaining canyons are major factors promoting grassland rather than forest vegetation (Roberts, 1992).

Annual precipitation of the Bunchgrass zone averages 250 millimetres, and is usually distributed bimodally in May-June and December-January (Nicholson et al., 1991). Grasslands of the Bunchgrass zone comprise approximately 1.3 million hectares, and are dominated at climax by widely-spaced bunchgrasses (Nicholson et al., 1991). Two subzones occur in the Bunchgrass zone — the Very Hot Subzone and the Very Dry Warm Subzone. These subzones occupy two distinct elevational bands between 300 and 1000 me-

Michael Pitt
Department of Plant Science
Faculty of Agricultural Sciences
University of British Columbia
2357 Main Mall, Room 248
Vancouver, B.C.
V6T 1Z4

Tracey D. Hooper
Department of Plant Science
Faculty of Agricultural Sciences
University of British Columbia
2357 Main Mall, Room 248
Vancouver, B.C.
V6T 1Z4

tres, and were formerly referred to as the lower and middle grasslands (Spilsbury and Tisdale, 1944; McLean, 1969; Nicholson et al., 1991).

Bluebunch wheatgrass (*Elymus spicatus [Agropyron spicatum]*) is the dominant and most productive species in the Bunchgrass Zone. Associated plant species vary, depending on the site and its grazing history, but may include needle-and-thread grass (*Stipa comata*), Sandberg's bluegrass (*Poa sandbergii*), prairie junegrass (*Koeleria macrantha*), arrow-leaved balsamroot (*Balsamorhiza sagittata*), and numerous other forbs (herbaceous plants other than grass) (Wikeem et al., unpubl.).

The Ponderosa Pine zone occupies nearly 800,000 hectares between 335 and 900 metres elevation, with annual precipitation ranging from 280 to 550 millimetres. Ponderosa pine (*Pinus ponderosa*), at climax, generally forms open stands. Bluebunch wheatgrass is often associated with the same plant species present in the Bunchgrass zone. Rough fescue (*Festuca campestris [F. scabrella]*) and Idaho fescue (*F. idahoensis*) often replace bluebunch wheatgrass, or co-dominate in the most southerly portions of the zone (Wikeem et al., unpubl.).

The Interior Douglas-fir zone extends over 4.8 million hectares, and supports large grassland communities. Elevations range from 300 to 1400 metres, and annual precipitation averages between 300 and 750 millimetres (Hope et al., 1991). Grassland associations are often dominated by rough fescue, bluebunch wheatgrass, or Idaho fescue (Wikeem et al., unpubl.), and were formerly classified as upper grassland (Tisdale, 1947; McLean and Marchand, 1968).

Uniqueness of British Columbian Grasslands

British Columbian grasslands are unique in Canada because they are dominated primarily by the bunchgrass, bluebunch wheatgrass, and many other plant species that occur only rarely east of the Canadian Rocky Mountains. Moreover, British Columbian grasslands represent the northern limit of extensive bunchgrass vegetation in North America, and are distinguished from bluebunch wheatgrass-dominated communities in Oregon and Washington by a greater proportion of plant species of northern than southern origin (Daubenmire, 1978).

British Columbian grasslands are also rare within Canada because a much smaller proportion than elsewhere has been converted to crop production. Other than low-elevation sites and areas adjacent to urban development, many B.C. grasslands remain extensive grazing ecosystems.

History of Grassland Disturbances in British Columbia

In the American Great Plains, between 45% and 85% of the major grassland vegetation types have been eliminated, while in the Canadian prairies, between 76% and 99% of the native grasslands have been lost, primarily to cultivation (Klopatek et al., 1979; World Wildlife Fund Canada, 1989). In addition, livestock grazing has occurred in virtually all grasslands of North America, so that unaltered grassland ecosystems are now extremely rare. In the Okanagan Region of British Columbia, less than 10% of the historical grasslands remain in a relatively natural state (Redpath, 1990).

Since the mid-1800s, B.C. grasslands have sustained development of the beef industry, which began following the discovery of gold on the Fraser River in 1858. Initially, grazing was confined almost exclusively to the grasslands, which soon became completely stocked and utilized. By 1900, widespread overstocking and season-long grazing by cattle, sheep and horses had reduced grazing capacity, and had also promoted an increase in weedy, sometimes introduced plant species. Since the 1940s, grassland range conditions have generally improved in British Columbia, primarily in response to better management and livestock reductions imposed by government. Depending on regional differences, as much as 50% of forage consumed by livestock on Crown ranges is now provided by forest grazing. Moreover, the numbers, duration and seasons of livestock grazing are now regulated on Crown land by the B.C. Ministry of Forests.

Currently, Crown lands under grazing permit or licence total about 8.3 million hectares, some of which may be on forest land also allocated for timber production (Ministry of Crown Lands, 1989). Figure 20-1 shows the Crown forest cover and range types used for grazing. An additional 168,000 hectares are used as community

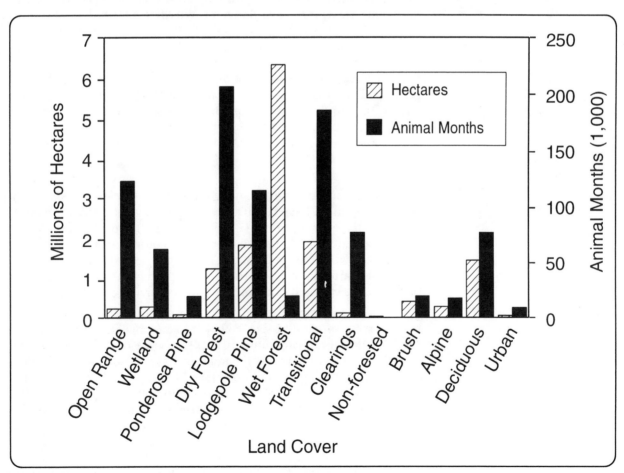

Figure 20-1. Use and Cover of Rangeland on Crown Forest Land in B.C. Source: Ministry of Crown Lands, 1989.

pasture land, and ranchers hold 206,000 hectares of improved pasture privately. The total area grazed in British Columbia exceeds 10,000,000 hectares, more than 1/10 of the province.

Beef production operations today typically involve spring, summer and fall grazing on Crown lands; cattle are moved to low-elevation, private pastures for winter. Before marketing, cattle are often taken to feedlots where they are fed hay, silage and grain grown on irrigated cropland.

In addition to livestock grazing, range re-seeding and off-road recreation have modified much of the remaining "undeveloped" grassland areas. Moreover, cultivation, urban development and hydro-electric power projects have caused outright, irreversible losses of native grasslands. Because of these combined influences and the relatively limited distribution of grasslands, "ancient" grasslands represent a much more endangered space in British Columbia than do "ancient" or old-growth forests.

Current Status of British Columbian Grasslands

Because grasslands have been so influenced by human activities, a relatively large number of wildlife species associated with grasslands are listed as threatened or endangered. In Canada, more than 1/3 of the birds and mammals on the 1988 COSEWIC (Committee on the Status of Endangered Wildlife in Canada) list are from the prairie provinces and most of these are associated with the prairie grasslands and parklands (World Wildlife Fund Canada, 1989). Similarly, 18 of the 64 species on British Columbia's Threatened or Endangered (Red) List are most closely associated with grasslands (British Columbia Ministry of Environment, 1993). It should be noted, however, that B.C. grasslands have been studied primarily in terms of livestock, wild ungulates, and forage production; consequently, the status of many grassland species remains unknown.

Because impacts on grasslands do not arouse public emotions like those associated with logging old-growth forests, grasslands in British Columbia have received comparatively little environmental attention (Hooper and Pitt, 1993). Less than 1% of B.C. grasslands have received protected status (personal communication in September, 1992, with T. Vold, Min. of Environment, Lands and Parks). It is also important to recognize that these "protected" grassland areas often permit various kinds of use, particularly recreation.

Approximately 11 grassland areas are included within ecological reserves, ranging in size from 4.2 to 884 hectares (British Columbia Ministry of Environment, 1992). Some of these reserves were established to protect features such as saline lakes, and include relatively small amounts of grassland. Examples of reserves that were established primarily to protect bunchgrass vegetation include the 257-hectare Big Creek Ecological Reserve in the Chilcotin-Cariboo Basin Ecosection. Grasslands considered to be ungrazed by livestock comprise 55% of the reserve, while forests and rocky outcrops comprise the remaining 45%. Thirty-six hectares of ungrazed ponderosa pine/bunchgrass vegetation are also contained within the Skihist Ecological Reserve northeast of Lytton. The Hayne's Lease Ecological Reserve near Osoyoos encompasses 101 hectares of Canada's most arid bunchgrass ecosystem.

The Junction Wildlife Management Area (WMA) in the Chilcotin-Cariboo comprises an area of about 4860 hectares near the confluence of the Fraser and Chilcotin Rivers. Concerns about the impacts of livestock grazing on California bighorn sheep (*Ovis canadensis californiana*) habitat led to establishment of the WMA, and to virtual exclusion of cattle from the site in 1973 (Mitchell and Prediger, 1975). The site is now being managed for biodiversity conservation (personal communication September, 1992, with H. Langin, Ministry of Environment, Lands, and Parks). Similarly, the Canadian Wildlife Service's Vaseux Lake National Wildlife Area in the south Okanagan was established for bighorn sheep.

Kalamalka Lake Grasslands Park, in the Okanagan, contains 890 hectares that served as spring grazing for the Coldstream Ranch until 1975. The park is now used for hiking, nature study, horse riding and other forms of recreation.

Additional grassland areas in the South Okanagan and East Kootenay Regions have been purchased by the Nature Trust of British Columbia, to be held in the public interest for conservation purposes (personal communication in December, 1992, with B. Brink, Professor Emeritus, University of British Columbia). Where appropriate, these conservation areas may include recreation and resource uses such as livestock grazing and agriculture.

Attributes of Grassland Ecosystems
Undisturbed Grasslands

Fairly detailed inventories of plant and animal species associated with grasslands are currently available for some regions, particularly the South Okanagan (for example, Hlady, 1990). Other than these inventories, however, very little work in British Columbia has investigated the complex relationships among grassland floristics, vegetation structure, grazing ungulates, insect populations, small mammals, birds, and reptiles and amphibians, particularly in climax grasslands.

Pitt and Wikeem (1990) investigated a big sagebrush (*Artemisia tridentata*)/bluebunch wheatgrass community near Penticton, and classified four plant groups (Spring Ephemeral, Summer Dormant, Summer Quiescent, Protracted Growth) that reflect adaptations to spatial and temporal distribution of soil moisture. Annual growth of each group occurs successively later during the growing season, and provides sequential peaks in forage and nutrient availability, which in turn, determines the variety of animal species that may be supported by these plant communities (Sauer and Uresk, 1976).

In British Columbia, there has been a great deal of recent interest in grassland microphytes (organic crusts of lichens, bryophytes, and cyanobacteria) (West, 1990). Under relatively pristine conditions, these organic crusts are typically well developed between the dominant vascular plants of arid grasslands (Daubenmire, 1970; Anderson and Rushforth, 1977). They stabilize soils (Anderson et al., 1982), enhance soil moisture retention (Brotherson and Rushforth, 1983), and may increase available soil nitrogen through fixation by the cyanobacteria (Cameron and Fuller, 1960). Microphytes may also inhibit germination and establishment of weedy plant species (Mack and Thompson, 1982), and of native grasses, forbs and shrubs (West, 1990).

Disturbed Grasslands

Although many human activities affect grassland biodiversity, most grassland research in British Columbia has focused on the impacts of cattle and wildlife grazing, particularly on botanical composition and forage production. Moreover, cattle grazing is widely believed to be a significant influence on grassland ecosystems (Hooper and Pitt, 1993). This section, therefore, will summarize the impacts of grazing on grassland biodiversity.

McLean and Tisdale (1972) reported on the recovery rate of range sites that had been protected from livestock grazing by exclosures established during the 1930s. They estimated that 20-40 years of full rest were required for grassland communities on rough fescue and Ponderosa pine/bluebunch wheatgrass sites to recover from poor to excellent range condition. Little change in the plant composition of grasslands that were in poor condition took place inside the exclosures in less than ten years following fencing. It took longer for the sites to progress from poor to fair condition than from fair to good condition. The primary plants that increased with protection were bluebunch wheatgrass and rough fescue. The main species to decrease included Sandberg's bluegrass, low pussytoes (*Antennaria dimorpha*) and rabbit-brush (*Chrysothamnus nauseosus*). After 25 years, average herbage yields inside the exclosures were 124% higher than yields outside the exclosures on Ponderosa pine sites, and 97% higher on rough fescue sites.

Cattle generally prefer to forage on large bunchgrasses rather than on smaller grasses and forbs. Bunchgrass vegetation grazed by cattle, therefore, includes lower proportions of bluebunch wheatgrass, rough fescue and Idaho fescue. Smaller perennial grasses, annual grasses, and forbs tend to increase on grazed sites, signifying downward trends in range condition and reduced habitat suitability for cattle (McLean and Marchand, 1968).

Native grazing ungulates (hoofed animals) may also affect botanical composition, forage production and plant vigour of grasslands. Wikeem and Pitt (1991) demonstrated that departures from climax caused by California bighorn sheep grazing may produce higher proportions of bluebunch wheatgrass and lower proportions of forbs than at climax. Like cattle, then, native ungulates can also become overstocked in relation to their food supply (Anderson and Scherzinger, 1975), which may alter grassland biodiversity.

No information is available in British Columbia regarding the role of grassland microphytes in nitrogen fixation, soil stabilization, impacts on soil moisture, or site resistance to weed invasion. Climax species, however, may be reduced or eliminated by livestock or wildlife trampling (Anderson et al., 1982; Pegua, 1970). Anderson et al. (1982) demonstrated that lichen cover often declined in grazed areas even when climax herbaceous plants still dominated the site.

Alternatively, some lichens may increase with grazing. MacCraken et al. (1983) speculated that *Parmelia chlorochroa* may be more abundant in grazed areas of Montana because of its positive correlation with bare ground, and negative association with vascular plant cover.

Hooper and Pitt (1993) concluded that the impacts of livestock grazing on wildlife are variable and difficult to assess. For example, birds that prefer to nest in tall vegetation, and/or require elevated singing perches, such as Western Meadowlarks (*Sturnella neglecta*), and Savannah Sparrows (*Passerculus sandwichensis*) may benefit from levels of grazing and browsing that leave large amounts of standing vegetation or stimulate lateral branching and bushier shrub growth by removal of terminal buds (Ryder, 1980). Sharp-tailed Grouse (*Tympanuchus phasianellus*) will abandon their dancing grounds and adjacent nesting areas when nesting and brooding cover is removed by grazing (Pepper, 1972). Conversely, grazing levels that reduce vegetation height and density may benefit species such as the Long-billed Curlew (*Numenius americanus*) which requires large, open areas of relatively short vegetation for communal predator detection and unhampered movements of feeding chicks (Allen, 1980). The Horned Lark (*Eremophila alpestris*), another short-grass nester, is unable to forage efficiently in tall grass, which can result in chicks starving to death in the nest (Cody, 1985).

Grazing can affect insect populations by altering the microclimate, living space, and food resources of insect habitats (Breymeyer and van Dyne, 1980). While these changes may lead to an increase in insect numbers, insect species generally decline under intense grazing pressures (Smith, 1940). For example, grasshoppers apparently benefit from grazing (Shotwell, 1958; Skinner, 1975), and generally are more common in heavily than lightly grazed areas (Smith, 1940; Kelly and Middlekauff, 1961). In contrast, populations of harvester ants (*Pogonomyrmex* spp.) are negatively associated with grazing intensity (Rogers et al., 1972).

Similarly, most of the butterfly species found on the Junction WMA are closely associated with their host plants, and are vulnerable to alteration of their habitat by grazing cattle. For example, the artemisia swallowtail (*Papilio machaon*) occurs only on tarragon (*Artemisia dracunculus*), and the buckwheat blue (*Euphilotes battoides subsp. glaucon*) is found only with parsnip-flowered buckwheat (*Eriogonum heracleoides var. angustifolium*) (Roberts, 1992).

Ecological overlap between domestic livestock and wild ungulates is a complex topic filled with uncertainty due to: (1) foraging differences between animal species; (2) changing patterns of livestock grazing, agriculture, and recreation; (3) variable annual weather patterns, such as winter snow pack; (4) plant community dynamics; and (5) difficulty in quantifying competition. Pitt (1982) reviewed dietary overlap between cattle and wild ungulates in British Columbia, and concluded that the greatest potential for direct competition exists on critical winter habitat, most notably low-elevation grasslands. Beyond this generalization, however, few local, site-specific conclusions regarding the magnitude of temporal and spatial dietary overlap between cattle and wild ungulates are currently possible. Nonetheless, cattle grazing may be detrimental to mammals that depend on climax vegetation, but may benefit species that occur more commonly at seral (successional) stages or prefer to forage on plant species that are more prevalent at seral stages.

Current Threats to Grassland Biodiversity

Forty-four British Columbian resource managers and professionals (ranchers, range agrologists, scientists and wildlife biologists) were interviewed by Hooper and Pitt (1993) regarding perceived threats to grassland biodiversity in the Chilcotin-Cariboo. Interviewees most commonly cited cattle grazing as their primary concern. Other factors, in decreasing order of perceived threats to grassland biodiversity, included prescribed burning, recreational use, forest encroachment, agricultural and urban development, rangeland re-seeding programs, road and trail development, alien plant and animal species introductions, predator control, and hunting.

This list may represent those factors of most topical or immediate interest to the interviewees, and does not necessarily identify the most significant impacts on biodiversity of B.C. grasslands. For example, in the Okanagan, urban development likely poses the greatest threat to grassland biodiversity. In the Chilcotin-Cariboo grasslands, biodiversity may be threatened most by forest encroachment, which has absorbed approximately 30% of the grasslands since 1960 (personal communication in September, 1992, with F. Knezevich, Ministry of Forests). Prescribed burning, although identified as a threat to grassland biodiversity by some interviewees, was recommended by others as a natural process for reversing or impeding forest and shrub encroachment (Hooper and Pitt, 1993). This recommendation for prescribed burning recognizes that fires undoubtedly occurred naturally in most British Columbian grasslands. Based on a review of historical fire frequency, Low (1988) estimated that grassland sites in the Kamloops and Okanagan regions burned, on average, once every seven to ten years.

Management Options

Regional or site-specific differences in threats to grassland biodiversity suggest that appropriate management options for maintaining biodiversity may also differ. Biodiversity management options most often emphasize either priority species or priority habitats.

Grassland Species of Concern

One approach to maintaining grassland biodiversity is to focus on priority species. The British Columbia Ministry of Environment (1993) identified 55 vulnerable (provincial Blue List) or endangered wildlife species (Red List) that occupy dry interior grassland habitats.

It should be noted that this list likely reflects current research associated with grassland biodiversity. For example, the Sprague's

Pipit (*Anthus spragueii*) was not red-listed until 1993 because it was not found in B.C. grasslands until 1991 (Hooper and Pitt, 1993). In contrast, the Common Poorwill (*Phalaenoptilus nuttallii*) and spotted bat (*Euderma maculatum*) were removed from the 1993 Red list because new information indicated that these species have larger populations and ranges than previously known. Sixty-six vascular plant species associated with the bunchgrass vegetation of British Columbia are considered threatened or endangered. The British Columbia Conservation Data Centre identifies 115 additional vascular plants as rare (personal communication in December, 1992, with G. Douglas, Conservation Data Centre). Despite these lists, however, it should be noted that many plant species naturally occur infrequently, and would be considered rare, even under pristine conditions (West, 1993). Additionally, some of these species, such as blue grama (*Bouteloua gracilis*), are distributed relatively widely outside B.C., and may not be immediately threatened.

Grassland Habitats of Concern

An alternative approach for maintaining grassland biodiversity is to focus on priority habitats. The areas or habitats of concern most often identified by Hooper and Pitt (1993), in decreasing order, included:

a) riparian areas or wetlands (including saline ponds) associated with grasslands;

b) shrub/grassland areas;

c) forest/grassland edge;

d) cliffs;

e) draws and ravines;

f) talus slopes;

g) sensitive soils or sites with poor soils stability; and

h) clay banks.

This list, like the ones for wildlife and plant species of greatest concern, may reflect current research or topical interest. Nonetheless, riparian sites in particular are likely to contribute significantly to grassland biodiversity. Although less than 1% of the western U.S. contains riparian vegetation, riparian habitat supports more species than any other type of habitat (Knopf and Samson, 1988). Similar patterns may exist in British Columbia. In the Chilcotin-Cariboo grasslands, riparian habitats support the greatest number of Blue- and Yellow-listed species and are the second most commonly used habitat by Red-listed species. Riparian areas, though, continue to be threatened by agricultural development, recreation, livestock grazing, and wood harvesting. In the Okanagan, these impacts have reduced riparian areas to 15% of their pre-European extent (Harper et al., 1992).

Shrub/grassland areas are also important habitats for maintaining biodiversity. Although these areas comprise only the sixth most commonly used wildlife habitat within the Bunchgrass and Interior Douglas-fir zones of the Chilcotin-Cariboo Basin Ecosection, they are used by more Red-listed species than any other habitat, and are the second most commonly used habitat by Blue-listed species (Hooper and Pitt, 1993).

Recommendations for Maintaining Grassland Biodiversity

Biodiversity is a complex issue involving organisms, genetics, communities, ecosystems, landscapes and human values, topics on which information is usually lacking (West, 1993). Within this uncertain context, we offer the following recommendations for managing and conserving grassland biodiversity in British Columbia.

Previous grassland management programs have tended to emphasize priority species such as bluebunch wheatgrass or big game. Similarly, there is a tendency for biodiversity proponents to concentrate on rare, threatened or endangered species (Harper et al., 1992). Too little information is available for species populations and distributions to provide for effective grassland biodiversity management based solely on priority species. Moreover, provincially-listed species of concern are usually those most well-publicized and studied. Placing management emphasis on those species potentially neglects the needs of other, perhaps equally endangered but certainly less well-known species. Grassland biodiversity programs, therefore, should adopt a habitat, rather than a priority-species management approach (Hooper and Pitt, 1993; Chapter 16, this volume). A combination of protected and well-managed habitat complexes will be more effective in conserving grassland species, processes and functions, than a species-specific approach.

The long-term success of biodiversity programs also depends on recognition of human activities as enduring and legitimate components of most ecosystems. While human influences may need to be excluded from some protected areas, it is neither feasible nor responsible for biodiversity conservation programs to exclude human activities from all sites. For example, it is virtually certain that British Columbian grasslands will continue to be grazed by livestock. If well managed, such grazing may be compatible with some grassland biodiversity goals, as it may provide or maintain a mosaic of successional stages that provides a mixture of grasses, forbs, and browse.

Unfortunately, the relationships among successional stage, livestock grazing, wildlife habitat and biodiversity are often obscured by the qualitative terms excellent, good, fair and poor, which are used to classify range condition. There is a temptation to assume that excellent range condition provides excellent habitat for all wildlife and plant species. In reality, these terms categorize the seral stages induced or maintained by livestock grazing and offer no direct information about other specific grassland values, such as biodiversity, forage production or optimal wildlife habitat. Less subjective terms like climax, late-, mid- and early-seral would more appropriately characterize these successional stages caused by livestock grazing (Pitt, 1984). Grassland managers should identify the specific successional stages needed to achieve explicit biodiversity goals, and manage livestock accordingly to maintain or produce those successional stages.

Optimal wildlife habitat depends more on a complement of plant species than on dominance by a few, key management species (Wikeem and Pitt, 1991). Nonetheless, such key species will likely remain important as indicators of desired plant communities, either seral or climax. Like bluebunch wheatgrass, microphytes may also be useful indicator species of climax grassland communities because of their apparent sensitivity to disturbances such as air pollution, fire, grazing and wind-blown silt deposition (West, 1990). Similarly, invertebrate species such as robber flies (Family *Asilidae*) may be useful indicators of successional and climax grassland communities (personal communication in September, 1992, with S. Cannings, Ministry of Environment Lands & Parks).

Ultimately, however, long-term maintenance of grassland biodiversity requires public, industry, and government involvement and cooperation. Public education and support are particularly important if grassland biodiversity programs are to be successful. Inventories, research, monitoring, and conservation also comprise essential components of grassland biodiversity programs. Inventories should identify types and distributions of all habitats associated with grasslands, relative abundance of wildlife and plant species, and types and extent of disturbances influencing these habitats and species. Research and monitoring should emphasize species/habitat relationships, and should document the short- and long-term impacts of grassland disturbances on biodiversity.

To a considerable degree, biodiversity is about the conservation of genetic resources (Wilson and Peter, 1988). We recommend, therefore, that grassland biodiversity programs conserve large, contiguous, representative areas to ensure that the genetic resources of grassland ecosystems will withstand disturbances and adapt to the inevitable changes in human activities and climatic patterns.

Acknowledgements

This chapter was based partially on a problem analysis (Hooper and Pitt, 1993) initiated by the Ministry of Environment, Lands and Parks (Fish & Wildlife Branch, Williams Lake) with assistance from the Habitat Conservation Fund. The purpose of the analysis was to identify habitat enhancement objectives and make recommendations for maintaining biodiversity of the Chilcotin-Cariboo grasslands. This chapter was reviewed by Bert Brink, Professor Emeritus, Department of Plant Science, University of British Columbia, and George Douglas, Conservation Data Centre.

References

Allen, J.N. 1980. The ecology and behavior of the long-billed curlew in southwestern Washington. *Wildlife Monograph* 73:3-67.

Anderson, D.C., K.T. Harper and S.R. Rushforth. 1982. Recovery of cryptogamic soil crusts in Utah deserts. *Journal of Range Management* 35:355-359.

Anderson, D.C. and S.R. Rushforth. 1977. The cryptogamic flora of desert soil crusts in southern Utah, U.S.A. *Nova Hedwigia* 28:691-729.

Anderson, E.W. and R.J. Scherzinger. 1975. Improving quality of winter forage for elk by cattle grazing. *Journal of Range Management* 28:120-125.

Breymeyer, A.I. and G.M. van Dyne. 1980. *Grasslands, Systems Analysis, and Management.* Cambridge University Press, New York, New York.

British Columbia Ministry of Environment. 1992. *Guide to Ecological Reserves in British Columbia.* B.C. Ministry of Environment, Parks Branch, Victoria, B.C.

British Columbia Ministry of Environment. 1993. *Red and Blue Lists.* B.C. Ministry of Environment, Wildlife Branch, Victoria, B.C.

Brotherson, J.D. and S.R. Rushforth. 1983. Influence of cryptogamic crusts on moisture relationships in Navajo National Monument, Arizona. *Great Basin Naturalist* 43:73-78.

Cameron, R.E. and W.H. Fuller. 1960. Nitrogen fixation by some algae in Arizona soils. *Proceedings - Soil Science: Society of America* 24:353-356.

Cody, M.L. 1985. Habitat selection and open-country birds. In: *Habitat Selection in Birds.* M.L. Cody (ed.). Academic Press. pp. 191-226.

Daubenmire, R. 1970. Steppe vegetation of Washington. Washington Agricultural Experiment Station Technical Bulletin 62. 131 pp. Washington State University Press, Pullman.

Daubenmire, R. 1978. Plant geography. *Academic Press.*

Graul, W.D. 1980. Grassland management practices and bird communities. In: *Management of Western Forest and Grasslands for Non-game Birds.* U.S.D.A. Forest Services General Technical Report INT-86, Ogden, Utah. pp. 38-47.

Harper, W.L., E.C. Lea and R.E. Maxwell. 1992. *Biodiversity Inventory in the South Okanagan.* B.C. Ministry of Environment, Lands and Parks, Wildlife Branch, Victoria, B.C.

Hebda, R.J. 1982. Postglacial history of grasslands of southern British Columbia and adjacent regions. In: *Grassland Ecology and Classification Symposium Proceedings.* A.C. Nicholson, A. McLean and T.E. Baker (eds). B.C. Ministry of Forests, Victoria. pp. 157-191.

Hlady, D.A. 1990. *South Okanagan Conservation Strategy: 1990 - 1995.* B.C. Ministry of Environment. Integrated Resource Management Branch, Victoria, B.C.

Hooper, T.D. and M.D. Pitt. 1993. *Problem analysis for Chilcotin-Cariboo grassland biodiversity.* Ministry of Environment, Lands and Parks, Fish and Wildlife Branch, Williams Lake, B.C.

Hope, G.D., W.R. Mitchell, D.A. Lloyd, W.R. Erickson, W.L. Harper and B.M. Wikeem. 1991. Interior Douglas-fir Zone. In: *Ecosystems of British Columbia.* D. Meidinger and J. Pojar (compilers and eds.). B.C. Ministry of Forest Special Report Series 6, Victoria, B.C. pp. 153-166.

Karasiuk, D., H. Vriens, J.G. Stelfox and J.R. McGillis. 1977. Study results from Suffield, 1976 In: *Effects of Livestock Grazing on Mixed Prairie Range and Wildlife within PFRA Pastures, Suffield Military Reserve.* J.G. Stelfox (comp.). Range-Wildlife Study Committee, Canadian Wildlife Service, Edmonton, Alta. pp. E33-E44.

Kelly, G.D. and W.W. Middlekauff. 1961. Biological studies of Dissosteira spurcata Saussure with distributional notes on related California species (Orthoptera: Acrididae). *Hilgardia* 30:395-424. In: *Rangeland Entomology.* Hewitt, E.W. Huddleston, R.J. Lavigne, D.N. Ueckert, and J.G. Watts. (eds.) 1974. Colorado State University, Range Science Department, Range Science Series No. 2G.

Klopateck, J.M., R.J. Olson, C.J. Emerson and J.L. Jones. 1979. Land-use conflicts with natural vegetation in the United States. *Environmental Science Division Publication No. 1333.* National Technical Information Service, U.S. Department of Commerce, Springfield, Virginia. 22 pp. (Original not seen; taken from Graul, 1980).

Knopf, F.L. and F.B Samson. 1988. Ecological patterning of riparian avifaunas. In: *Streamside Management; Riparian Wildlife and Forestry Interactions.* K.J. Raedeke (ed.). College of Forest Resources and Institute of Forest Resources, University of Washington.

Low, D.J. 1988. Effects of prescribed burning on non-target wildlife species associated with fire prone ecosystems in the southern Interior of British Columbia. *Wildlife and Range Prescribed Burning Workshop Proceedings.* M.C. Feller and S.M. Thomson (eds.). University of British Columbia, Faculty of Forestry. Vancouver, B.C. pp. 185-196.

MacCraken, J.G., L.E. Alexander and D.W. Uresk. 1983. An important lichen of southeastern Montana rangelands. *Journal of Range Management* 36:35-37.

Mack, R.N. and J.N. Thompson. 1982. Evolution in steppe with few large, hoofed animals. *American Naturalist* 119:757-773.

McLean, A. 1969. *Plant Communities of the Similkameen Valley, British Columbia and Their Relationships to Soils.* Washington State University Ph.D. Thesis. University Microfilms, Inc., Ann Arbor, Michigan.

McLean, A. and L. Marchand. 1968. *Grassland Ranges in the Southern Interior of British Columbia.* Canadian Department of Agriculture Publication, 1319, Ottawa, Ontario.

McLean, A. and E.W. Tisdale. 1972. Recovery rate of depleted range sites under protection from grazing. *Journal of Range Management* 25:178-184.

Ministry of Crown Lands. 1989. *B.C. Land Statistics.* Province of B.C.

Mitchell, H.B. and G.W. Prediger. 1975. *Junction Wildlife Management Area: A Management Plan 1975-1980.* B.C. Ministry of Environment, Fish and Wildlife Branch, Williams Lake, B.C.

Nicholson, A., E. Hamilton, W.L. Harper and B.M. Wikeem. 1991. Bunchgrass zone. In: *Ecosystems of British Columbia.* D. Meidinger and J. Pojar (eds.). B.C. Ministry of Forests Special Report Series 6, Victoria, B.C. pp. 125-137.

Pegua, R.E. 1970. Effect of reindeer trampling and grazing on lichens. *Journal of Range Management* 23:95-97.

Pepper, G.W. 1972. *The Ecology of the Sharp-tailed Grouse During Spring and Summer in the Aspen Parklands of Saskatchewan.* Department of Natural Resources and Wildlife Report No. 1. (Original not seen; information taken from Karasiuk, Vriens, Stelfox and McGillis, 1977).

Pielou, E.C. 1991. *After the Ice Age; The Return of Life to Glaciated North America.* University of Chicago Press, Chicago Il.

Pitt, M.D. 1982. *East Kootenay Problem Analysis. Interactions Among Grass, Trees, Elk and Cattle.* B.C. Ministry of Forests, Victoria, B.C.

Pitt, M.D. 1984. *Range condition and trend assessment in British Columbia.* B.C. Ministry of Forests Research Report RR84004-HQ, Victoria, B.C.

Pitt, M.D. and B.M. Wikeem. 1990. Phenological patterns and adaptations in an *Artemisia/Agropyron* plant community. *Journal of Range Management* 43:350-358.

Redpath, K. 1990. Identification of relatively undisturbed area in the South Okanagan. In: *Biodiversity Inventory in the South Okanagan.* W.L. Harper, E.C. Lea, and R.E. Maxwell (eds.). 1992. Canadian Wildlife Service, Delta, B.C.

Roberts, A. 1992. *A Report on the Ecology of the Junction Wildlife Management Area.* Ministry of Environment, Lands and Parks, Fish & Wildlife Branch, Williams Lake.

Rogers, L., R. Lavigne and J.L. Miller. 1972. Bioenergetics of the western harvester ant in the shortgrass plains ecosystem. *Environmental Entomology* 1:763-768. (Original not seen; information taken from Breymeyer and van Dyne, 1980).

Ryder, R.A. 1980. Effects of grazing on bird habitats. In: *Management of Western Forests and Grasslands for Non-game Birds.* R. M. DeGraff and N.G. Tilghman (eds.). U.S. Forests Service General Technical Report INT-86. pp. 51-64.

Sauer, R.H. and D.W. Uresk. 1976. Phenology of steppe plants in wet and dry years. *Northwest Science* 50:133-139.

Shotwell, R.L. 1958. *The Grasshopper Your Sharecropper.* University of Missouri Agricultural Experimental Station Bulletin 714. (Original not seen; information taken from Ryder, 1980).

Skinner, R.M. 1975. *Grassland Use Pattern and Prairie Bird Populations in Missouri.* (Original not seen; information taken from Ryder, 1980).

Smith, C.C. 1940. The effect of overgrazing and erosion upon the biota of the mixed-grass prairie of Oklahoma. *Ecology* 21:381-397.

Spilsbury, R.H. and E.W. Tisdale. 1944. Soil plant relationships and vertical zonation in the southern interior of British Columbia. *Scientific Agriculture* 24:395-436.

Tisdale, E.W. 1947. The grasslands of the southern interior of British Columbia. *Ecology* 28:346-365.

West, N.E. 1990. Structure and function of microphytic soil crusts in wildland ecosystems of arid to semi-arid regions. In: *Advances in Ecological Research,* Vol. 20. M. Began, A.H. Fitter and A. MacFadyen (eds.). Academic Press Ltd. pp. 179-223.

291

West, N.E. 1993. Biodiversity of rangelands. *Journal of Range Management* 46:2-13.

Wikeem, B.M., A. McLean, A. Bawtree and D. Quinton. Unpublished. *An Overview of the Forage Resource and Beef Production on Crown Land in British Columbia.*

Wikeem, B.M. and M.D. Pitt. 1991. Grazing effects and range trend assessment on California bighorn sheep range. *Journal of Range Management* 44:466-470.

Wilson, E.O. and F.M. Peter (Eds.) 1988. *Biodiversity.* National Academy Press. Washington, D.C.

World Wildlife Fund Canada. 1989. *Prairie Conservation Action Plan: 1989-1994.* World Wildlife Fund Canada, Toronto, Ontario.

Chapter 21
Threats to Biodiversity in the Strait of Georgia

Ecosystem Diversity in the Strait of Georgia

British Columbia's variable terrestrial physiography is reflected in its marine environments. Estuaries, fjords, shoreline and benthic habitats in the Strait of Georgia harbour many very different marine ecosystems which contribute to high species diversity.

Highly productive estuaries, such as those of the Fraser, Squamish, Cowichan, Nanaimo and Campbell Rivers, support rich, detritus-based communities of invertebrates, fish and waterfowl. From about half to 90% of the area, and presumably the primary production, of these estuaries has been lost to agricultural and industrial development. However, since about 1980, management plans have limited further habitat losses in all of the major estuaries. Two of them—the Campbell and Fraser River Estuaries—have regained some of their lost habitat through intensive restoration efforts (Levings, 1991). The largest tidal flat, Boundary Bay, is a highly productive eel grass community which supports shorebirds of "hemispheric" importance (Butler and Campbell, 1987). It is largely intact.

The numerous deep (300 metre and more) fjords at the northern end of the Strait are less productive than the estuaries in surface waters (Stockner et al., 1979) and, consequently, in the deep bottom waters, which are cold and frequently hypoxic. However, like the many narrow passages of the Gulf Islands, the rocky sides of these fjords (particularly near their mouths where tidal flow creates constant currents) provide habitat for very diverse attached invertebrate communities (Levings et al., 1983; Chapter 6, this volume).

There is currently no inventory of shoreline habitats and the amount lost to uses such as shoreline construction, marinas, wharves, and log booming in the Strait as a whole, but one estimate for Howe Sound is 40% (Ministry of Environment, 1981). Log booming has a variety of effects, from reduction in growth of geoduck clams, a large and long-lived species (Noakes and Campbell, 1992) to nearly complete obliteration of benthic communities (Colodey and Wells, 1992). Marinas and wharves are sources of fecal contamination (shellfish harvesting is automatically prohibited within 125 metres of such facilities) and heavy metal contamination (from antifouling paints used on boat hulls; Saavedra and Ellis, 1990). These activities have altered an unquantified amount of intertidal and shallow habitat, with unknown effects on biodiversity of species in these habitats.

The deep central portion of the Strait is home to six distinct communities of macrobenthos: the Osteochella/Sipuncula, Paracaudina/Compsomyax, Brisaster/Lucina/Aphrocallistes,

Lee E. Harding
Canadian Wildlife Service
P.O. Box 340
Delta, B.C.
V4K 3Y3

Glycera/Macoma, Glycymeris/Chlamys, and Acila/Nemocardium communities (Levings et al, 1983). The characteristics of the habitats supporting these different communities are, however, less well known than those of the shoreline.

Assessment of Ecosystem Health in the Strait of Georgia

Waldichuk (1983) provided the first overall assessment of the condition of the Strait. His analysis was based on sources of pollutants, concentrations of "critical" pollutants in sediments and tissues, and some direct, toxic effects of these pollutants. He concluded that the Strait, as a whole, did not yet exhibit an alarming degree of pollution, and the assimilative capacity of the Strait was not exceeded, except in local nearshore areas. Subsequent reviews (Kay, 1989; Wells and Rolston, 1991) have shown that, except for localized areas of contamination and habitat disturbance, there are few serious threats to marine environmental quality along most of Canada's west coast. However, the Strait of Georgia has a much higher concentration of pollution sources (Figure 22-1), foreshore uses and other localized disturbances than other, more remote ecosystems. The purpose of this chapter is to provide an

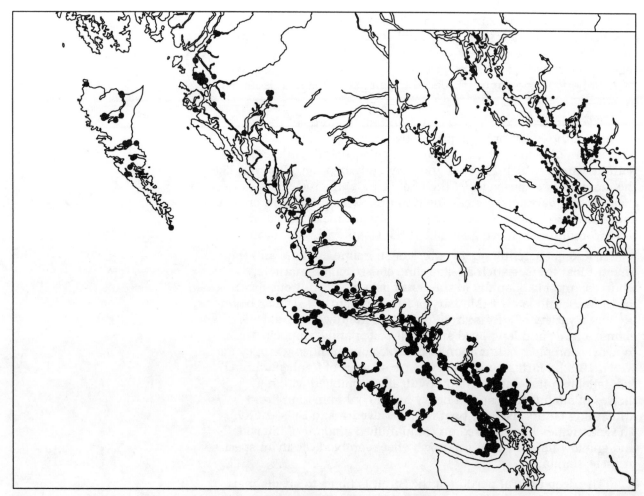

Figure 21-1. Effluent Discharge Locations on the B.C. Coast and the Strait of Georgia (Inset). Source: Environment Canada Marine Pollution Source Database.

update of Waldichuk's 1983 assessment, from the point of view of marine ecosystem health. The framework for this analysis is that proposed by Rapport (1989), who discussed human health as a metaphor for ecosystem health. He suggested that the following attributes of marine ecosystems could be used to diagnose ecosystem pathology: primary productivity, nutrients, species diversity, instability, contaminants, disease prevalence and size spectrum. This framework will be followed in this analysis.

Primary Productivity and Nutrients

Harrison (1985) and Clifford et al. (1988) showed that even the large nutrient sources of the heavily agricultural, urban and industrial Fraser River and City of Vancouver have only a local effect, if any, on nutrients and primary production. Primary production of plant biomass in the Strait and the zooplankton community that it supports fluctuates within normal ranges (Harrison et al., 1984; Waldichuk, 1983). It is, however, elevated in local embayments, such as in Vancouver Harbour's Port Moody Arm, a warm, shallow inlet with urban runoff (Stockner and Cliff, 1976), and in the receiving waters of six pulpmills that discharge into the Strait (Stockner and Costella, 1976; Colodey et al., 1990). Moreover, Macdonald et al. (1991) found that terrestrial carbon increased towards the sediment surface in a core from the north central Strait of Georgia, and was correlated with increasing nitrogen. They believe the additional carbon and nutrients to result from pulp mill effluent. If so, this discovery would imply a widespread, low level impact of pulp mill effluent throughout the Strait of Georgia. This hypothesis is strengthened by data on dioxin distributions discussed in the Contaminants section below.

Species Diversity

By piecing together evidence from the relatively few studies in the Strait of Georgia, a rough picture of changes in benthic infaunal (ocean floor dwelling invertebrate), marine floral, fish, bird and marine mammal diversity can be obtained.

Benthic infaunal communities have been analyzed at a number of sites in Boundary Bay (Burd et al, 1987) and Vancouver Harbour (Burd and Brinkhurst, 1990). Burd and Brinkhurst (1990) found that, except in the industrialized eastern end of Vancouver Harbour, benthic communities were diverse and apparently healthy in comparison with reference sites on the west coast of Vancouver Island (Brinkhurst, 1987) and the north coast (Brinkhurst and Kathman, 1987).

Hawkes (1991) notes that, while the west coast of Canada has one of the richest and most diverse marine floras in the world, its environment is very different from 200 years ago. Elimination of sea otters (*Enhydra lutris*) by Russian hunters from Alaska during the 1700s would have allowed the sea urchin population to expand, with consequent depression of benthic algal populations. He observes that, although no marine plants are known to have been extirpated from the province in recent times, a few species, including some in the Strait of Georgia, are rare and in need of protection.

Ketchen et al. (1983) showed that commercial landings of spiny dogfish (*Squalus acanthus*), ling cod (*Ophiodon elongatus*), sole (various species, especially English sole [*Parophrys vetulus*]) and shrimp (various species, especially pink shrimp [*Pandalus borealis*]) have declined steadily in recent decades. Walters and Cahoon (1985) provided evidence of decreasing diversity of salmon stocks, which are presumed to be genetically distinct: 90% of the chinook (*Oncorhynchus tshawytscha*), coho (*O. kisutch*), pink (*O. gorbuscha*) and chum (*O. keta*) salmon escapements are now produced from only half as many streams as in 1950. The decreases in genetic diversity have been most severe in the Strait of Georgia, where Riddel (Department of Fisheries and Oceans, personal communication, March, 1992) estimates that, excluding Fraser River stocks, one third of spawning populations have been nearly or completely eliminated. However, the catches of coho and, to a lesser extent, chinook have remained quite high as a result of hatchery production. From 1976 to 1990, 76% of the spawning returns of B.C. salmon stocks have been stable or increasing, and 24% have declined (Province of British Columbia and Environment Canada, 1993). Canadian catches of salmon, in general, are close to historic high levels. Pink, chum and sockeye (*O. nerka*) from the Fraser River spend at least one month in the Strait, so conditions in the Strait itself appear to be adequate at least for the current salmon populations (personal communication in April, 1993, with R. Beamish, Department of Fisheries and Oceans).

Other factors which may affect the diversity of salmon stocks are predation and the carrying capacity of the Strait of Georgia. Predation by spiny dogfish (which are abundant in the Strait), may be a factor in the observed decline in the percentage of smolts that return from the ocean as adult salmon (Beamish et al., 1992). The most abundant commercial species in the Strait of Georgia is Pacific hake (*Merluccius productus*). As many as 100,000 tonnes of hake now support a fishery of about 10,000 tonnes. Pacific hake in the Strait are known to feed on fish, but the extent of their predation on salmon is unknown. It is possible, however, that their abundance affects the Strait's carrying capacity for other species (personal communication in April, 1993, with R. Beamish, Department of Fisheries and Oceans).

Among fish-eating birds, Bald Eagle (*Haliaeetus leucocephalus*) populations and both Pelagic (*Phalacrocorax pelagicus*) and Double-crested Cormorants (*P. auritus*) populations in the Strait of Georgia are stable or increasing (personal communication in April, 1992 with J. Elliott, Canadian Wildlife Service). Reproduction of alcids and perhaps other fish-eating birds in B.C. fluctuates with populations of Pacific sand lance (*Ammodytes hexapterus*) which, in turn, are controlled by sea-surface temperatures (Kaiser and Forbes, 1992). Hence, their populations are affected more by natural factors than by human disturbance of marine ecosystems.

Although populations of seals (northern elephant seal [*Mirounga angustirostris*] and habrour seal [*Phoca vitulina*]), sea lions (California sea lion [*Zalophus californianus*] and Stellar sea lion [*Eumetopias jubatus*]), and killer whales (*Orcinus orca*) are stable or increasing (personal communication in April, 1993, with T. Smith, Department of Fisheries and Oceans), humpback whales (*Megaptera novaeanglia*), which were abundant enough in the Strait to support a commercial fishery for some 80 years, have been absent since 1908 (Ketchen et al, 1983).

While some species have declined or been extirpated from the Strait, others have been introduced. These include Japanese oysters (*Crassostrea gigas*) and Manila clams (*Venerupis japonica*), a number of small snail species (introduced with Japanese oyster spat), a Japanese seaweed, wood boring worms and crustaceans (Waldichuk and Smiley, 1991; and Chapter 17, this volume), and some polychaete worm species (Fournier and Levings, 1982). California sea lions, a smaller relative of the native Stellar sea lions, have become resident in the Strait in recent decades (Ketchen, 1983), and are apparently expanding their range northward in response to spatial shifts in herring (*Clupea harengus*) stocks (personal communication in April, 1993, with T. Smith, Department of Fisheries and Oceans).

Instability

Analysis of the degree of instability gives some indication of whether physical and ecological structures are likely to be permanently altered. Some changes in the physical features of the Strait of Georgia can be discerned from long term data. Freeland (1991) found that sea surface temperatures from all stations in British Columbia show a significant warming trend of about 1° per century. This trend strongly corresponds to the average Northern Hemisphere surface air temperature increases shown in Figure 24-11. Freeland (1991) noted that, if climate change continues, the frequency of red tides and other toxic algal blooms might also increase because the organisms that cause them favour warmer water.

Within the Strait, temperature and possibly water column structure appear to have been affected by changes in estuarine circulation (in which high salinity bottom water is entrained by the low-salinity surface plume, forcing deep water renewal) resulting from altered weather patterns. Beamish (in prep.) noted that discharge from the Fraser River, which contributes 80% of the runoff received by the Strait of Georgia, began declining after 1976 when precipitation and snowpack levels in the interior fell (see Chapter 24). He showed that the annual temperature cycle of bottom water was relatively uniform until 1976, when it was disrupted. The disruption lasted until 1980, after which a new cycle having warmer annual bottom water temperatures and less annual fluctuation became established. The timing of the change in this annual pattern of bottom water temperature coincided with other changes in climate and oceanography, including the drop in discharge from the Fraser River.

Ecological response to the instability of physical variables is difficult to measure without a focused monitoring program, but some surrogate indices are available. Beamish (in prep.) states that,

> It is clear that the climate change that began in 1976 was a major event all along the west coast of North America that was associated with increases in primary and secondary production on a large scale. Associated with these changes were major changes in fish abundance.

Increases in primary production are suggested by outbreaks of toxic algal species. Both the frequency and intensity of paralytic shellfish poison outbreaks seem to have increased in recent years (Chiang, 1991), although the data do not lend themselves to rigorous analysis. This possible increase suggests a corresponding rise in dinoflagellate production, though its spatial distribution does not indicate a direct relationship to anthropogenic nutrient sources. Blooms of other algae species which are toxic to farmed salmon have caused millions of dollars of losses in some years, and very little damage in others (Black, 1991). These blooms involve a number of species and are highly variable in space and time. Although the available data are not really adequate for rigorous statistical analysis, there is a general sense among professionals in the field that these events have been increasing in frequency and intensity in recent years (Forbes, 1991). If so, the most likely cause would be the warmer sea-surface temperatures, which would increase the frequency of ideal combinations (for toxic algae) of temperature, sunlight and nutrients.

Overall, there are subtle but large scale indications of ecological instability in the Strait of Georgia, but their significance is unknown. The apparent increase in toxic algal blooms, if confirmed, could have severe consequences for several marine resource sectors. Similarly, changes in fish abundance could affect harvesting patterns.

Contaminants

Contaminants were reviewed by Kay (1989) and Wells and Rolston (1991). In general, they found that while "hot spots" exist in sediments and in marine organisms near specific outfalls, sediments and tissues collected even short distances from outfalls have normal background levels of trace metals and undetectable amounts of most organic contaminants.

Although a wide range of organic contaminants are present in the marine food chains of inshore and continental shelf areas, some of them have declined. For example, examination of eggs collected from seven seabird species at colonies on the B.C. coast from 1971 to 1990 showed that mean residue levels of most organochlorine pesticides have fallen since the early 1970s (Elliott and Noble, 1992; Elliott et al., 1992). At a cormorant colony on Mandarte Island in the Strait of Georgia, all measured organochlorine pesticides and PCBs have decreased since the early 1970s, when they were controlled or banned. In offshore areas, only DDE (a metabolite of DDT) is declining; other organochlorine contaminants persist at low but stable levels. There was no evidence of impaired productivity in the seabird populations surveyed.

However, new evidence suggests that the distribution and, possibly, effects of contaminants are more widespread than previously thought. MacDonald et al. (1991) found rising levels of contaminants in the more recent layers of a sediment core from the central Strait of Georgia. They interpreted this increase as evidence for widespread (but low level) contamination of the Strait as a whole, rather than merely localized contamination from point sources as had previously been supposed. In 1983, the Canadian Wildlife

298

Service discovered dioxins in Great Blue Heron (*Ardea herodias*) eggs at certain colonies in the Strait of Georgia, the highest levels being at a colony that fed in tidal flats near a pulp mill at Crofton (Elliott et al., 1989). Another colony near Crofton, but located by a lake where the herons fed nearly exclusively on freshwater organisms, had negligible levels, suggesting that the route of exposure was through the marine food web of the Strait of Georgia. During 1988-1990 approximately 90,000 hectares of nearshore habitats near pulp mills and associated industries were closed to fishing because of dioxin contamination in sediments, crabs, shrimp, bivalves and fish (Harding and Pomeroy, 1990). Elevated levels were also found in double-crested and pelagic cormorants and in some fish-eating waterfowl. Advisories were issued against consumption of certain species of waterfowl. Elliott and Noble (1992) noted that

> Double-crested cormorant eggs from Mandarte Island in the Strait of Georgia, distant from industrial pollutant sources, were also contaminated with PCDDs and PCDFs, indicating widespread contamination during the 1980s of the marine food chains of the Strait.

Annually monitored dioxin and furan levels increased until 1987 and then appeared to decline somewhat, particularly at the Crofton site, following improved organochlorine control at coastal mills (Whitehead et al., 1992). Although stable or increasing population trends of herons and cormorants suggest that these contaminants are not affecting productivity, they do cause sublethal toxicity to developing heron embryos (Bellward et al., 1990; Hart et al., 1991). Dioxins, along with PCBs, also show up in Bald Eagle eggs (personal communication in April, 1992 with J. Elliot, Canadian Wildlife Service) and in both grey (*Eschrichtius robustus*) and killer whales (Muir et al., 1991). Elevated dioxin and furan levels in marine invertebrates, fish, seabirds and cetaceans suggests widespread uptake of low levels of these contaminants through marine food webs (Norstrom et al., 1991). Pulp mills, the dominant source of dioxins (Cretney and Crewe, 1991; Norstrom et al., 1991), have greatly reduced their discharge of organochlorines pursuant to new provincial and federal legislation. Other sources of contaminants may include atmospheric deposition, ocean current importation, migratory seabird and marine mammal importation, and continued local uptake in marine ecosystems.

Disease Prevalence

Like contaminants, disease seems to often occur in hot spots. It manifests itself in various forms, from liver lesions in sole to lower breeding success in bald eagles.

In Vancouver Harbour (Goyette and Boyd, 1989), the Fraser River estuary (E.V.S Consultants Ltd., 1990), and the waters near four pulp mills in the Strait of Georgia (Brand and Goyette, 1991), prevalence of ideopathic liver lesions in English sole varies from 30-75%. At some stations, these neoplasms were associated with biochemical indicators of sublethal toxicity (Brand, 1991). By contrast, fish from reference areas in Satellite Channel (southwest Georgia Strait), Loughborough Inlet and Rivers Inlet (mainland north of Georgia Strait) had neither preneoplastic nor neoplastic lesions.

All populations of six species of marine gastropods in the Strait of Georgia exhibited imposex, a morphological disorder (Saavedra and Ellis, 1990) symptomatic of poisoning by tributyl tin (TBT) poisoning. TBT has been used as an antifoulant on boat hulls and, from about 1985 to 1988, on salmon net pens. Imposex was more prevalent near marinas and wharves. By comparison, snails at reference sites outside the Strait had none of these disorders.

For a short time after rapid expansion of the aquaculture industry in 1986, oysters grown near several salmon farms began showing classic symptoms of tributyl tin poisoning, including mortality. These effects disappeared when federal regulations eliminated the use of tributyl tin on salmon net pens and controlled its use on boat hulls (Harding and Kay, 1988). However, marine snails have not been re-surveyed since these controls were imposed.

Heron chicks from a site (Crofton) contaminated with dioxins and furans from pulp mill effluent were smaller, suffered more from edema and had elevated levels of enzymes that indicate exposure to certain organic contaminants, compared to those from less contaminated sites (Bellward et al., 1990; Hart et al., 1991). A few bill deformities have been found in Strait of Georgia cormorant chicks in a ratio of 13.3 to 10,000 birds, compared to a clean colony ratio of 0.6 per 10,000 birds (personal communication in April, 1992 with J. Elliot, Canadian Wildlife Service). This condition is usually fatal. More study is needed to confirm the ratio and determine if it is increasing or decreasing. Its relation, if any, to dioxins or other contaminants in the Strait of Georgia is unknown.

Preliminary data shows that there are significant regional differences in bald eagle breeding success on the southern B.C. coast (personal communication in April, 1992, with J. Elliott, Canadian Wildlife Service). The number of young produced per active nest is relatively high in the Fraser Valley (Hope to Mission) and on Vancouver Island and the Gulf Islands from Nanoose to Crofton. Productivity is lower around Powell River, in Johnstone Strait and in Clayoquot and Barkley Sounds (west coast of Vancouver Island). Differences in biological productivity and food availability during the breeding season are likely the primary controlling factors. However, breeding success is also poor at nests near pulpmills at Powell River and Crofton. Levels of PCDDs, PCDFs, PCB and DDE are quite high in the eggs from some nests. It is currently too early to determine if there is any cause-effect relationship between these contaminants and productivity (personal communication in April, 1992, with J. Elliot, Canadian Wildlife Service). During 1989 to 1991, 65 sick or dead Bald Eagles brought to veterinarians or wildlife rehabilitators were tested for lead poisoning, which they get from feeding on dead or wounded ducks shot by hunters (Elliott et al., 1992). Preliminary results indicated that 36% had been exposed to lead and 14% had been poisoned by it. These results have contributed to efforts to ban lead use by hunters in certain areas.

A stranded Whale and Dolphin Program of B.C. has recently been established to investigate cetacean strandings. In B.C., 24 strandings were reported for 1987 and 28 for 1988 (Stacey et al., 1989). Causes of death were given as starvation of juveniles following separation from their mothers, parasite infection, net entanglement, neonatal mortality (in killer whales), and gastritis. One whale, a male false killer whale (*Psuedorcas crassidens*) that beached on Denman Island in May, 1987, had very high levels of pesticides and heavy metals, including the highest level of liver mercury reported in any cetacean worldwide (Baird et al., 1989). This false killer whale was the first to be found in Canada.

Size Spectrum

Theory on size spectrum relates productivity to size of organisms, particularly as it relates to conversion of energy from lower trophic levels to higher ones (predators): the more productive a system is, the larger organisms it can support (Platt and Denman, 1978; Ware, 1978). Environments with degraded primary productivity have, on average, smaller organisms. Conversely, harvest of larger organisms can lead to increases in the smaller organisms on which the larger ones prey. Both effects result in overall smaller size of organisms. It may be ecologically significant that, in a variety of predacious taxonomic groups in the Strait of Georgia, the largest species have been eliminated or are declining: humpbacks among the whales, ling cod among the bottom fish, chinook among the salmon. It is possible that the loss or reduction of these and other commercially fished species may have altered the size spectrum of organisms in the Strait of Georgia, although directed studies of this phenomenon in the Strait of Georgia are lacking. Parsons (1992) documented several cases elsewhere in the world where continued removal of large predators by fishing altered the trophic structure in marine waters similar to the Strait of Georgia. Such a mechanism may be involved in the changing proportions of hake, pollock (*Theragra chalcogramma*), spiny dogfish and salmon. If a new equilibrium under the control of external factors, such as altered estuarine circulation or continued fishing, is established among species, one could say that a shift has occurred in ecosystem state. In effect, a new ecosystem may have been created, and an old one lost. Such hypotheses await testing by studies of secondary production and biomass partitioning among species.

Conclusions

The analysis of marine ecosystem health comes down to whether ecological processes or ecosystem structures have been altered, and, if so, whether the changes are desirable. Evidence leads to the conclusions below.

1. Excessive stimulation of primary productivity by nutrients (eutrophication) is not occurring, except possibly on a very small local scale. However, increasing carbon levels in sediments suggest that processes that can lead to eutrophication are present. Industrial development has reduced the area of productive marshes in the major estuaries, where productivity is probably stable or slightly increasing due to marsh rehabilitation projects.

2. There is no evidence for declining diversity of pelagic (phytoplankton and zooplankton communities) or benthic (infauna) ecosystems in the main part of the Strait of

Georgia. Benthic communities in reference areas (sites not directly exposed to pollution sources) in the Strait are structurally sound. Perhaps one third of the salmon stocks (and, hence, the genetic diversity of the species) has been nearly or completely eliminated. Important ecological components have changed, primarily as a result of fish harvesting and the introduction of new species.

3. There are no strong indices of ecological instability. However, the apparent increase in temporally and spatially variable toxic algal blooms, and possibly increasing frequency and intensity of paralytic shellfish poisoning warrant monitoring.

4. Many areas of low to moderate metal and organic contamination exist in nearshore habitats associated with specific pollution sources, such as municipal and industrial outfalls and ocean dump sites. Bioaccumulation of contaminants is occurring, and being magnified in successive trophic levels. Major sources of organic contaminants (pesticides, PCBs, dioxins, tributyl tin) have been eliminated or controlled, but the contaminants persist in marine ecosystems throughout the Strait of Georgia.

5. Indications of disease prevalence, which may or may not be attributable to contaminants, include imposex in marine snails (high frequency), bill deformation in cormorant chicks (low frequency, but above background), and ideopathic lesions in bottom fish (confined to areas near pollution sources). There is no evidence that these effects at the individual or population levels of biological organization have affected ecosystem structure, but there is also no room for complacency.

6. Removal of large predators in several groups, including whales, bottom fish and salmon, may have altered the size spectrum of marine ecosystems, with unknown effects.

Based on these conclusions, I would say that, while the Strait of Georgia is diverse and biologically productive, its ecosystem has changed with European and Asian settlement. New processes that exert ecosystem control, such as restructured estuaries, altered weather patterns and large scale fish harvest, have been added to the natural processes that shaped the original ecosystem of the Strait of Georgia. Some structural changes, such as loss of humpback whales and changes in fish abundance, are evident. These changes will continue, with unknown consequences, unless action is taken to curtail them.

Acknowledgements

J. Elliott of the Canadian Wildlife Service, C. Levings and R. Beamish of Fisheries and Oceans Canada reviewed the manuscript. An earlier draft was also reviewed by Martin Pomeroy of Environmental Protection Service, Environment Canada.

Baird, R.W., K.M. Langelier and P.J. Stacey. 1989. First records of False Killer whales (*Psuedorcas crassidens*) in Canada. *Can. Field Nat.* 103(3):368-371.

Beamish, R.J., B.L. Thomson and G.A. McFarlane. 1992. Spiny dogfish predation on chinook and coho salmon and the potential effects on hatchery-produced salmon. *Trans. Am. Fish. Soc.* 121(4): 444-455.

Bellward, G.D., R.J. Norstrom, P.E. Whitehead, J.E. Elliott, S.M. Bandiera, C. Dworschak, T. Chang, S. Forbes, B. Cadario, L.H. Hart and K.M. Cheng. 1990. Comparison of polychlorinated dibenzodioxin levels in hepatic mixed-function oxidase induction in Great Blue Herons. *J. Toxicol. & Environ. Health* 30:33-52.

Black, E.A. 1991. B.C. Ministry of Agriculture, Fisheries and Food studies on the management of salmon farming algae problems. In: Pacific coast research on toxic marine algae. *Can. Tech. Rep. Hydrog. and Ocean Sci.* No. 135.

Brand, D. and D. Goyette. 1991. Ideopathic lever lesions from sole inhabiting Vancouver Harbour and from areas near the vicinity of pulp mills. In: *Proceedings: 17th Annual Aquatic Toxicity Workshop.* November 5-7, 1990. P. Chapman, F. Bishay, E. Power, K. Hall, L. Harding, D. McLeay, M. Nassichuk and W. Knapp (eds.). *Can. Tech. Rep. Fish. and Aquat. Sci.* No. 1774.

Brand, D.G. 1991. Neoplasia and biomarkers in fish and bivalves collected near pulpmills. In: *Proceedings: the 12th Annual Meeting of the Society of Environmental Toxicology and Chemistry.* November 3-7, Seattle WA.

Brinkhurst, R.O. and R.D. Kathman. 1987. Benthic studies in Alice Arm, B.C. during and following cessation of mine tailings disposal 1982 to 1986. *Can. Tech. Rep. Hydrog. Ocean Sci.* No.89.

Brinkhurst, R.O. 1987. Distribution and abundance of macrobenthic infauna from the continental shelf off southwestern Vancouver Island, British Columbia. *Can. Tech. Rep. Hydrog. Ocean Sci.* No. 85.

Burd, B.J. and R.O. Brinkhurst. 1990. Vancouver Harbour and Burrard Inlet benthic infaunal sampling program, October, 1987. *Can. Tech. Rep. Hydrog. Ocean Sci.* No. 122.

Burd, B.J., D. Moore and R.O. Brinkhurst. 1987. Distribution and abundance of macrobenthic infauna from Boundary and Mud Bays near the British Columbia/U.S. border. *Can. Tech. Rep. Hydrog. Ocean Sci.* No. 84.

Butler, R.W. and R.W. Campbell. 1987. *The Birds of the Fraser River Delta: Populations, Ecology and International Significance.* CWS Occasional Paper No. 65. Ottawa.

Chiang, R. 1991. Paralytic shellfish management program in British Columbia. In: Pacific coast research on toxic marine algae. (Introduction) Forbes, J.R. (ed.). 1991. *Can. Tech. Rep. Hydrog. and Ocean Sci.* No. 135.

Clifford, P.J., P.J. Harrison, K. Yin, M. St. John, A.M. Waite and L.J. Albright. 1988. Plankton and nutrient dynamics in the Fraser River plume, 1988. *U.B.C. Manuscript Report* No. 53.

Colodey, A.C. and P.G. Wells. 1992. Effects of Pulp and Paper Mill Effluents on Estuarine and Marine Ecosystems in Canada: A Review. *J. Aquatic Ecosystem Health* 1:201-226.

Cretney, W.J. and N.F. Crewe. 1991. History of dioxin and furan deposition in Howe Sound, B.C. from sediment cores. In: *Proceedings: 12th Annual Meeting of the Society of Environmental Toxicology and Chemistry.* November 3-7, Seattle WA.

Elliott, J.E. and D.G. Noble. 1992. Chlorinated hydrocarbon contaminants in marine birds of the temperate North Pacific, 1992. In: *Ecology and Conservation of Marine Birds of the Temperate North Pacific.* K. Vermeer and S. Causey (eds.). In press.

Elliott, J.E., D.G. Noble, R.J. Norstrom, P.E. Whitehead, M. Simon, P.A. Pearce and D.B. Peakall. 1992. *Patterns and Trends of Organic Contaminants in Canadian Seabird Eggs 1968-1990.* Canadian Wildlife Service.

Elliott, J.E., K. Langelier, A.M. Scheuhammer, P.H. Sinclair and P.E. Whitehead. 1992. Incidence of lead poisoning in Bald Eagles and lead shot from waterfowl gizzards from British Columbia. *Canadian Wildlife Service Progress Note* No. 200, June, 1992.

Elliott, J.E., R.W. Butler, R.J. Norstrom and P.E. Whitehead. 1989. Levels of polychlorinated dibenzodioxins and polychlorinated dibenzofurans in eggs of great blue herons (*Ardea herodias*) in British Columbia, 1983-1987: possible impacts on reproductive success. *Canadian Wildlife Service Progress Note* No. 176, April, 1988, Ottawa.

E.V.S. Environmental Consultants Ltd. 1990. *Iona Deep Sea Outfall Environmental Monitoring, 1990.* Report for Greater Vancouver Regional District, Vancouver, B.C. Vol. I and II.

Forbes, J.R. 1991. Introduction. *In:* Pacific coast research on toxic marine algae. J.R. Forbes (ed.). 1991. *Can. Tech. Rep. Hydrog. and Ocean Sci.* No. 135.

Fournier, J.A. and C.D. Levings, 1982. Polychaetes recorded near two pulp mills on the coast of northern British Columbia: a preliminary taxonomic and ecological account. *Syllogeus* 40:1.

Freeland, H. 1991. Sea surface temperature observations off Vancouver Island. In: Pacific coast research on toxic marine algae. J.R. Forbes (ed.). 1991. *Can. Tech. Rep. Hydrog. and Ocean Sci.* No. 135.

Goyette, D. and J. Boyd. 1989. Distribution and environmental impact of selected benthic contaminants n Vancouver Harbour, B.C. *Environment Canada Regional Program Report* 89-02.

Harding, L.E. and W.M. Pomeroy. 1990. Dioxin and furan levels in sediments, fish and invertebrates from fishery closure areas of coastal British Columbia. *Environment Canada Regional Data Report* 90-09.

Harding, L.E. and B. Kay. 1988. *Levels of Tributyl Tin (TBT) in Sediments and Oyster (Crassostrea gigas) Tissue.* Environmental Protection, Conservation and Protection, File Report (unpubl).

Harrison, P.J. 1985. Monitoring cutrients and plankton in the cicinity of the Iona Island sewage outfall. *Environment Canada Regional Manuscript Report* 85-04.

Harrison, P.J., J.D. Fulton, G. Miller, C.D. Levings, F.J.R. Taylor, T.R. Parsons, P.A. Thompson and D.W. Mitchell. 1984. A bibliography of the biological oceanography of the Strait of Georgia and adjacent inlets, with emphasis on ecological aspects. *Can. Tech. Rep. Fish and Aquatic Sci.* No. 1293.

Hart, L.E., K.M. Cheng, P.E. Whitehead, R.M. Shah, R.J. Lewis, S.R. Ruschkowski, R.W. Blair, D.C. Bennett, S.M. Bandiera, R.J. Norstrom and G.D. Bellward. 1991. Dioxin contamination and growth and development in Great Blue Heron embryos. *J. Toxicol. and Environ. Health* 32:331-344.

Hawkes, M.W. 1991. Conservation status of marine plants and invertebrates in British Columbia: are our marine parks and reserves adequate? In: *Proceedings: A Public Symposium on B.C.'s Threatened and Endangered Species.* Sept. 28-29, 1991, Vancouver.

Kaiser, G.W. and L.S. Forbes. 1992. Climatic and oceanographic influences on island use in four burrow-nesting alcids. *Ornis scandinavica* 23:1-6.

Kay, B.H. 1989. Pollutants in British Columbia's marine environment. *State of the Environment Fact Sheet* 89-2. Ottawa.

Ketchen, K.S., N. Bourne and T.H. Butler. 1983. History and present status of fisheries for marine fishes and invertebrates in the Strait of Georgia, British Columbia. *Can. J. Fish. Aquat. Sci.* 40(7):1095-1119.

Levings, C.D. 1991. Strategies for restoring and developing fish habitats in the Strait of Georgia-Puget Sound Inland Sea, Northeast Pacific Ocean. Proceedings: Environmental Management and Appropriate Use of Enclosed Coastal Seas - EMECS '90. *Mar. Poll. Bull.* 23:417-422.

Levings, C.D., R.E. Foreman and V.J. Tunnicliffe. 1983. Review of the benthos on the Strait of Georgia and contiguous fjords. *Can. J. Fish. Aquat. Sci.* 40:1120-1141.

Macdonald, R.W., D.M. Macdonald, M.C. O'Brien, and C. Gobeil. 1991. Accumulation of heavy metals (Pb, Zn, Cu, Cd), carbon and nitrogen in sediments from Strait of Georgia, B.C., Canada. *Marine Chemistry* 34:109-135.

Ministry of Environment. 1981. *Howe Sound Management Plan.*

Muir, D.C.G., C.A. Ford, B. Rosenberg, M. Simon, R.J. Norstrom and K. Langelier. 1991. PCBs and other organochlorine contaminants in marine mammals from the Strait of Georgia. In: *Proceedings: 12th Annual Meeting of the Society of Environmental Toxicology and Chemistry.* November 3-7, Seattle WA.

Noakes, D.J. and A. Campbell. 1992. Use of geoduck clams to indicate changes in the marine environment of Ladysmith Harbour, British Columbia. *Envirometrics* 3(1):81-97.

Norstrom, R.J., M. Simon, C. Macdonald, P.E. Whitehead, J.E. Elliott, D.C.G. Muir, C. Ford and K. Langelier. 1991. Food Chain Transfer and Sources of PCDDs and PCDFs in the Strait of Georgia Marine Ecosystem. In: *Proceedings: 12th Annual Meeting of the Society of Environmental Toxicology and Chemistry.* November 3-7, Seattle WA.

Parsons, T.R. 1992. The removal of marine predators by fisheries and the impact on trophic structure. *Mar. Poll. Bull.* 25(1-4):51-54.

Platt, T. and K. Denman. 1978. The structure of pelagic marine ecosystems. *Rapp. P.-v. Reun. Cons. int. Explor. Mer* 173:60-65.

Province of British Columbia and Environment Canada, 1993. *State of the Environment Report for British Columbia.*

Rapport, D.J. 1989. What constitutes ecosystem health? *Perspectives in Biology and Medicine* 33(1):120-132.

Saavedra Alvarez, M.M. and D.V. Ellis. 1990. Widespread neogastropod imposex in the Northeast Pacific: Implications for TBT contamination surveys. *Mar. Poll. Bull.* 21(5):244-247.

Stacey, P.J., R.W. Baird and K.M. Langelier. 1989. Stranded whale and dolphin program 1988. *Wildlife Veterinary Report* 2(1): 10-11.

Stockner, J.D. and D.D. Cliff. 1976. Phytoplankton recession and abundance in Howe Sound, British Columbia: A coastal marine embayment under stress. *Fish. Mar. Serv. Res. Dev. Tech. Rep.* 658:24 pp.

Stockner, J.G. and A.C. Costella, 1976. Field and laboratory studies on effects of pulpmill effluent on growth of marine phytoplankton in coastal waters of British Columbia. *Env. Prot. Serv. Rep.* PR-76-9:60 pp.

Stockner, J.G., D.D. Cliff and K.R.S. Shortreed. 1979. Phytoplankton ecology of the Strait of Georgia, British Columbia. *J. Fish. Res. Board Can.* 36:657-666.

Waldichuk, M. 1983. Pollution in the Strait of Georgia: A review. *Can. J. Fish. Aquat. Sci.* 40:1142-1167.

Waldichuk, M. and B. Smiley. 1991. *Exotic Introductions into B.C. Marine Waters.* Unpublished Information Note. DFO, Sidney, B.C.

Walters, C.J. and P. Cahoon, 1985. Evidence of decreasing spatial diversity in British Columbia salmon stocks. *Can. J. Fish. Aquat.* Sci. 42:1033-1037.

Ware, D.M. 1978. Bioenergetics of pelagic fish: theoretical change in swimming speed and ration with body size. *J. Fish. Res. Res. Board Can.* 35:220-228.

Wells, P.G. and S.J. Rolston, 1991. *Health of Our Oceans: A Status Report on Canadian Marine Environmental Quality.* Environment Canada, Dartmouth and Ottawa.

Whitehead, P.E., R.J. Norstrom and J.E. Elliott. 1992. Dioxin levels in eggs of Great Blue Herons (*Ardea herodias*) decline rapidly in response to process changes in a nearby kraft pulp mill. Extended abstract. *Dioxin '92*, Tampere, Finland.

Chapter 22
Urban Biodiversity

Urban environments suffer tremendous losses in biodiversity through a loss of natural habitat. Some 20-30% of the land surface is commonly paved, and much of the remainder is covered by buildings. Also, native plant species are removed and replaced by exotics. These two major factors cause a loss of ecosystem diversity, and result in extensive local extinctions of native species (see Thomas, 1984; Heath, 1981). With the world population expected to almost double to 10 billion people by the year 2050, and most of the population growth expected to occur in cities, losses in urban biodiversity are an important issue (Brown and Jacobson, 1987).

Furthermore, the expansion of cities into the natural environments along their periphery is a major factor contributing to global losses in biodiversity. Cities are relatively new to the biosphere, with the first appearing 5,000 years ago. The modern urban landscape with its extensive blacktop and lawns only became common after the second World War due to the prevalence of the automobile and rapid population growth. Urban impacts on the natural landscape are great. For example, about 2% of the entire land base of the United States is composed of lawns (Garber, 1987).

Urban areas are synonymous with a loss of biodiversity (Murphy, 1988). A number of forces are at work besides the direct removal of natural habitats. Reductions in biodiversity can also occur from air or water–borne pollutants, the overdrafting of local aquifers, and changes in groundwater flows caused by storm drains. Pets and livestock may prey on native animals, and spread diseases and parasites among more vulnerable native species. Introduced animals such as the European Starling (*Sturnus vulgaris*), House Sparrow (*Passer domesticus*), Norway rat (*Rattus norvegicus*), and house mouse (*Mus musculus*), may become more abundant than native species. Introduced plant species, such as the scotch broom (*Cytisus scoparius*) and purple loosestrife (*Lythrum salicaria*), can outcompete and displace the local flora. All these factors can disrupt local ecosystem functions, or cause them to collapse.

Despite their highly artificial composition, urban areas are not biological deserts. The city can be a complex habitat suitable for many species, with its buildings creating unique environments of temperature, moisture, and acidity (Oke, 1976). Gardens, yards, parks, and recreation areas establish significant areas of green space with linkages which can support a surprising array of plant and animal life (Laurie, 1979). Species and population diversities can be high, although in many cases they involve introduced species, and they rarely form fully functioning ecosystems.

In British Columbia, the two best known urban centres are those of Greater Victoria and the Lower Mainland. They serve as examples of the state of and future trends in urban biodiversity in British Columbia.

307

Biodiversity in Greater Victoria

It is difficult to measure biodiversity and no single descriptive statistic can accurately assess the complexity of the ecosystems, species, and gene pool characteristics which contribute to biodiversity (Pielou, 1973). However, in urban areas one indirect general indicator of biodiversity is a series of land use categories which reflect the broad vegetation characteristics of a site. It is reasonable to expect that animal diversity would be correlated with plant diversity and that these land use categories therefore reflect biodiversity as a whole. The spatial distribution of these land use categories can be mapped, as is shown for Greater Victoria in Figure 22-1. This map covers the more densely populated areas of the Capital Regional District—Victoria, Oak Bay, Saanich, View Royal, Esquimalt, Colwood, Langford, and Metchosin.

The land use categories range from Class 1 to Class 6. Included in Class 1 are the remaining unique and intact ecosystems like ecological reserves and nature sanctuaries which have a level of biodiversity very similar to other natural environments removed from the influence of the city. There is an ecological reserve at Ten Mile Point protecting an intertidal area with 42 species of algae and 55 species of invertebrates. Trial Islands contain 28 rare vascular plants, including the golden paintbrush *(Castilleja levisecta)*, Nuttall's quillwort *(Isocetes nuttalli)* and Macoun's meadow foam *(Limnanthes macounii)*. The Oak Bay Islands ecological reserve protects nesting seabird colonies and uncommon meadow communities. The Swan Lake-Christmas Hill Nature Sanctuary contains wet

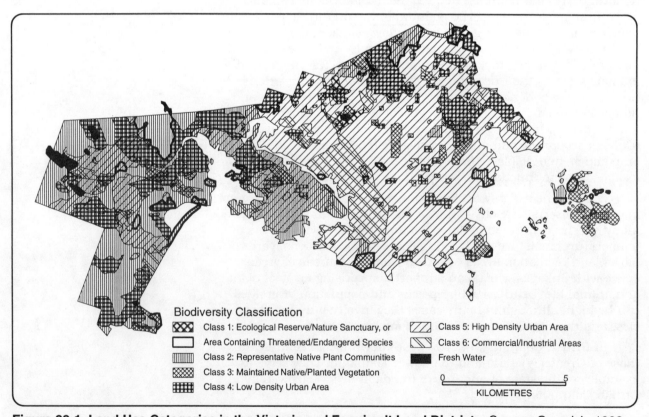

Biodiversity Classification

- Class 1: Ecological Reserve/Nature Sanctuary, or Area Containing Threatened/Endangered Species
- Class 2: Representative Native Plant Communities
- Class 3: Maintained Native/Planted Vegetation
- Class 4: Low Density Urban Area
- Class 5: High Density Urban Area
- Class 6: Commercial/Industrial Areas
- Fresh Water

0   5
KILOMETRES

Figure 22-1. Land Use Categories in the Victoria and Esquimalt Land Districts. Source: Czernick, 1993.

meadows, a Douglas-fir *(Pseudotsuga menziesii)* forest, a large eutrophic lake, and a tributary of the Colquitz River (Czernick, 1993).

Much of the native vegetation remaining in the Capital Regional District is contained in the regional parks. The Capital Regional District has 17 parks covering 3,000 hectares. Almost half of this area is in East Sooke Park. The Capital Regional District itself covers an area of 2,031 square kilometres (1,491 in Sooke).

The most disturbed sites are found in Class 5 (high density urban) and Class 6 (commercial/industrial). These sites would have little, if any, biodiversity. As might be expected, the map shows that there is little undisturbed habitat remaining. However, around the periphery, large areas are still covered by representative native plant communities (Class 2). In the Greater Victoria area there are 10 native plant communities recognized by McMinn et al. (1973), and they are listed in Table 22-1. They are becoming more and more fragmented by the expanding city. Scattered throughout the housing (Classes 4 and 5) are significant patches of green space, consisting of native or planted vegetation under management in parks and

Scientific Name	English Name
Quercus garryana	Garry oak
Quercus garryana-Arbutus menziesii	Garry oak-Arbutus
Arbutus menziesii-Pseudotsuga menziesii	Arbutus-Douglas-fir
Arbutus menziesii-Pseudotsuga menziesii-Pinus contorta	Arbutus-Douglas-fir-lodgepole pine
Gaultheria shallon-Berberis aquifolium	Salal-Oregon grape
Polystichum minitum-Gaultheria shallon	Sword fern-Salal
Polystichum minitum	Sword fern
Populus balsamifera-Pyrus fusca-Salix sitchensis	Black cottonwood-Crabapple-willow freshwater wetland saltwater wetland

Table 22-1: Native Plant Communities Found Around Greater Victoria. Source: McMinn et al. 1973.

sanctuaries (Class 3). Unfortunately, the Class 5 and 6 lands are concentrated along the foreshore, where natural ecosystems are often the most productive.

These observations demonstrate that, within the City of Victoria, there is a tremendous loss of natural habitat, and a loss of ecosystem diversity as a consequence. However, species and population diversities persist in the Class 1 and 2 lands, and they can still be surprisingly high in the Class 3, 4 and 5 lands. In Victoria, these latter three classes represent 924 species and subspecies of plants. Five hundred and ninety-one of these species and subspecies are indigenous and 333 are introduced (Szazawinski, 1973). More than 250 species of birds also occur here (for comparison, Stanley Park in Vancouver has some 230 bird species). Some 50 species of birds (of which 6 are introduced) are typical of highly developed urban areas and can be found in the Class 5 and 6 lands.

The Garry Oak Community

In some cases, the remaining ecosystem diversity in a city may be sufficiently intact to pursue protection and restoration of the native plant communities. Greater Victoria is an excellent example. It is located in the part of the Coastal Douglas-fir Zone with a cool summer, which gives it a Mediterranean climate unique to only 1% of Canada. According to Robert Prescott-Allen (Young, 1990), Victoria is in the heart of one of the most endangered ecosystems in North America. It has a unique combination of trees, with Garry oak (*Quercus garryana*) and arbutus (*Arbutus menziesii*) growing along with the Douglas-fir and broadleaf maple (*Acer macrophyllum*). This ecosystem is predominant on southeastern Vancouver Island in dry coastal regions and down the west coast as far south as California. Many wildflowers occur in this community, especially those with bulbs. Common are the common camas *(Camassia quamash)* and Hooker's onion *(Allium acuminatum)*.

The Garry oak is only one of several threatened communities in the area. Table 22-2 lists selected others, plus individual plant species endangered in the Victoria region (B.C. Conservation Data Centre). These plant species represent potentially important populations, which may well contribute adaptations and genetic diversity to the future survival of their species.

Biodiversity in the B.C. Lower Mainland

Large, relatively intact ecosystems can exist in urban areas, though usually along the periphery. The Fraser River estuary and coast mountains outside Vancouver are renowned for their biodiversity. The estuary is, for example, an important stopover point for birds migrating along the Pacific Flyway from as far away as Alaska in the north to Chile and Argentina in the south. Because of its role in the Pacific Flyway, many people believe that the estuary should be recognized under the United Nations Ramsar Convention on the Conservation of Wetlands of International Importance.

There are serious threats to natural ecosystems in the Greater Vancouver Regional District (GVRD) from population growth and urban sprawl. In response to this concern the Greater Vancouver Regional District and others are working to create a Green Zone that identifies and in some way protects remaining green space (GVRD, 1992a).

The extent and value of natural ecosystems in the Lower Mainland were recently reviewed under the Liveable Region Strategic Plan (GVRD, 1992b, 1992c). Approximately half of the land area within the Greater Vancouver Regional District was considered to be important in maintaining ecological functions. The north shore lands contained within the Greater Vancouver Water District and the estuary of the Fraser Delta formed the major portion of the remaining natural ecosystems. However, other important areas included river valleys, wetlands, forests, and bogs. Fifty-five rare vascular plants found in the Lower Mainland were identified, based on Straley et al. (1985) (GVRD, 1992c). Wildlife at risk was also identified. There are five endangered or threatened species (Red-listed),

Species or Community		Global Rank	Local Rank
Scientific Name	English Name		
Quercus garryana-Bromus sterilis	Garry oak-Barren brome	X	1
Abies grandis-Mahonia nervosa	Grand fir-Oregon grape	X	1
Abies grandis-Tiarella trifoliata	Grand fir-foam	X	1,2
Pseudotsuga menziesii-Gaultheria shallon	Douglas fir-Salal	X	1,2
Alnus rubra-Carex obnupta	Red alder-Slough sledge	X	1,2
Quercus garryana-Holodiscus disco	Garry oak-ocean spray	X	2
Thuja plicata-Oemleria cerasiform	Western redcedar-Indian plum	X	2
Montia howellii		1	1
Castilleja levisecta		1	
Aster curtis	White-top Aster	3	2
Juncus kelloggii		3	1
Callitriche marginata		4	1
Centaurium muhlenbergii		4	1
Orthocarpus castillejoides		4	1
Sliene scouleri grandis		4	1
Psilocarphus tenellus	slender woolly-heads	4	1
Orbanche corymbosa mutabilis		4	2
Progne subis		5	1
Sanicula arctopoides	Footsteps of spring	5	1
Sanicula bipinnatifida	Purple snake-root	5	1
Balsamorhiza deltoidea	Deltoid balsam root	5	1
Lotus formosissimus		5	1
Trifolium depauparatum		5	1
Montia dichotoma		5	1
Ranunculus alismaefolius		5	1
Salix lemmonii		5	1
Orthocarpus bracteosus		5	1
Alopecurus carolinianus		5	1
Cardamine angulata		5	2
Montia fontana	Blinks	5	2
Ranunculus californicus	California buttercup	5	2
Orthocarpus hispidus		5	2
Psilocarphus elatior		X	1

Table 22-2: Examples of Threatened or Endangered Species or Plant Communities in the Victoria Region with Global and Local Rarity Rankings. Source: B.C. Conservation Data Centre. The rank of 1 represents the most threatened or endangered, and 5 the least. The Victoria region has high local rankings for many species, two of which also have a high global ranking.

and 18 sensitive or vulnerable species (Blue-listed), based on an assessment by BC Environment. The endangered or threatened species are: long-eared Keen's myotis *(Myotis keenii)*, Townsend's big-eared bat *(Plecotus townsendii)*, Spotted Owl *(Strix occidentalis)*, Marbled Murrelet *(Brachyramphus marmoratus)*, and sharptail snake *(Contia tenuis)* (GVRD, 1992c).

Altogether, the GVRD (1992b) study considered the natural areas of Greater Vancouver to form four large, interrelated ecosystems: North Shore Systems, Coastal/Intertidal Systems, Fraser River Systems, and Fraser Lowland Systems. A map showing the locations of these ecosystems is contained in the GVRD (1992b) study.

The North Shore System forms a 50 kilometre uninterrupted band of western hemlock *(Tsuga heterophylla)*, western red cedar *(Thuja plicata)*, and Douglas-fir forest from Howe Sound to the Pitt River. It contains the Capilano, Seymour, and Coquitlam watersheds, which are estimated to be 40% old growth forest (trees of more than 140 years).

The Coastal/Intertidal Systems include the many marshes and mudflats around Burrard Inlet (Maplewood Mudflats, Port Moody Mudflats, Mossom Creek, and Noons Creek in particular), the Fraser River foreshore, and Roberts and Sturgeon Banks. Fed by the rich supply of silt from the Fraser River, they play an important role as rearing and nursery habitat for salmon *(Oncoryhnchus spp.)*, act as a spawning ground for Pacific herring *(Clupea harengus)*, support many shellfish species, and provide food for migrating birds.

The Fraser River System is the lower part of the Fraser River watershed, which drains an area of 233,000 square kilometres. About 20 million tonnes of silt are brought to the Lower Mainland each year. Each year about four million salmon and trout *(Salmo spp.)* use the Fraser River as a migration route.

The Fraser Lowland System has shallow water tables and contains bogs and wetlands. Burns Bog, in particular, is an "exceptional ecological treasure" (Hebda, 1991). At about forty square kilometres, Burns Bog is one of the largest urban wildernesses in the world, as well as the largest domed peat bog on the west coast of the Americas. Its sphagnum moss/acid bog based plant community is the southernmost location for subarctic plant species, such as cloudberry *(Rubus chamoemorus)*, crowberry *(Empetrum nigrum)*, and velvetleaf blueberry *(Vaccinium myrtilloides)* (Hebda, 1991). The bog also provides a haven for such species as the threatened Sandhill Crane *(Grus canadensis)*. Streams in the Fraser Lowland System are used by salmon and trout. The open fields are important for hawks, owls and waterfowl. The eastern part of the Fraser Lowland System also contains extensive forest habitat which supports a diverse number of birds and mammals.

The areas in which these ecosystems were identified by the GVRD (1992b, 1992c), are primarily on the periphery of urban development or in parks. Within the urban areas of Greater Vancouver fairly natural ecosystems can be found in the larger parks, such as Stanley Park, Pacific Spirit Regional Park, Central Park, and Burnaby Lake Regional Park. The Greater Vancouver Regional

District park system includes 16 parks and one park reserve, together covering 9,404 hectares (an additional 456 hectares may soon be added).

However, the study suggested that smaller significant natural ecosystems occur within the urban areas and outside of parks as well. They are more difficult to locate and require more detailed study to assess their value. Until recently urban biodiversity has not been valued and has been poorly protected. The current realization that nature in cities contributes to the quality of urban life is changing attitudes. There is a movement to identify, protect, and enhance remaining natural urban environments.

The beauty and solitude afforded by urban green space are important to people. However, such areas also play an important role as wildlife corridors, nurseries of biodiversity, and moderators of the local environment, particularly air quality, wind and water movement, and water storage and filtration (Watson et al., 1981; P. Whitfield, Environment Canada, unpubl. data). Although urban green spaces individually appear small and insignificant, collectively they have a major impact on a city and its surrounding wilderness areas, and can represent a surprisingly high degree of biodiversity. Urban ravines are an excellent example—they act as wildlife corridors and have a rich plant and animal life.

Ravines—Unique Natural Urban Environments in the Lower Mainland of B.C.

Steep slopes have in many cases spared ravines and canyons from development. These areas can act as refugia for the original native vegetation of an area. Well known examples of such areas are Rock Creek Park in Washington, D.C., and Fairmont Park in Philadelphia (Murphy, 1988). The ravines and canyons represent hidden community assets by providing relatively high biodiversity in the heart of residential development.

A study of urban ravines in the Lower Mainland of B.C. (Schaefer et al., 1992), documented 140 ravines with a total area of over 15 square kilometres within urban limits. About one-third of the ravines occur in parks. The remainder are mainly in private or municipal land (and therefore may be destroyed by development), with a few being a mixture of park land and some other category. The ravines occur along the escarpments of the Burrard Peninsula, the Fraser River, and Boundary and Mud Bays. A few also occur on the flatlands along major creeks.

Some of the ravines were only selectively logged in the past and retain a few of the original trees. Several of the ravines contain stands of very large Douglas-fir, western red cedar, and western hemlock. Individual examples greater than 0.7 metres in diameter are common, and some are more than one metre in diameter, despite the fact that just over the tops of the banks might be a thriving, high density urban community. The map of an urban ravine shown in Figure 22-2 illustrates how it is indeed an isolated environment.

313

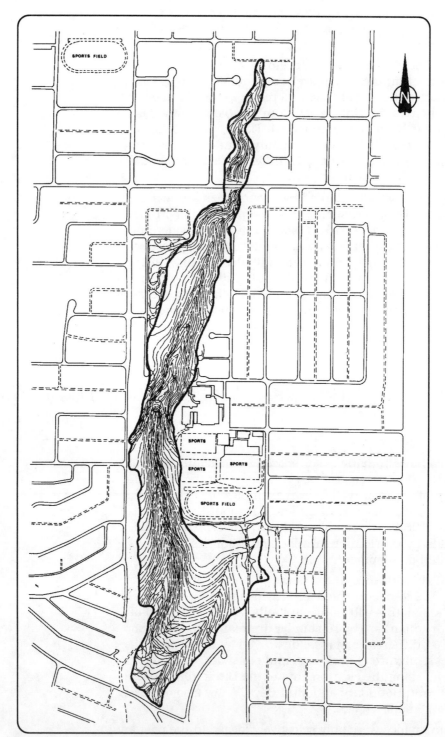

Figure 22-2. An Urban Ravine in Greater Vancouver, British Columbia. Source: Schaefer et al., 1992.

In several urban ravines, the herb layer has already changed in composition, with only sword ferns *(Polystichum munitum)*, remaining dominant. The canopy is sufficiently closed to exclude salal *(Gaultheria shallon)* and Oregon grape *(Berberis* spp.), clearly indicating an early climax community sere (stage in plant community development).

Along the floor of many ravines by the stream or creek would be a rich community of skunk cabbage (*Lysichiton kamtschatcense*), salmonberry (*Rubus spectabilis*), and mature broadleaf maple adorned with licorice fern (*Polypody* spp.) in the spring. The rich insect life associated with the bark of mature broadleaf maples, the berries of the salmonberry and red elderberry (*Sambucua racemosa*) in the understory, and the standing snags scattered along the ravine floor all contribute to making ravines excellent wildlife habitat.

The ravine creates a unique stream environment. Figures 22-3a and 22-3b show changes in several stream characteristics as the water passes through a ravine. In this example, the ravine is about one kilometre long. The top of the ravine is site 6, and the bottom site 1. Site 5 is at the mouth of a large storm drain. The shade afforded by a ravine significantly lowers the temperature of the water from urban runoff, making it more suitable for fish (most creeks in the area are part of the salmonid enhancement program). The water is also oxygenated by the turbulence caused by the high proportion of rocks and cobbles in the ravine (Schaefer et al., 1992). All things considered, although ravines exist as isolated fingers of green space in the city, they represent tremendous value in terms of biodiversity.

Threats to Urban Biodiversity

Urban biodiversity is not always lost through development. Fragmentation effects may ultimately be just as serious at the level of the microhabitat (Wilcox and Murphy, 1985). In cities, it is common to remove the understory plants in parks and housing developments. The understory contributes greatly to moderating temperature, humidity, light availability, and wind exposure. It is also an important food source and habitat for many bird species (Dean, 1976; Laurie, 1979).

Biodiversity can also be threatened by local storm drain systems which use the creeks. Chemical pollutants can enter these creeks from street runoff. The potential for such chemical pollution is high when industry occurs in the watershed. For example, during the course of the ravine study (Schaefer et al., 1992) a sudsing agent was accidentally introduced into the storm drain system of Byrne Creek Ravine, causing a metre of foam to form on the creek. There were also many instances when the creek was filled with silt due to construction in the water-

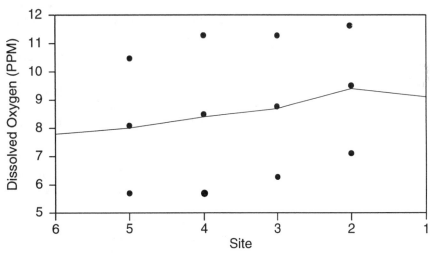

Figure 22-3a. Water Quality Results for Byrne Creek: Changes in Dissolved Oxygen Levels by Site. Site 6 is at the top of the ravine; Site 1 is at the bottom of the ravine; points indicate mean and ± one standard deviation. Source: Schaefer et al., 1992.

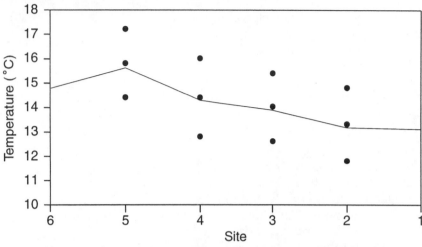

Figure 22-3b. Water Quality Results for Byrne Creek: Changes in Temperature by Site. Site 6 is at the top of the ravine; Site 1 is at the bottom of the ravine; points indicate mean and ± one standard deviation. Source: Schaefer et al., 1992.

shed. Any salmon or trout eggs in the shallows would have been quickly smothered.

Changes in drainage patterns in the community also affect biodiversity. For example, increased blacktop in the watershed contributes to floods in a ravine during periods of rain—less water percolates into the ground and more is collected into the storm drains emptying into the creek (Watson et al., 1981; P. Whitfield, Environment Canada, unpubl. data). In the summer, when the creek relies on groundwater flows, the water levels may drop because the reduced infiltration due to blacktop has lowered groundwater supplies. The low water levels directly lead to a loss of suitable habitat for fish, and the reduced volume of water means that there is less turbulence and thus less dissolved oxygen, indirectly threatening fish life.

Protecting Urban Biodiversity

Preserving biodiversity in urban areas is very difficult. There are many agencies and interests involved. The land is usually privately owned and expensive. Politicians usually favour development over protection. Also, conservation agencies usually do not have a presence in urban areas. Local natural history clubs and other citizens groups, if informed of the value of biodiversity, can play an important role in applying political pressure, and even inspiration, to protect natural ecosystems in urban areas in spite of these difficulties.

Even narrow rows of trees and shrubs should be protected because they may represent valuable wildlife corridors between large patches of habitat. They may appear to be insignificant to the health of local ecosystems, but their destruction may lead to species losses, and the compounded effects of secondary extinction events (Gilbert, 1989).

Biodiversity can also be protected by reducing the number of exotic species planted and, instead, encouraging the use of native vegetation in landscaping. Organizations such as the Urban Wilderness Gardeners already promote this practice (Savage, 1987). An additional benefit to using native vegetation is that it requires fewer pesticides and fertilizers, and less water. Native vegetation can also be transplanted to help restore a disturbed site, particularly if the native vegetation would have been destroyed by development anyway. Such a transplant occurred this year in Victoria when a meadow community was taken from Broadmead to the Christmas Hill-Swan Lake Nature Sanctuary (Cowley, 1992).

Other options for preserving biodiversity in urban areas involve private landowners. A program could be implemented to evaluate the biodiversity of an owner's site (O'Connell and Noss, 1992). Decisions on development could then be made with this consideration in mind.

Owners of private land may also be required to maintain biodiversity. Tree protection bylaws are an option that is being considered in Burnaby, B.C. Under such bylaws an owner wishing to remove a tree on private property must seek the approval of the municipality. The District of Coquitlam has classified all ravines as ecologically sensitive lands, has included them in the district's green space inventory, and requires a permit for someone to cut a tree on private property.

Finally, a particularly important site may be protected by purchasing it from the owner and placing it in a land trust (Elfring, 1989).

Whatever the means of protection chosen, it is imperative that communities take steps to preserve urban biodiversity before it is overwhelmed by the relative uniformity of development.

Acknowledgements

Sarah Groves, Minicipality of Burnaby, reviewed the manuscript.

References

Brown, L. and J. Jacobson. 1987. *The Future of Urbanization: Facing the Ecological and Economic Constraints*. Worldwatch Paper #77, Worldwatch Institute, Washington, D.C.

Cowley, J. 1992. Conservationist and developer cooperate to move meadow in Broadmead. *Victoria Naturalist* 48:4.

Czernick, G. 1993. A biodiversity map for Greater Victoria. In: *Biodiversity in Greater Victoria*. V.H. Schaefer (ed.). Environmental Studies, University of Victoria. Victoria, B.C.

Dean, P.B. 1976. Wildlife needs and concerns in urban areas. In: *Ecological (Biophysical) Land Classification in Urban Areas*. E.B. Wiken and G.R. Ironside. Fisheries and Environment, Canada.

Elfring, C. 1989. Preserving land through local land trusts. *Bioscience* 39:71-74.

Garber, S.D. 1987. *The Urban Naturalist*. John Wiley and Sons Inc., Toronto, Ontario.

Gilbert, O.L. 1989. *The Ecology of Urban Habitats*. Chapman and Hall, New York, New York.

Greater Vancouver Regional District. 1992a. *Creating Greater Vancouver's Green Zone Conference: Proceedings*. Development Services, Burnaby, B.C.

Greater Vancouver Regional District. 1992b. *Greater Vancouver's Ecology. Volume 1. Summary Report*. Development Services, Burnaby, B.C.

Greater Vancouver Regional District. 1992c. *Greater Vancouver's Ecology. Volume 2. Technical Report*. Development Services, Burnaby, B.C.

Heath, J. 1981. *Threatened Rhopalocera (Butterflies) in Europe*. Council of Europe, Strasbourg, France.

Hebda, R.J. 1977. *The Paleoecology of a Raised Bog and Associated Deltaic Sediments of the Fraser River Delta.* PhD Thesis, Department of Botany, University of British Columbia. Vancouver, B.C.

Hebda, R.J. 1991. Burns Bog: vegetation and future. *Discovery* 20(1):13-16.

Laurie, I.C. (ed.). 1979. *Nature in Cities.* John Wiley and Sons Inc., Toronto, Ontario.

McMinn, R.G., S. Eis and E. Oswald. Native vegetation. In: *An Inventory of the Land Resources and Resource Potentials in the Capital Regional District.* C.V. Stanley-Jones and W.A. Benson (eds.). 1973. pp. 59-75.

Murphy, D. 1988. Challenges to biodiversity in urban areas. In: *Biodiversity.* E.O. Wilson (ed.). National Academy Press, Washington, D.C.

Oke, T. 1976. The significance of the atmosphere in planning human settlements. In: *Ecological (Biophysical) Land Classification in Urban Areas.* E.B. Wiken and G.R. Ironside (eds.). Fisheries and Environment, Canada.

O'Connell, M.A. and R.F. Noss. 1992. Private land management for biological conservation. *Environmental Management* 16(4):435-450.

Pielou, E.C. 1973. *Ecological Diversity.* Wiley-Interscience, New York, New York.

Savage, J. 1987. Greening the city. *Canadian Heritage* Aug/Sept:31-35.

Schaefer, V.H., M. Aston, M. Bergstresser, N. Gray, J. and P. Malacarne. 1992. *Urban Ravines (Vol. 1). Byrne Creek Ravine—A Case Study. (Vol. 2). B.C. Lower Mainland Urban Ravines Inventory.* Douglas College Institute of Urban Ecology, New Westminster, B.C.

Schaefer, V.H. (Ed.) 1993. *Biodiversity in Greater Victoria.* Environmental Studies, University of Victoria, Victoria, B.C.

Straley, G.B., R.C. Taylor and G.W. Douglas. 1985. The Rare Vascular Plants of B.C. *Syllogeus,* No. 59.

Szazawinski, A.F. 1973. *Flora of the Saanich Peninsula.* B.C. Provincial Museum Handbook.

Thomas, J.A. 1984. The conservation of butterflies in temperate countries: Paste efforts and lessons for the future. In: *The Biology of Butterflies.* R.I. Vane-Right and P.R. Ackery (eds.). Academic Press, London.p p. 333-353.

Watson, V.J., O.L. Loucks and W. Wojner. 1981. The impact of urbanization on seasonal hydrologic and nutrient budgets of a small North American watershed. *Hydrobiologia* 77:87-96.

Wilcox, B.A. and D.D. Murphy. 1985. Conservation strategy: The effects of fragmentation on extinction. *American Naturalist* 125(6):879-887.

Young, C. 1990. Your own abode. *Monday Magazine* 16 (48):1.

Lee E. Harding
Canadian Wildlife Service
P.O. Box 340
Delta, B.C.
V4K 3Y3

About 250 species of birds breed in North America and winter in the forests of the tropics. These "neotropical migrants" include many of British Columbia's songbirds. Evidence that songbirds throughout North America are declining comes from the annual Breeding Birds Survey and from meteorological radar records. The annual Breeding Birds Survey, organized by dedicated amateur bird watchers and professional ornithologists, has been conducted by volunteers throughout North America since 1965. The North American report on results from 1965 to 1979 gave no indication that neotropical migrants were decreasing, but the report on results from 1978 to 1987 documented declines in 75% of the species of forest-dwelling long-distance migrants (Robbins, 1989; cited by Terborgh 1992). Other studies have corroborated these results (Collins and Wendt, 1989; Terborgh, 1987).

Gauthreaux (1992) used radar records from stations along the coast of the Gulf of Mexico to monitor flights of small birds on their spring migration from their tropical wintering areas. Large numbers of birds migrating north from South and Central America and the Antilles wait on the Yucatán Peninsula for good weather, and then head north to the gulf coast, where they are detectable by radar. Gauthreaux reported that only half as many waves of migratory birds crossed over the Gulf in the late 1980s as in the 1960s.

Such dramatic declines throughout North America must include British Columbia's migratory bird population; unfortunately, there are very little hard data with which to assess population trends in forest birds. In British Columbia, breeding bird surveys are conducted at 25 locations near cities where there are interested volunteers, leaving most of the province unsurveyed. Nevertheless, even with these limited data, Erskine et al. (1990) have found statistically significant, long term (1968-1988) declines in Northern Flickers (*Colaptes auratus*), Swainson's Thrushes (*Catharus usulatus*), Yellow Warblers (*Dendroica petechia*), Chipping Sparrows (*Spizella paserina*), and Dark-eyed Juncos (*Junco hyemalis*) in southern B.C. Savard (1991) discussed this issue for the Fraser River Basin. He noted that over 60% of the passerines (perching birds or songbirds) breeding within the Fraser Basin winter in the southern U.S., Central America or South America. Some examples are most species of warblers and vireos, Eastern Kingbirds (*Tyrannus tyrannus*), Western Wood Peewees (*Contopus sordidulus*), Black-headed Grosbeaks (*Pheucticus melanocephalus*), Western Tanagers (*Piranga ludoviciana*) and Swainson's Thrush (*Catharus ustalatus*). The B.C. provincial government has red-listed the Cape May Warbler (*Dendroica tigrina*), Connecticut Warbler (*Oporornis agilis*), Canada Warbler (*Wilsonia canadensis*), Yellow-Breasted Chat (*Icteria virens*), Yellow-billed Cuckoo (*Coccyzus americanus*), and Purple Martin (*Progne subis*), all of which breed in the tropics.

There are at least two reasons for declines in neotropical migrant bird numbers: tropical deforestation of their winter range and fragmentation of summer breeding habitat. At current rates of tropical forest loss, estimated at between 142,000 and 200,000 square kilometres per year, the last hectare of tropical forest will fall between 2030 and 2045 (Terborgh, 1992). Diamond (1986) expressed "direct and particular concern" about the "imminent loss of winter habitat for 90 species of bird that breed in Canada and migrate to Latin America for the winter." He noted that, "some of the species concerned are known to be of potential economic importance as controllers of forest insect pests." Their decline may be relevant to recent forest insect pest irruptions noted in Chapter 19.

Fragmentation of summer breeding habitat combines with tropical deforestation to affect neotropical migrant populations more than other bird species (such as resident or short distance migrants). Terborgh (1992) has shown that the long-distance migrants tend to arrive on the breeding ground later and to depart earlier. Hence, they are able to make fewer nesting attempts and are more vulnerable to nest predation and parasitism (when one bird lays its eggs in the nest of another species). Also, many long-distance migrants have smaller clutch sizes and are, therefore, less able to recover from population pressures. Terborgh (1992), Wilcox and Murphy (1985) and others have shown that forest habitat fragmentation on breeding ranges greatly increases nest predation and parasitism, and is responsible, in combination with tropical deforestation, for large declines in these species in North America, particularly in the Southeastern United States.

Though no such studies have been carried out in British Columbia, many of the factors are the same. For example, the main species associated with nest parasitism, the Brown-headed Cowbird (*Molothrus ater*), occurs everywhere except in the northeast corner of British Columbia (Figure 23-1). Terborgh (1992) points out that, since cowbirds forage alongside cattle, scavenging seeds from their dung, they have benefitted from "the creation of open agricultural habitats in formerly forested sections of the country." Cowbirds also eat insects attracted to or stirred up by the grazers (Godfrey, 1966). Terborgh (1992) took some comfort in the "expansive virgin forests of the Northwest" as the last bastion of habitat for neotropical migrants. British Columbia contains some of the largest tracts of virgin forest remaining in North America. However, habitat fragmentation (Chapter 18, this volume) already affects more than a third of British Columbia, roughly coinciding with the area occupied by the Brown-headed Cowbird. While there is no information on the effects of cowbird parasitism in British Columbia, the fact that cowbirds are widely distributed over large parts of the province in fragmented ecosystems is cause for concern. It shows that the same features of altered environments that contribute to songbird declines elsewhere in North America are present in British Columbia.

Given that some British Columbian songbirds are known to be declining, a question that needs to be asked is: what is the role of these birds in maintaining forest ecosystem health? For example, will the loss of these species reduce consumption of insects and cause an increase in the spread of forest pest infestations? What other birds or other insect predators will expand to fill the ecological

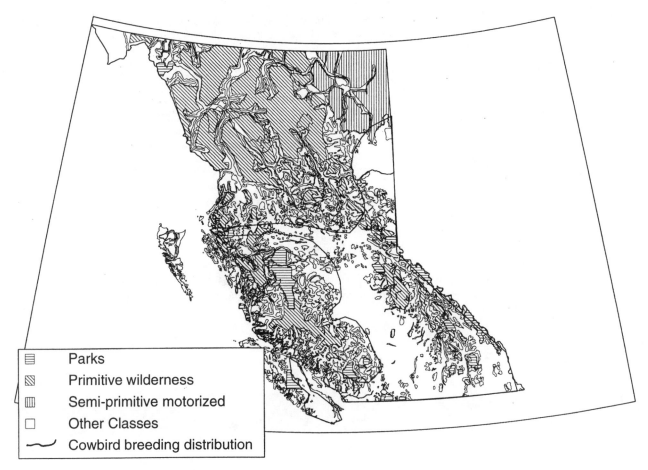

Figure 23-1. Breeding Distribution of the Brown Headed Cowbird in British Columbia Overlaid on a Base Map of Roaded and Unroaded (Primitive and Pristine) Areas. This bird's known breeding distribution closely matches the area of most severe ecosystem fragmentation, represented by the white areas on the base map. Source: draft map prepared for *Birds of British Columbia Vol. 3* (in prep.), provided by R. Wayne Campbell; Base Map from Vold, 1992.

The legend reads:
- Parks
- Primitive wilderness
- Semi-primitive motorized
- Other Classes
- Cowbird breeding distribution

niche vacated by the birds? Will avian and mammalian bird predators, such as Gyrfalcons (*Falco rusticolus*) and pine martens (*Martes americana*, member of the weasel family), decline because of the loss of their prey species? The Canadian Wildlife Service has recently expanded its nongame bird research program to obtain a better understanding of the degree and consequences of loss of neotropical migrant and other forest birds.

Acknowledgements

Kathy Martin, Canadian Wildlife Service, reviewed the manuscript.

References

Collins, B.T. and J.S. Wendt. 1989. *The Breeding Bird Survey in Canada, 1966-1983: Analysis of Trends in Bird Populations.* Canadian Wildlife Service Technical Report. No. 75. Ottawa, Ontario.

Diamond, A.W. 1986. *An Evaluation of the Vulnerability to Changes in Neotropical Forest Habitats.* Canadian Wildlife Service Regional Report, Ottawa, Ontario.

Erskine, A.J., B.T. Collins and J.W. Chardine. 1990. *The Cooperative Breeding Bird Survey in Canada, 1988.* Progress Notes No. 188. Canadian Wildlife Service, Ottawa, Ontario.

Gauthreaux, S.A. Jr. 1992. J. M. Hagan, D.W. Johnston (eds.). *Proceedings: Symposium on Ecology and Conservation of Neotropical Migrant Land Birds.* Smithsonian Institution Press.

Godfrey, W.E. 1966. *The Birds of Canada.* National Museum of Canada Bulletin No. 203, Biological Science No. 73.

Savard, J.P. 1991. Birds of the Fraser Basin in sustainable development. In: *Perspectives on Sustainable Development in Water Management: Towards Agreement in the Fraser River Basin.* A.H.J. Dorcey (ed.). University of British Columbia. pp. 189-215.

Terborgh, J. 1987. *Where Have All The Birds Gone? Essays on the Biology and Conservation of Birds That Migrate to the American Tropics.* Princeton University Press.

Terborgh, J. 1992. Why American songbirds are vanishing. *Scientific American* 266(5):98-104.

Wilcox, B.A. and D.D. Murphy. 1985. Conservation strategy: The effects of fragmentation on extinction. *American Naturalist* 125(6):879-887.

Chapter 24
Atmospheric Change in British Columbia

Atmospheric environmental changes in British Columbia may result from two separate but related phenomena: climate change and depletion of the stratospheric ozone layer. Climate change, the main feature of which is global warming, is caused by the accumulation of gases like water vapour, carbon dioxide, and methane, which trap the heat of the sun next to the earth in much the same way that the glass of a greenhouse does. While too much ozone in the lower atmosphere contributes to air pollution, too little of it in the upper atmosphere (the stratosphere), allows too much ultraviolet radiation to reach the surface of the earth. Much more is known about climate change, so we will discuss this phenomenon and its potential impact on British Columbia first, and then close with a brief look at ozone depletion.

Lee E. Harding
Canadian Wildlife Service
P.O. Box 340
Delta, B.C.
V4K 3Y3

Eric Taylor
Environment Canada
700-1200 West 73rd Avenue
Vancouver, B.C.
V6P 6H9

Climate change

Gases that contribute to the greenhouse effect include: carbon dioxide, methane, and nitrous oxide. These gases occur naturally, but human activities such as the burning of fossil fuels and changing land uses are leading to increased concentrations of these gases. Significant, worldwide increases in these greenhouse gases have been well documented. Atmospheric carbon dioxide is now 25% higher than in pre-industrial times, and methane has fully doubled in concentration (Houghton et al., 1992). In recent decades, the concentration of carbon dioxide has increased at unprecedented rates, as shown in Figure 24-1, and the other greenhouse gases have similarly increased worldwide (CCPB, 1991).

In 1989, an Intergovernment Panel on Climate Change, established under the auspices of the World Meterological Organization and the United Nations Environment Program, published a major report documenting climate change (Houghton et al., 1992). Most atmospheric scientists now agree that greenhouse gas emissions will continue to increase for the next several decades. Carbon dioxide is currently expected to double within the next few decades (CCPB, 1991). The magnitude and speed of climate change are still uncertain, particu-

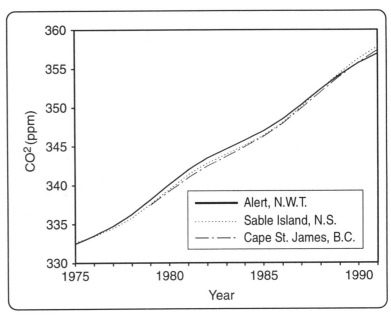

Figure 24-1. Increases in Carbon Dioxide Concentrations at Three Canadian Sites (1975-1990). Source: CCPB, 1991.

323

larly since global conventions limiting greenhouse gas emissions may slow the process somewhat. But even if global reductions in emissions of greenhouse gases were achieved today, their concentrations in the atmosphere would still increase until well into the next century because of the long lag times in the physical, chemical and biological processes controlling them. This increase in greenhouse gases would likely cause surface warming, increased global precipitation (mainly because warmer air holds more moisture), higher sea levels, and changes in snow and ice cover. These effects will be more pronounced toward middle and high latitudes, as in Canada.

Primary Effects of Climate change

Predictions

The primary effects of climate change will be on temperature and precipitation. Global Circulation Models (GCM) are computer simulations of future temperature and precipitation regimes under various scenarios of greenhouse gas levels. The GCM developed and used by Environment Canada is among the world's most sophisticated; its predictions for temperature and precipitation changes in different areas of British Columbia, based on a doubling of carbon dioxide, are shown in Figures 24-2 - 24-5. This degree of change is expected around the middle of the next century, although the timing is uncertain. McBean et al., 1991, considering the results of several GCMs, also using a scenario of doubled carbon dioxide levels, suggested the following general effects on British Columbia's climate:

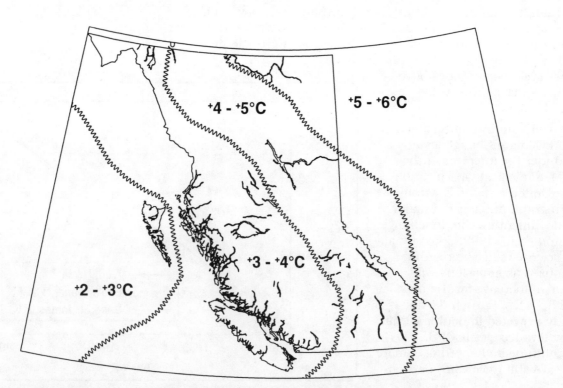

Figure 24-2. Global Climate Model Prediction for Winter Temperature Changes Under a Carbon Dioxide Doubling Scenario. Source: CCPB, 1991.

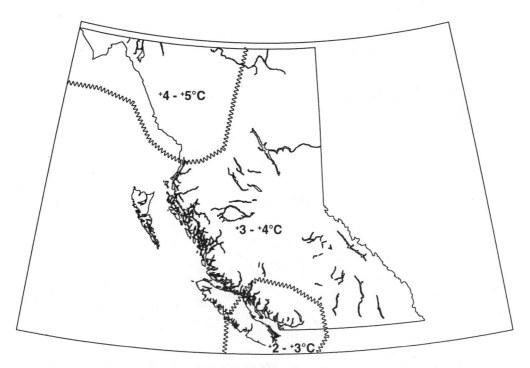

Figure 24-3. Global Climate Model Prediction for Summer Temperature Changes Under a Carbon Dioxide Doubling Scenario. Source: CCPB, 1991.

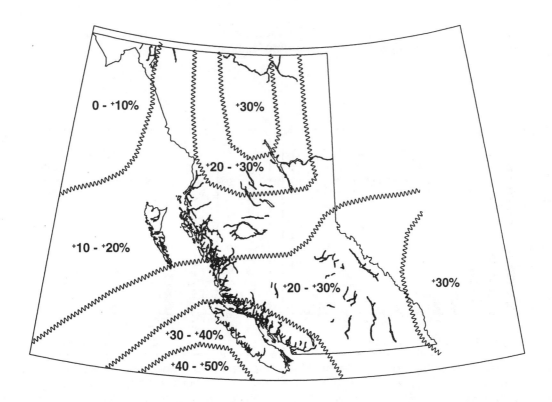

Figure 24-4. Global Climate Model Prediction for Winter Precipitation Changes Under a Carbon Dioxide Doubling Scenario. Source: CCPB, 1991.

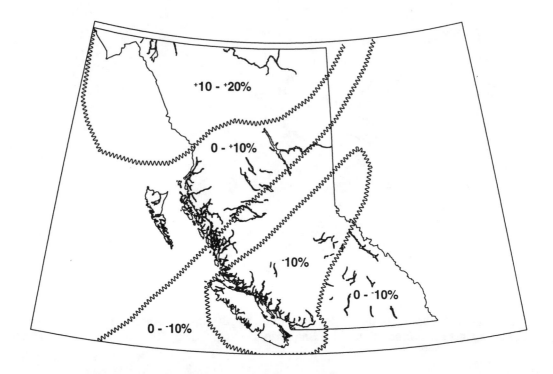

Figure 24-5. Global Climate Model Prediction for Summer Precipitation Changes Under a Carbon Dioxide Doubling Scenario. Source: CCPB, 1991.

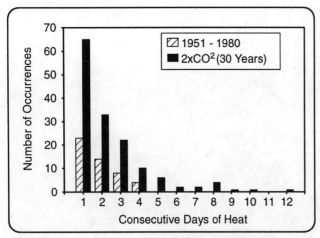

Figure 24-6a. Vancouver Region Heat Waves. Heat waves are defined as those consecutive days with a July temperature greater than 20° Celsius. With a doubling of carbon dioxide over the next 30 years, the occurence of heat waves is predicted to increase as shown by the black bars. Source: Hengeveld, 1989.

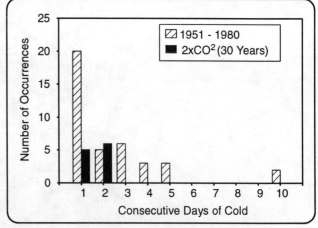

Figure 24-6b. Vancouver Region Cold Spells. In this figure, cold spells are defined as those consecutive days with a January temperature less than -2.8° Celsius. With the doubling of carbon dioxide over the next 30 years, the occurrence of cold spells is predicted to decrease as shown by the black bars. Source: Hengeveld, 1989.

326

o   a winter temperature increase of 7°C, with an uncertainty of ±3°C, though this increase would probably be 1-2°C lower in the coastal areas;

o   a summer temperature increase of 4°C, with an uncertainty of ±3°C;

o   a winter precipitation increase of about 0.5 millimetres per day in winter, with possibly double that along the coastal mountains; and

o   a summer precipitation increase which would be smaller and might even be negative in places like the interior.

These are continental-scale models which, unfortunately, do not yet have sufficient resolution to precisely predict the many variations arising from British Columbia's complex topography.

Such global annual averages do not take into account the extremes of weather on smaller spatial and temporal scales. Hengeveld (1989) modelled climate change in the Pacific region of Canada, based on a doubling of carbon dioxide. He predicted that the frequency and duration of heat waves (consecutive days of greater than 20°C) in the Vancouver region will increase, and that the frequency and duration of cold spells (consecutive days of less than -2.5°C) will decrease (Figures 24-6a and b). Extreme events such as these may elicit more profound ecological responses than subtle average changes.

Observations

Natural variations in long term temperature and precipitation patterns can be caused by: periodic events such as El Niño, a warming of the eastern equatorial Pacific Ocean which occurs every 4-5 years; unusual events, such as the eruption of Mount Pinatubo in the Phillipines, which has had a worldwide cooling effect; and unknown periodic and non-periodic geologic, atmospheric and oceanographic processes (Gullet and Skinner, 1992). Climate change due to greenhouse gases, if it occurs, may either compound or reverse current trends, depending on the direction of current change. Recent observation of long term changes in weather patterns, whatever their cause, are relevant to a discussion of biodiversity because of the profound effect that climate has on ecosystems.

Average global temperatures rose 0.3°C to 0.6°C during the last century, and more sharply during the past two decades (Houghton et al., 1992). Gullett and Skinner (1992) observed increasing temperature trends across Canada. They concluded that, though the causes of these trends cannot be definitely established, they "are consistent with predictions of warming resulting from a human-induced buildup of greenhouse gases—and, indeed, most of the world's leading scientists see this as the most important single factor." It is important, however, to remember that changes in temperature and precipitation may be caused by a number of factors in addition to—or even other than—climate change.

Though Gullett and Skinner (1992) considered the Canada-wide temperature trends significant, they found that B.C.'s average annual temperature increases were not statistically significant. The largest increases were in the northern mountains, followed by the southern mountains and lastly by the coast, where the temperatures are moderated by the ocean. Moore (1991) also found that annual temperatures in northern B.C. (Fort St. James) warmed during this century.

However, annually and areally averaged data often do not detect seasonal and local changes. For example, seasonal (winter and spring) warming trends are evident in data from the southern and northern interior, though not from the coast. Taylor (1990) spatially averaged monthly temperature data recorded since 1916 for six southern interior climate stations (Summerland, Kamloops, Princeton, Cranbrook, Golden and Kaslo). He found that winter temperatures (January through March) have warmed by about 2.3°C, significantly more than the annual temperature increase of about 0.7°C (Figures 24-7a and b). Raphael (1993) found that temperatures in Prince George have increased over the period of record (1943-1991), with the greatest warming occurring in spring (Figure 24-8). On the coast, however, Taylor (unpubl. data) found no significant changes in temperatures, measured at Vancouver.

Precipitation also shows long term trends. When analyzed by season, precipitation in the southern interior (Summerland) appears to have decreased slightly during fall months and increased at other seasons since 1916 (Taylor, unpubl. data). However, precipitation on the south coast, measured at Vancouver over the last four decades, has definitely been increasing and is most apparent in winter (Figure 24-9). McBean and Thomas (1991) noted that, "In many respects, the most important weather element in terms of its impact on ecosystems and human activities is precipitation. .... Changes to available water, in regions where there are already water supply problems, may be catastrophic."

Secondary Effects of Climate change

Predictions

The secondary effects of climate change will be on hydrology and may include rising sea levels and sea surface temperatures, and changes in ocean currents, river hydrology, snowpack depths, and rates of glacial movement.

The sea level is predicted to rise 0.65 metres by 2100 (Houghton et al., 1992). Dunn (1989) suggested that, in British Columbia, such a rise would probably cause coastal erosion and flooding, permanent inundation of low gradient, intertidal marshes, and increased intrusion of salt water into estuaries.

Changes in river hydrology, snow packs and glacial melt rates due to any climate change would be difficult both to predict and to detect because they depend on the climate regimes in each part of each river basin. For example, increased winter precipitation in the warmer parts of a basin, or in coastal mountains, may fall as rain and increase winter runoff, leaving a lower snowpack which would result in lower summer runoff. The annual precipitation average may not change, but the ecological consequences of increased flooding in winter and lower water flows in summer could be profound. Conversely, the same precipitation increase in the colder

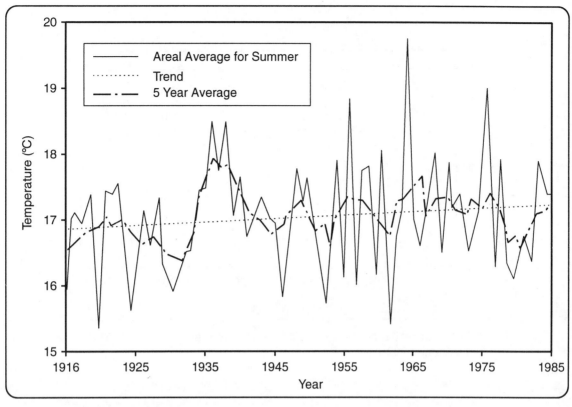

Figure 24-7a. Summer Temperature in the Southern Interior (1916-1985). Increasing trends are statistically significant. Source: Taylor, 1990.

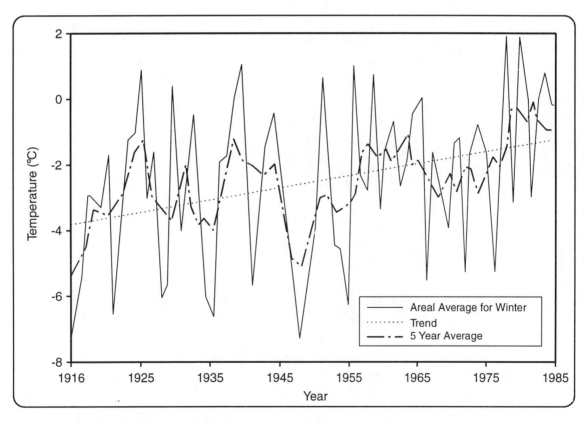

Figure 24-7b. Winter Temperature in the Southern Interior (1916-1985). Increasing trends are statistically significant. Source: Taylor, 1990.

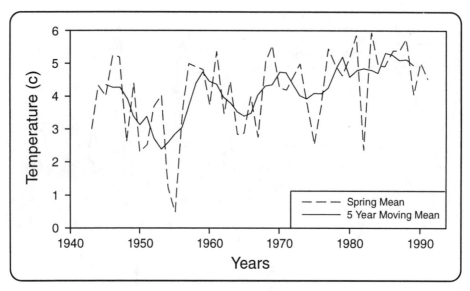

Figure 24-8. Spring Temperature Trends at Prince George (1943-1991).
Source: Raphael, 1993.

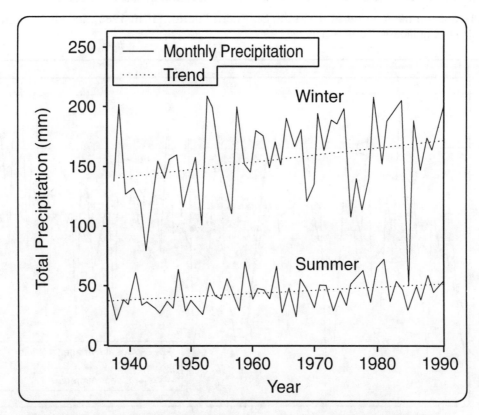

Figure 24-9. Total Monthly Precipitation on the Coast (1937-1990). The increasing winter trend is statistically significant. Source: E. Taylor, Environment Canada, unpubl. data.

parts of the basin may fall as snow and add to the snowpack, thereby increasing summer runoff. Similarly, increasing summer temperatures may cause some moderate climate glaciers to retreat more rapidly, while increases in winter precipitation, falling as snow, may cause others in colder regions to advance.

Oerlemans and Fortuin (1992) have shown that, because they receive significantly more rain and less snow as the temperature increases, glaciers in wetter regions are more sensitive to climate change that those in drier regions. Oerlemans and Fortuin (1992) conclude that, overall, "even a significant increase in precipitation cannot compensate for increased melting, and further shrinkage of glaciers...must occur in a warmer climate."

Martinec et al. (1992) modelled the effects of temperature and precipitation changes on seasonal flow in the Illecilleweat River basin in the Columbia Mountains of British Columbia, an area of very heavy snowfall and numerous glaciers. They predicted that, while precipitation increases during the snow accumulation periods will increase the total flow of melted water if significant warming occurs, temperature increases will cause the timing of melted water runoff to advance from summer to early spring (Figure 24-10). Moore (1991) made a similar prediction for the Fraser River basin: in the southern interior, flow increases would be at least partially offset by increased evaporation. These effects may worsen summer water supply problems in areas of existing water shortage, and may stimu-

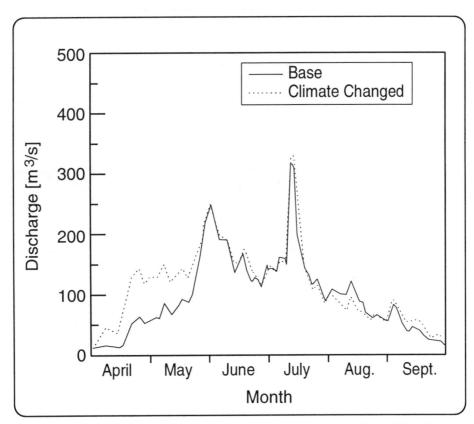

Figure 24-10. Computer Projection of Shifts in Meltwater Runoff in the Illecillewaet Basin in the Columbia Mountains of British Columbia, based on a scenario of 3° Celsius increase in annual temperature. Source: Martinec et al. (in press.).

late pressure to increase water storage in reservoirs. Places like the Okanagan which lack significant water storage capacity would be hit hardest by such climate changes. On the coast, the higher winter and spring water flows would likely cause more landslides and floods, particularly from "rain on snow" events (CCPB, 1991).

Observations

As with temperature and precipitation, the observed hydrological changes may be due to factors other than climate change. For example, on the west coast, very small long term sea level changes on the order of millimetres per year have been attributed to geological plate movements, rather than to climate change (personal communication on June 9, 1992, with F. Stephenson, Institute of Ocean Sciences). On the other hand, Freeland (1991) found a trend in increasing sea surface temperatures that parallels temperature increases in the northern hemisphere (Figure 24-11). Though this trend is only marginally significant because the data is highly variable, it is worth noting that virtually all sea surface temperature measurements since about 1977 have been above normal for the 60 year period of record. This sort of uncertainty over cause and effect means that considerable time and careful study will be needed to determine to what degree hydrological changes are attributable to climate change.

Slaymaker (1991) put some thought into how to detect the effects of climate change on the water balance. He determined that the water balance should be modelled and the following parameters included: precipitation, snow cover, available soil moisture,

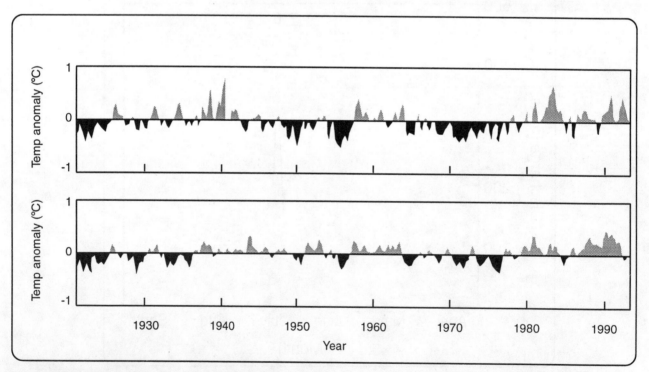

Figure 24-11. Comparison of Sea Surface Temperature Anomalies at Race Rocks on the British Columbia Coast (upper panel) with Northern Hemisphere Air Temperature Anomalies (lower panel).
Source: Howard Freeland, Institute of Ocean Sciences, Sidney, B.C.

evapotranspiration, storm events (magnitude, frequency, duration), surface runoff, glacier ice, and groundwater. Hydrological impacts to look for, he thought, would include changes in water quantity, water quality, magnitude of peak events (both high and low flows), and freshet flows. To date, no such analysis has been completed for most parts of B.C.

However, several hydrologists (Kite, 1992; Leith, 1991; Moore, 1991; Coulson, 1989) have analyzed time series snowpack and/or water flow data for B.C. rivers. Unfortunately, they all looked at annually averaged data, which could mask the seasonal differences that may be more significant under climate change scenarios. Kite (1992) found no trends, but the others found trends which were more or less consistent with climate change predictions. For example, Leith (1991) and Coulson (1989) found that snowpack and water flows have tended to increase in northern stations and decrease in southern interior stations. Moore (1991) found that snowpacks and water flows in the Fraser Basin have been decreasing since about 1976.

Since the 1800s, glaciers in the Coast Mountains, the Rockies and the Purcells have generally been receding (Osborn and Luckman, 1988). The rapid rates of recession that prevailed from 1920-1950 have decreased markedly in the last few decades, and some glaciers have advanced (Osborn and Luckman, 1988). Whether these recent changes are in any way related to climate change is unknown.

Ecosystem Effects of Climate Change

Climate change may occur faster than ecosystems can adapt. Models predict that the yearly temperature change will be the equivalent of a shift of several thousand meters southward in latitude (CCPB, 1991). A study in eastern North America showed that trees migrated northward at the end of the last glaciation at a rate of 100-400 metres per year (Davis, 1981; Johnson and Adkisson, 1986)—roughly 10% as fast as the predicted rate of temperature change over the next century. The mountainous character of much of this province will mean that the trees will not have as far to migrate because they can move upslope as well as northward. However, such barriers to natural migration as freeways, farms and housing development were not impediments 12,000 years ago. During previous migrations, birds such as jays and nutcrackers carried seeds of the nut-bearing trees northward, but these birds do not readily cross open land (Perry and Borchers, 1990). Therefore, changing climate will not immediately result in movement of whole ecosystems, but only of those generalized and opportunistic components able to rapidly colonize new environments, like the introduced species discussed in Chapter 17.

Franklin et al. (1991) have modelled the likely consequences of average annual temperature increases of 2.5°C and 5.0°C on forests on the east (dry) and west (moist) sides of the Cascade Range of central Oregon (Figure 24-12). Their predictions may be applicable to the Cascade Range in British Columbia in the area around Manning Provincial Park east to Keremeos (dry) and west to Hope (moist). In general, they show a diminution of forest and alpine zones, and creation and expansion of shrub and grassland zones.

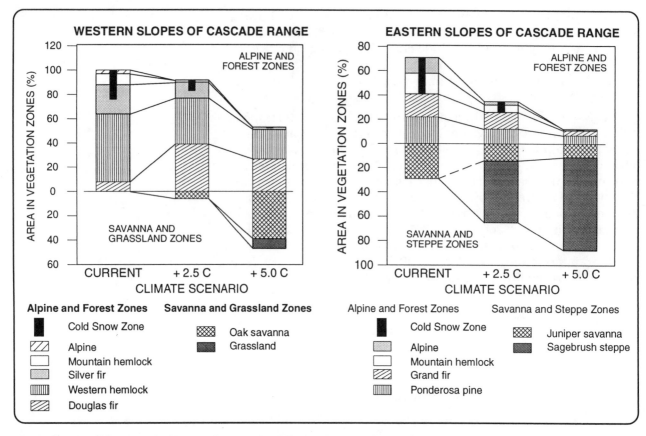

Figure 24-12. Changes in Vegetation as the Climate Warms in the Cascade Range. Source: Franklin et al, 1991.

Similarly, Rizzo and Wiken (1989) predicted that a doubling of carbon dioxide levels would cause the boreal forest of the B.C.'s northeast corner and the Peace River district to shift to grassland. And Woo et al. (1992) predict that the zone of discontinuous permafrost, which coincides roughly with the boreal forest zone, will move northward out of the province entirely.

Levy (1992) felt that climate change would adversely affect salmonids such as trout and salmon. He noted that, as ectotherms (cold-blooded organisms), fish are sensitive to changes in water temperature. While overall fish production may increase from the effects of global warming, fish distributions would likely change, species composition would shift and invasions by exotic species would alter British Columbian fish communities. Salmon are cold-water species (as opposed to warm-water species such as bass and catfish) and can spawn only in a narrow range of temperatures. Moreover, because salmon migrate to spawning grounds and lay eggs in gravel in shallow water, they are very sensitive to water level changes, including both increases (as may occur in winter) which would increase the scouring of gravels where eggs are deposited, and decreases (predicted for late summer and autumn) which may expose spawning gravel and make upstream migration difficult. Levy (1992) concluded that most salmonid species would be affected negatively by climatic warming, particularly those which rely on freshwater habitats near the southern margin of their geographical

334

range for juvenile rearing. These effects would compound existing temperature and water level changes caused, for example, by hydro-electric dams and by agricultural and urban water withdrawals.

Some components of ecosystems may be more affected by the climatic extremes than by the gradual changes resulting from climate change. Heat waves and consequent drying through evapotranspiration will adversely affect plants that require cool, moist conditions, of which there are many in British Columbia, some endangered or threatened (Chapter 10, this volume). Fewer cold spells will mean that many plants and animals vulnerable to freezing, including a wide variety of insect pests, will survive where they have not before. Borden (1991) gave three examples: the pinewood nematode (*Bursaphenelenchus xylophilus*), the mountain pine beetle (*Dendroctonus ponderosae*), and the watermould fungus (*Phytophthora cinnamomi*). The pinewood nematode can only kill trees south of the 20°C isotherm, which will progress north and upslope as temperatures warm. The mountain pine beetle is limited by the climate of the Pacific Northwest to one generation per year, but will probably manage two generations with only a slightly earlier spring. Borden (1991) mentioned that, "It is possible we are seeing this effect already, for in British Columbia in the last 15 years, the beetle has frequently violated the low and moderate hazard zones established for this species." Pollard (1988) also notes that there have been new infestations of mountain pine beetles in the Nass and Skeena Valleys, hitherto low hazard areas. The watermould fungus requires a currently rare combination of soil moisture and soil temperatures, which may occur much more frequently if the climate warms. Bergvinson (1988) studied the effect of climate on the introduced green spruce aphid (*Elatobium abietinum*) and concluded that climatic changes brought about by the 'greenhouse' effect may increase the aphid outbreak frequency. These examples support the hypothesis that the trends in forest insect pest outbreaks discussed in Chapter 19 may be at least partly related to the recent winter warming and summer drying trends.

Many authors (Pollard, 1991a; Franklin et al., 1991) have emphasized the uncertainty in such predictions, and the certainty that there will be surprises. Not all the changes caused by climate change will be bad for British Columbian ecosystems. Pollard (1991b) and others have pointed out that increased carbon dioxide levels may stimulate plant growth, and areas not stressed by lack of water will become more productive, provided that catastrophic disturbances can be managed. Such changes as more degree-days, longer frost-free periods, northward-shifting isotherms, and less winter snow accumulation will benefit many plant and animal species, as well as fortuitously-located farmers, ranchers and tree farm operators. Climatic conditions suitable for Douglas-fir, a tree species most valued for its lumber, are likely to expand. Relative to adjacent regions—Pacific Northwest forests to the south of British Columbia where many tree species are already at the southern limits of their physiological tolerance for temperature and water stresses, and the sagebrush/grasslands of the Great Basin—British Columbia is in a favoured position to weather the coming changes. The challenge for society and its government institutions will be to adapt to changes in resource abundance and distribution while maintaining ecological diversity.

Ozone Depletion

Depletion of the stratospheric ozone layer is caused by anthropogenic (human-caused) pollutants, such as chlorofluorocarbons (CFCs), and natural disturbances, such as volcanic eruptions. Ozone (O_3)forms when oxygen molecules (O_2) are dissociated by ultraviolet radiation and the resulting oxygen atoms combine with other oxygen molecules. Chlorine atoms from CFC compounds break up the ozone molecule by taking one of the oxygen atoms, leaving oxygen and chlorine monoxide (ClO). Then the chlorine monoxide reacts with a free oxygen atom (formed by the photodissociation of another ozone molecule) and liberates more oxygen and the chlorine atom, which can initiate the cycle again. Because each chlorine atom can go on breaking down ozone molecules again and again, ozone depletion will continue for long after we have ceased discharging CFCs. The ozone layer insulates the earth from the sun's ultraviolet radiation, and even a small loss is serious for plants and animals, including humans, that are damaged by this radiation.

Predictions

Exposure to ultraviolet radiation has a range of effects on vertebrates, including skin cancer, immune deficiency and a variety of eye diseases such as cataracts (UNEP, 1991a). It may also cause physiological stress to terrestrial and marine plants (UNEP, 1991a); hence the concern that global primary productivity of the earth's polar seas may be affected (Toon and Turco, 1991). Since marine phytoplankton are a major sink for atmospheric carbon dioxide, any reduction in these plants would decrease the uptake of carbon dioxide and so exacerbate the greenhouse effect of climate change (UNEP, 1991a).

Borden (1991), discussing the potential effects of ozone depletion on British Columbia forests, notes that ultraviolet radiation may reduce the photosynthetic capacity of trees and other plants, disrupt growth, cause gene mutations, and kill cells. He suggested that these effects, plus the additional stress from lack of water and temperature, may put forest trees at very high risk for pest infestations.

Observations

Ozone depletion progresses during the late winter and spring and recovers in summer and fall. Independent satellite and ground-based measurements have confirmed that atmospheric (mainly stratospheric) ozone has decreased in both the southern and northern hemispheres at middle and high latitudes, mainly in later winter and spring (UNEP, 1991b). The recent trends in stratospheric ozone levels over Edmonton, the nearest monitoring station to B.C. with a long term record, show the same pattern (Figure 24-13). A recent study has definitively shown that ultraviolet radiation is increasing over Canada as a result of ozone thinning (Kerr et al., 1993).

336

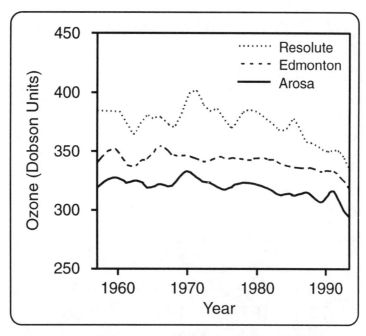

Figure 24-13. Total Ozone Levels over: Resolute Bay, Northwest Territories, Canada; Edmonton, Alberta, Canada; and Arosa, Switzerland (1955-1993). Source: J. Kerr, C. McElroy and D. Wardle, Atmospheric Environment Service, Environment Canada.

There is no information on the effects of increased ultraviolet radiation on wildlife and other fauna in B.C. However, Elwood et al. (1985) and Holman et al. (1986) showed that exposure to the ultraviolet radiation in sunlight increases the risk of malignant melanoma in humans and Gallagher et al. (1989) showed that chronic exposure increases the risk in British Columbia. Gallagher et al. (1990) found significant increases in three kinds of skin cancer (basal cell carcinoma, squamous cell carcinoma and melanoma) in British Columbians between 1973 to 1987 (Figures 24-14 and 24-15). Melanoma, a relatively uncommon cancer, has been increasing in British Columbians faster than all other cancers (McBride et al., 1989). These cancers are characterized by a long latency period, which suggests that they resulted from exposure to ultraviolet radiation many years ago, long before depletion of the ozone layer was significant. Any increases as a result of ozone depletion would tend to exacerbate these recent trends. All weather offices in Canada now report daily levels of ultraviolet radiation through Environment Canada's Ultraviolet Advisory Program.

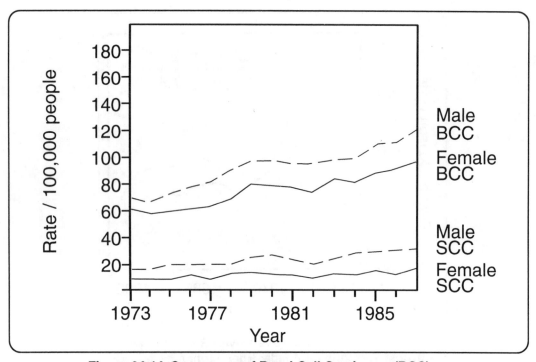

Figure 24-14. Occurrence of Basal Cell Carcinoma (BCC) and Squamous Cell Carcinoma (SCC): Age-Standardized Rate per 100,000 population. Source: Gallagher et al., 1989.

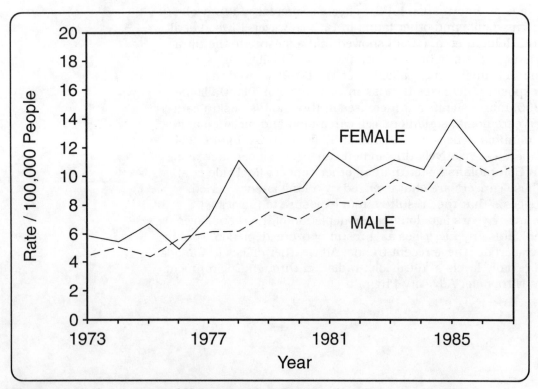

Figure 24-15. Occurrence of Melanoma: Age-Standardized Rate per 100,000 population. Source: Gallagher et al., 1989.

Acknowledgements

Information on climate change was provided by the following people: Henry Hengeveld of the Atmospheric Environment Service (Downsview, Ontario), April Ingraham, and Eric Taylor of the Atmospheric Environment Service (Vancouver, B.C.); Cliff Raphael of the College of New Caledonia (Prince George); Ralph T. Roberts and Albert Rango of the U.S. Department of Agriculture (Beltsville, MD); and Gordon McBean of the University of British Columbia. The manuscript was reviewed by: Rick Williams of the British Columbia Ministry of Agriculture, Fisheries and Food; Hal Coulson of the British Columbia Ministry of Environment, Lands and Parks; Phil Comeau, Jim Crover and Evelyn Hamilton of the British Columbia Ministry of Forests; Cliff Raphael of the College of New Caledonia; Douglas Pollard of the Pacific Forestry Centre; Richard Hebda of the Royal British Columbia Museum; and A.C. Allaye-Chan and Gordon McBean, formerly of the University of British Columbia (now with Environment Canada).

References

Bergvinson, D. 1988. *The Green Spruce Aphid, Elatobium abietinum, (Homoptera: Aphididae): A Review of its Biology, Control and Status in British Columbia.* Unpublished M.Sc. Thesis, Simon Fraser University, Burnaby, B.C.

Borden, J.H. 1991. *Ozone Depletion, Global Warming and the Potential Impact of Forest Pests.* President's Lecture Series, Simon Fraser University, Burnaby, B.C.

Canadian Climate Program Board (CCPB) 1991. *Climate Change and Canadian Impacts: the Scientific Perspective.*

Coulson, H. 1989. The impact of climate variability and change on water resources in British Columbia. *Symposium on the Impacts of Climate Variability and Change on British Columbia.* Environment Canada. pp. 41-50.

Davis, M.B. 1981. Quaternary history of the stability of forest communities. In: *Forest Succession.* D.E. Reichle (ed.). New York: Springer Verlag.

Dunn, M.W. 1989. Sea level rise and implications to coastal British Columbia: an overview. *Symposium on the Impacts of Climate Variability and Change on British Columbia.* pp. 59-76.

Elwood, J.M., R.P. Gallagher, J. Davidson and G.B. Hill, 1985. Sunburn, suntan and risk of cutaneous malignant melanoma: Western Canada Melanoma Study. *Br. J. Cancer* 51: 543-549.

Franklin, J.F., F.J. Swanson, M.E. Harmon, D.E. Perry, T.A. Spies, V.H. Dale, A. McKee, W.K. Ferrell, J.E. Means, S.V. Gregory, J.D. Lattin, T.D. Schowalter and D. Larsen. 1991. Effects of global climate change on forests in northwestern North America. *Northwest Environmental Journal* 7: 233-254.

Freeland, H. 1990. Sea surface temperature observations off Vancouver Island. In , Pacific coast research on toxic marine algae. J.R. Forbes (ed.). *Can. Tech. Rep. Hydrog. and Ocean Sci.* No. 135.

Gallagher, R.P, J.M. Elwood and C.Paul Yang 1989. Is chronic sunlight exposure important in accounting for increases in melanoma incidence? *Int. J. Cancer* 44: 813-815.

Gallagher, R.P., B. Ma, D.I. McLean, C. Paul Yang, V. Ho, J.A. Carruthers and L.M. Warshawski 1990. Trends in basal cell carcinoma, squamous cell carcinoma and melanoma of the skin from 1973 through 1987. *J. Amer Academy of Dermatology* 23(3): 413-421.

Gullet, D.W. and W.R. Skinner. 1992. The state of Canada's climate: temperature change in Canada 1985-1991. *Environment Canada SOE Report 92-2.*

Hengeveld, H. 1989. Future climate scenarios for Pacific Canada. In: *Proceedings of the Symposium on the Impacts of Climate Variability and Change on British Columbia.* E. Taylor and K. Johnstone (eds.). Environment Canada, Atmospheric Environment Service Report PAES-89-1, Vancouver, B.C.

Holman, C.D.J., B.K. Armstrong and P.J. Heenan 1986. Relationship of cutaneous malignant melanoma to individual sunlight-exposure habits. *J. Nat. Cancer Inst.* 76: 403-414.

Houghton, J.T., B.A. Callander and S.K. Varney (eds.). 1992. *Climate change 1992. Supplementary report to the IPCC Scientific Assessment.* Intergovernmental Panel on Climate Change (IPCC), World Meteorological Organization/United Nations Environment Program, Cambridge University Press.

Johnson, A.H. and C.S. Adkisson. 1986. Air lifting the oaks. *Natural History* 95:40-47.

Kerr, J.B. and C.T. McElroy, 1993. Evidence of large upward trends of ultraviolet-B radiation linked to ozone depletion. *Science* 262: 1032.

Kite, G.W. 1992. Detecting climatic change in hydrometeorological time series. In: *Workshop on Using Hydrometric Data to Detect and Monitor Climatic Change.* April 8-9, 1992, Saskatoon, Saskatchewan.

Leith, R.M. 1991. Patterns in snowcourse and annual mean flow data in British Columbia and the Yukon. In: *Using Hydrometric Data to Detect and Monitor Climatic Change (Proceedings of NHRI Symposium No.8, April, 1991).* NHRI, Saskatoon, Saskatchewan.

Levy, D.A. 1992. *Potential impacts of global warming on salmon production in the Fraser River watershed.* Canadian Technical Report of Fisheries and Aquatic Science 1889.

Martinec, J., A. Rango and R. Roberts. 1992. (In prep.) *Rainfall—Snow Melt Peaks in a Warmer Climate.* U.S.D.A. Hydrology Laboratory, Beltsville, Maryland.

Mathewes, R.W. 1985. Paleobotanical evidence for climatic change in southern British Columbia during late-glacial and holocene time. In: *Climatic Change in Canada 5, Critical Periods in the Quaternary Climatic History of Northern North America.* C.R. Harrington (ed.). National Museums of Canada, National Museum of Natural Sciences, Syllogeus Series, no. 55, pp. 397-422.

Mathewes, R.W. and M. King. 1989. Holocene vegetation, climate and lake level changes in the Interior Douglas-fir Biogeoclimatic Zone, British Columbia. *Canadian Journal of Earth Sciences* 26:1811-1825.

McBean, G.A. and G. Thomas. 1991. Regional climate change for the Pacific Northwest. Presented to the *Symposium on the Implications of Climate Change for Pacific Northwest Forest Management.* Seattle, WA., October 22-25, 1991.

McBean, G.A., O. Slaymaker, T. Northcote, P. Leblond and T.S. Parsons. 1991. *Review of Models for Climate Change and Impacts on Hydrology, Coastal Currents and Fisheries in British Columbia.* Canadian Climate Centre Report No. 91-11.

McBride, M., B. Ma and P.R. Band, 1989. *Projected Cancer Frequency in British Columbia and Regions, 1986, 1991, 1996, 2001.* Monograph No. 2, Cancer Control Agency of British Columbia.

Meidinger, D. and J. Pojar (eds.). 1991. *Ecosystems of British Columbia.* B.C. Ministry of Forests, Victoria, B.C.

Moore, D.R. 1991. Hydrology and water supply in the Fraser River Basin. In: *Water in Sustainable Development: Exploring our Common Future in the Fraser River Basin.* A.H.J. Dorcey, J.R. Griggs (eds.). Westwater Research Centre, University of British Columbia.

Neilson, R.P., G.A. King, R.L. DeLevice, J. Lenihan, D. Marks, J. Dolph, B. Campbell and G. Glick. 1989. *Sensitivity of Ecological Landscapes and Regions to Global Climate Change.* U.S.A. Environmental Protection Agency, Corvalis, Oregon.

Oerlemans, J. and J.P.F. Fortuin. 1992. Sensitivity of glaciers and small ice caps to greenhouse warming. *Science* 248:115-117.

Osborn, G. and B.H. Luckman. 1988. Holocene glacier fluctuations in the Canadian cordillera (Alberta and British Columbia). *Quaternary Science Reviews* 7: 115-128.

Perry, D.A. and J.G. Borchers. 1990. Climate change and ecosystem response. *Northwest Environmental Journal* 6:293-313.

Pollard, D.F.W. 1988. Impacts of climate change on the forest in British Columbia In: *Symposium on the Impacts of Climate Variability and Change on British Columbia.* December 14, 1988, Vancouver, B.C.

Pollard, D.F.W. 1991a. Forestry in British Columbia: planning for the future climate today. *Forestry Chronicle* 67(4):336-341.

Pollard, D.F.W. 1991b. Climate change as a current issue for the Canadian forest sector. *The Environment Professional* 13:37-42.

Raphael, C. 1993. Temperature trends at Prince George, British Columbia, 1943-1992. *Western Geography* Volume 3, 1992, University of Victoria, B.C. (In press)

Rizzo, B. and E.B. Wiken. 1989. Assessing the Sensitivity of Canada's Ecosystems to climate change. Presented at the *European Conference on Landscape—Ecological Impact of Climate Change,* Lunteren, The Netherlands, December 3-7, 1989.

Slaymaker, O. 1991. Hydrological change in British Columbia. In: *Review of Models for Climate Change and Impacts on Hydrology, Coastal Currents and Fishing in British Columbia.* G.A. McBean, O. Slaymaker, T. Northcote, P. Leblond and T.S. Parsons (eds.). Canadian Climate Centre Report 91-11.

Straley, G.B., R.L. Taylor and G.W. Douglas. 1985. *The Rare Vascular Plants of British Columbia.* National Museum of Canada, Syllogeus 59, Ottawa, Ontario.

Sucoff, E. 1969. Freezing of conifer xylem and the cohesion-tension theory. *Physiologia Plantarum* 22:424-431.

Taylor, E. 1990. *Warm and Cool Periods in the Southern Interior of British Columbia.* Environment Canada Report PAES-90-3.

Toon, O.B. and R.P. Turco, 1991. Polar stratospheric clouds and ozone depletion. *Scientific American* 264(6): 68-75.

United Nations Environmental Programme (UNEP). 1991a. *Environmental Effects of Ozone Depletion: 1991 Update.* Panel report pursuant to Article 6 of the Montreal Protocol on substances that deplete the ozone layer.

United Nations Environmental Programme (UNEP). 1991b. *Science Assessment of Stratospheric Ozone.* Nairobi.

Woo, M.K., A.G. Lewkowicz and W.R. Rouse. 1992. Response of the Canadian permafrost environment to climatic change. *Physical Geography* 13(4): 287-317.

Chapter 25
The Future of British Columbia's Flora[1]

The possible effects of global warming on British Columbia's flora cannot be discussed in their entirety because we know little about many areas. My work on ecological responses to past climate change (Hebda 1982, 1983, in press) has led me to conclude that, with global warming, the vegetation and climate of the Pacific Northwest may become like that of the Xerothermic (dry-warm) Interval of the Holocene (Mathewes and Heusser, 1981; see also Chatters, 1989; Chatters et al., 1991; Houghton et al., 1990). During the Xerothermic Interval, 10,000 to 7,000 years ago, southern British Columbia was about 2°C warmer and drier than at present and, as many researchers have shown, the distribution of species was then quite different from today (for examples, see: Hebda, 1982 and 1983; Heusser, 1983; Mathewes and King, 1989). In this discussion, then, I will focus on five broad vegetation assemblages where impacts will be significant and predictions can be made with a considerable degree of confidence. The five categories are: alpine tundra (equivalent to the Alpine Tundra Biogeoclimatic Zone), wetlands, dry interior, shoreline vegetation, and forests.

Richard Hebda
Royal British Columbia Museum
675 Belleville Street
Victoria, B.C.
V8V 1X4

Alpine Tundra

Alpine tundra occupies approximately 150,000 square kilometres, 16 percent of the province's land area. Many plant species grow in no other zone. Large areas of alpine tundra occur along the Coast Mountains, the Rocky Mountains, and in the northern half of the province. Alpine habitats may be continuous or nearly so in the north, where the treeline is low, and in the highest mountains throughout the province. However, many alpine areas, especially in the south, are discontinuous, small, and isolated. These areas are most at risk from global warming.

With increasing temperatures, the treeline may be expected to move higher up mountain slopes (Clague and Mathewes, 1989). Since alpine zones are topographically top-limited, even a slight rise of the timberline would dramatically reduce alpine habitat size. This "cone" effect would work as follows. On a conical peak 2,300 metres high with 25° slopes, the alpine zone today begins at about 2,000 metres. A 100 metre rise of timberline would reduce the alpine area by 56%.[2]

Consequently, the area of distribution of many species will be reduced and, perhaps, some populations will be isolated, leading to potential increases in speciation. Unfortunately, because there have been few paleoecological studies in alpine sites (see Clague and

---

[1] Based on a paper presented at a colloquium, *British Columbia native plants, their current status and future*. Botany Department, University of British Columbia, May 12, 1990.

[2] Surface area of a cone= $\pi r^2 \times \sqrt{r^2 + h^2}$.

Mathewes, 1989), we know little about how the alpine flora and zone were affected by the warmer and presumably drier Xerothermic climate. Notably, many species survived the Xerothermic Interval intact to give us the alpine flora that we know today. Some alpine and subalpine species even survived on the Brooks Peninsula, in one of the most moist and equable climatic regimes in British Columbia (Hebda, in press).

Nevertheless, we must expect some loss of populations and, perhaps, of species if global warming occurs. B.C.'s Alpine Tundra Biogeoclimatic Zone contains 106 taxa listed as rare by Straley et al., (1985). Though many of these taxa also occur in other zones, they are often found only in openings in the zonal vegetation. These taxa, especially those of southern B.C. alpine habitats, are at the greatest risk from global warming.

Wetlands

For at least four reasons, global warming will likely have the greatest impact on wetland taxa and ecosystems, including lakes, ponds, swamps, bogs, fens, and marshes. First, any change in moisture availability results in a change in wetland hydrology and, hence, wetland character. Unlike alpine tundra species, wetland species cannot always migrate along a climatic gradient—they are physiographically limited. Therefore, a change in hydrological regime often results in wholesale changes in plant community structure and complete elimination of certain habitats and species. In some cases, hydrological change might be sufficient to cause disappearance of the wetland.

Second, even a small decline in moisture supply at the margins results in a proportionately large loss of, or change in, wetland habitat. For example, in a circular wetland with a radius of 2 kilometres, a decline of moisture sufficient to shrink the wetland's radius to 1.6 kilometres results in a loss of 36% (πr^2) of the original wetland.

Third, many, if not most, wetland species tolerate only small changes in nutrient flux, nutrient concentration, and water levels (particularly the duration of inundation). We observe the effects of this specificity in the wild by the concentric pattern of species distribution around wet zones. These species exist on an environmental tightrope, unable to compete in the terrestrial realm and squeezed against an inhospitable aquatic environment. Changes in any of the elements of the wetland gradient, including its steepness, could easily eliminate their niche along the gradient. Further, they cannot easily escape because habitat opportunities are not continuous.

Fourth, global warming will increase existing water supply conflicts among agricultural and urban users, with consequent effects on wetlands. Increasing human populations will likely demand more from small neighbouring wetlands, lowering their water levels or digging them out. Also, drying wetlands may be seen as excellent sites for agriculture under droughty conditions. On a community or regional scale, water consumption may drastically reduce water levels in large wetland reservoirs or reduce the ground water table. The latter effect has already been noticed in places such

344

as Grand Forks in the southern interior, where, according to local residents, water levels in wells are lower than ever before in living memory. Already, many water sources are polluted, forcing people to look elsewhere for clean water.

With global warming, then, the area covered by wetland and aquatic environments may be expected to shrink. My work (Hebda, 1982) and that of Rolf Mathewes and students (Mathewes and King, 1989) at Simon Fraser University reveals that, especially in southern and central interior B.C., lakes, ponds and wetland systems shrank or dried up completely during the warmest part of the early Holocene (about 10,000-8,000 years ago). Finney Lake in the Hat Creek Valley provides an excellent example. Today, the lake covers about 15 hectares and has a mean depth of 3 metres and a maximum depth of 5 metres. During the early Holocene, there was little or no water in the lake—its level was 10 metres (Holocene sediments included) lower and its volume 95-100% smaller. Obviously, the lake could not have supported the aquatic flora it does today. At best, it might have harboured alkaline or salt-tolerant taxa in the bottom of the basin, like much smaller and shallower basins do today.

The moisture loss does not have to be nearly so extreme to have a significant impact. It is my experience that many interior lakes, like Finney Lake, have a wide, shallow, littoral platform. At the lakeward edge of the platform the bottom drops off. During the winter season the lake is full to the limits of the landward edge of the platform. As the lake level drops during the summer, ephemeral species grow along a marginal community gradient forming concentric vegetation zones. Many uncommon species thrive in this setting. Under slightly drier conditions, this zone may revert largely to terrestrial habitat and the peculiar ephemeral habitat disappear or be greatly reduced. The stress on such marginal zones is increased by cattle in the interior, whose hooves turn the soil into mush, providing sites for weedy species to colonize. With changing climate, the native species of this zone may be decimated or extirpated, enabling weeds, such as dandelions (*Taraxacum officinale*) and *Tragopogon spp.* to thrive. Other species, such as the rare hairy water clover (*Marsilea vestita*), would possibly disappear. Scarlet ammania (*Ammania coccinea*) and rotalla (*Rotalla ramosior*) could likewise be extirpated from British Columbia.

The changes that organic wetlands (peatlands) may experience are recorded in wetland deposits along the coast of B.C. Banner et al. (1988) examined organic sediment records from several wetlands in the Pacific Temperate Wetland Region and the Pacific Oceanic Wetland Region of B.C.'s coast (National Wetlands Working Group, 1988). These records reveal major changes in organic sediment type and, hence, environment of deposition and plant habitat and community composition since deglaciation. A slimy, humic horizon, indicative of a drier climate, usually occurs in the mid to lower part of the sequence. At Bear Cove Bog near Port Hardy, this horizon occurs during the Xerothermic Interval of the early Holocene (Hebda, 1983). The development of the *Sphagnum*-dominated wetland found in today's cooler, wetter climate is a mid to late Holocene phenomenon. I believe that many, if not most, of the changes in organic wetland sequences on the coast are primarily the result of climatic change and related hydrological and, hence, community changes.

On the coast of British Columbia, the last 5,000 years has seen paludification (wetland expansion into non-wetland areas), rather than loss of wetlands due to forest encroachment (Quickfall, 1987). We may well experience the decline of paludification if the climate warms as predicted. In general, I expect warming to precipitate a shift from acidic, rain water fed wetland systems to less acidic and even alkaline ground water fed systems, resulting in a change from bogs and bog woodlands to fens, swamps and marshes. Certainly, the conditions favourable to *Sphagnum* growth and peat accumulation would be reduced, eventually causing a major decline in the acidic bog and bog forest presently so typical of the Pacific Oceanic Wetland Region, as well as a major reduction in appropriate habitat for acid-requiring species. The changes would be most profound in the regions with the greatest summer moisture deficits, such as the Coastal Douglas-fir Biogeoclimatic zone of southwest British Columbia. In the interior, fens and boggy fens may shift to alkaline marshes, especially at low to mid elevations. Species of hyperacid bog environments or boreal species surviving in southern bog enclaves, such as cloudberry (*Rubus chamaemorus*) in Burns Bog, Fraser Delta, would be seriously affected, while those of hypermaritime acidic habitats would be at less risk. Even with a major net reduction in moisture availability, the hyperoceanic settings may remain moist and cool enough for these species to survive. Studies on the Brooks Peninsula (Hebda, in press) reveal that moist acidic subalpine-alpine communities survived during the Xerothermic Interval. Within those communities were preserved several of our rarest species such as, Calder's lovage (*Ligusticum calderi*) and Queen Charlotte avens (*Geum schofieldii*). However, if temperature increases and moisture decreases are greater than they were during the Xerothermic, we may lose these botanical treasures too.

How can we plan for and mitigate changes in our wetland flora? First, we must develop a functional classification system and a predictive model for wetlands. Several models are available. In Banner et al. (1988), I briefly mentioned a hydrological gradient approach wherein wetlands could be classified according to three interdependent aspects of their water supply: 1) water source; 2) water flow rate; and 3) water table position. Wetland communities and species distributions could then be plotted on a multi-dimensional coordinate axis system. By inspecting the arrangement of communities or species on the gradient, it should be possible to predict in which direction a wetland would change with a change in one of the hydrological gradient variables. Because nutrient flux and concentration depend on the water table, this axis would probably be the most critical indicator of global warming.

Interior wetlands particularly need to be studied and classified. One of the great omissions of the National Wetlands Working Group's 1988 report, *The Wetlands of Canada*, was the absence of a description of montane wetlands, which are most important in southern British Columbia. We know very little about them and they are the ones most likely to suffer the greatest change. These wetlands need to be described and classified and the occurrence of species noted.

Dry Interior

We can expect the greatest shifts in vegetation distribution as a result of global warming in the dry interior of B.C. Here, we can anticipate disappearance of large tracts of forest and the upward and northward expansion of open vegetation, rather than simple species replacements, as in more moist, forested regions.

Reference to the paleoecological record of the southern interior reveals major up-slope shifts in the lower tree line (forest to grassland ecotone). Hebda (1982) estimated that the tree line may have been at, at least, about 1,500 metres above sea level during the warmest and driest part of the Xerothermic Interval. Today, the Ponderosa Pine and Bunchgrass Biogeoclimatic Zones cover only 10,500 square kilometres of B.C. A mean annual temperature increase of 2°C - 5°C could easily convert the vegetation of most of the Interior Douglas-fir zone to that more characteristic of a Ponderosa Pine or Bunchgrass Biogeoclimatic zone. Such a change would add 48,000 square kilometres to the area of these arid zones. Indeed, at the northern end of the range of Ponderosa pine and bunchgrass, large areas of logged-over second growth pine may convert to open habitats in the near future. Thus, we could see as much as a four hundred percent increase in the most arid vegetation types.

Such a change might look very positive for species of the dry interior, but the rare ones may suffer from competition with the many adventive weeds, such as knapweeds (*Centaurea spp.*), that have found their way into this habitat. It may well be that these vigorous competitors disperse to form a new kind of hybrid interior vegetation. Such will be the case particularly if open and woodland areas are continuously overgrazed and raw soil turned up, providing seedbed. On south facing slopes, the grave potential exists for the adventive-native hybrid grassland developed at lower elevations to merge with dry alpine habitats and facilitate the upward spread of the weedy scourges of the overgrazed lowlands.

The many rare species of the driest southern interior sagelands may spread if sufficient populations are retained to form effective dispersal centres. Protection from urban and agricultural development and provision of corridors or pathways for natural dispersal, especially northward, would augment their chances. We may also see the northward spread of arid land species from the Great Basin of Washington and Oregon, enriching our native flora.

Distinct zones of arid vegetation may be expected to develop. The most recent report on biogeoclimatic zones (Meidinger and Pojar, 1991) recognized two distinct arid land units: Ponderosa pine and bunchgrass. Initially, they were not differentiated, partly because the climatic gradients in the areas where they occur were so steep that the zones tended to merge together. With increasing warmth and drought, the gradient along the warm dry end of the climate spectrum will likely become gentler and species will find their appropriate position. We may see the development of: an upper elevation zone dominated by grasses and forbs, patches of which occur at the upper limits of today's open arid vegetation; a Ponderosa pine zone; and a well-developed arid or semi desert steppe dominated by big sagebrush (*Artemisia tridentata*). Certainly, big sagebrush was much more abundant in the southern interior of B.C. during the Xerothermic Interval than it is today (Hebda, 1982).

The new vegetation communities may enable yet other, unguessed-at species to flourish. The fossil record reminds us that species behaviour is individualistic and often unpredictable. New combinations should be expected and, indeed, the postglacial fossil record shows they are the norm when major climatic shifts occur.

Shoreline Vegetation

The predicted 0.3 to 1.1 metre rise in sea level (Houghton et al., 1990) will significantly affect shoreline habitats and flora. Higher water will drown low-lying coastal plant communities and substantially modify shoreline morphology, resulting in the redistribution and, possibly, the loss of some species.

The greatest impact will be on estuaries, all of which will be inundated to some degree. Though inundation will mostly occur gradually, catastrophic inundation and erosion may be expected when winter storms roar across the deltaic platform. Similarly, changes in distributary channels will occur suddenly through a process called channel avulsion, in which, at high-water phase, a channel suddenly finds an easier path to the ocean.

The effects of inundation will likely be severest on those estuaries and coastal wetlands that are bounded by urban or rural land. Where there are dikes, the full force of inundation will be concentrated outside the dikes, damaging if not destroying seaward plant communities. In the case of estuaries, such as the Fraser Delta, which have sufficient sediment supply, the estuarine part of the flood plain may build up (aggrade) quickly enough to maintain the estuary at sea level (Williams and Hebda, 1991). In other cases, when the landward side is not overly developed, species may be able to disperse up the floodplain and re-establish populations. But when the estuary is small and constrained or highly developed, there is no physiographic continuity for the wetland to re-establish further up the gradient. Consequently, the estuarine systems or wetlands may simply be squeezed out, though some species may persist along the steep gradient of the dike face or natural banks.

Therefore, any rare or endangered species in small and constrained or highly developed estuaries could face extirpation. Henderson's checker-mallow (*Sidalcea hendersonii*), for example, would be at considerable risk. One way such extirpations can be mitigated is through the exclusion of natural estuarine expanses or embayments from diked land. These embayments must contain sufficient area and appropriate elevational gradient to accommodate upslope dispersal and re-establishment of estuarine communities.

In other shoreline environments, a rapidly rising sea level or, more specifically, a catastrophic storm surge might destroy a population of rare species. Almost all the populations of California bayberry (*Myrica californica*) in B.C. could be eliminated in this way. Two taxa of sand verbena (*Abronia latifolia* and *Abronia umbellata subsp. acutalata*) could similarly be affected. More common species would likely reoccupy newly created shore habitats.

Historically speaking, sea-level change has been a normal process, with rapid submergence and emergence occurring during the late Pleistocene and early Holocene in B.C. (Clague et al., 1982). There was a relatively major sea level rise of about 10 metres along the southern B.C. coast between 7,000 - 5,000 years ago. Paleoecological studies reveal that some shoreline taxa (such as saltwort [*Salicornia spp.*]) survived, but we do not know how many were lost or decimated. The last 5,000 years have seen relatively stable shorelines and well developed habitat zones. Notably, the predicted rise will occur at about twice the rate (1 metre in 100 years) experienced between 7,000 and 5,000 years ago (1 metre in 200 years; Williams and Hebda, 1991).

Forests

Tree distribution and, consequently, forest composition will likely change with global warming. The forests on valley sides in the Southern Interior may be expected to be confined to the upper slopes. In arid regions of central and southern B.C., we can expect: 1) reduction of lowland forests and expansion of upland forests; 2) general northward migration of forest zones, with gradually increasing change in their structures; 3) increasing fire frequency and the predominance of seral lodgepole pine forests, especially in central B.C. (see Mathewes, 1985). Some species, such as the moisture-requiring western red cedar (*Thuja plicata*), will diminish. Only in the last 5,000 years has this species become a major forest element (Hebda and Mathewes, 1984), and its tenure may be shortlived. In the Victoria area, during the relatively droughty summer of 1988, for example, large healthy cedar trees died, presumably because of an insufficient supply of nutrient-rich water. Episodes of drought will likely eliminate some species from large parts of their range.

An important point to consider is the organic content of the forest soil. Large organic deposits in moist forests cool the soil, reduce moisture loss and erosion, and store carbon. Consequently, they moderate climatic effects and the release of carbon dioxide, thereby helping to slow the rate of change of vegetation and species distribution as warming takes effect. Such organic-rich soils must be protected from disruption and desiccation as much as possible. Doing so might mean treading as lightly as possible when logging an area, and perhaps leaving a partial canopy (through selective logging) to protect the organic component of the soil from decomposition or erosion.

General Effects and Summary

In general, global warming may be expected to affect our native flora in the ways listed below.

1. Species distribution and vegetation composition will change, a natural process that is well-demonstrated by studies of vegetation history.

2. The change associated with global warming will be as rapid as or, likely, more rapid than, any change since the last glaciation.

3. Species will behave individualistically and new biogeoclimatic zones will arise.

4. Weedy adventive species will become more abundant and find permanent niches in new vegetation communities.

5. Some rare species will expand their populations, whereas other will decline and disappear.

6. Moisture-and acid-loving species may decline in number, and drought resistant and alkaline-tolerant species increase.

7. Wetland and southern alpine species are potentially at greatest risk.

8. Re-equilibration of the flora in new vegetation units will be a lengthy process because climatic change will continue for many years, as will changes in species distribution.

9. Loss of populations will likely be sudden rather than gradual, precipitated by climatic extremes or their effects. In general, this suddeness will lead to impoverishment of local and regional floras before the re-establishment of new species.

Recommendations

1. We must establish a long term project to inventory our flora and identify populations of rare plant species, especially in the south.

2. We must take steps to protect rare plant populations, so that they can serve as the reservoir from which new genetic strains adapted to the changed conditions can arise. The fossil record shows that rare plants have served this function before—western red cedar is a good example of a once uncommon plant that became abundant as the climate became cooler and wetter. Extant and new ecological reserves would harbour this vital genetic potential.

3. We must begin a program of ecological gradient conservation, with a particular focus on climatic and hydrological gradients, so that species can disperse along them as conditions change.

4. We must complete paleoecological studies of major biogeoclimatic regions, gathering information on potential changes in species range, vegetation, hydrology and landscape process. This information may help us to de-

velop a conservation strategy which ensures that native
flora persist and new vegetation types arise from as broad
a base of native species as possible.

In conclusion, we need to change our conservation philosophy.
By all means, we must continue to preserve or set aside tracts of
land to act as reservoirs of native species and vegetation, not only
for economically valuable species, but especially for rare taxa and
rare vegetation assemblages. However, we must also look at conserving the continuity of the natural setting from a much broader perspective. Rather than only trying to save or reserve small parcels, we
must begin with the assumption that all of the provincial area
merits a conservation strategy. Further disruptions to the natural
environment must be sufficiently important to merit withdrawals
from our inherited portfolio of life forms and assemblages. For,
unlike a financial investment portfolio, we cannot buy and sell to
meet our long term goals. When the life-form portfolio loses a species, it is gone forever and we will never know, or have the
opportunitiy to benefit from, what it might have become.

Acknowledgements

The manuscript was reviewed by: Rick Williams of the British Columbia Ministry of
Agriculture, Fisheries and Food; Hal Coulson of the British Columbia Ministry of
Environment, Lands and Parks; Phil Comeau, Jim Crover and Evelyn Hamilton of
the British Columbia Ministry of Forests; Cliff Raphael of the College of New
Caledonia; Douglas Pollard of the Pacific Forestry Centre; and A.C. Allaye-Chan and
Gordon McBean of the University of British Columbia.

References

Banner, A., R.J. Hebda, E.T. Oswald, J. Pojar and Trowbridge. 1988. Wetlands of
Pacific Canada. In: *National Wetlands Working Group*. Wetlands of Canada.
Ecological Land Classification Series, No. 24. Sustainable Development Branch,
Environment Canada, Ottawa, Ontario and Polyscience Publications Inc., Montreal,
Quebec. pp. 305-346.

Barnosky, C.W. 1984. Late Pleistocene and early Holocene environmental history of
southwestern Washington State, U.S.A. *Canadian Journal of Earth Sciences* 21:619-
629.

Chatters, J.C. 1989. The hypsithermal peleohydrology as an analog of greenhouse
effects on the Columbia River basin, Washington. *Geological Society of America.
Abstracts with Programs* 21(6).

Chatters, J.C., D.A. Neitzel, M.J. Scott and S.A. Shankle. 1991. Potential impacts of
global climate change on Pacific Northwest Spring Chinook Salmon (*Oncorhynchus
tshawytscha*): an exploratory case study. *Northwest Environmental Journal* 7(1):71-
92.

Clague, J.J. and R.W. Mathewes. 1989. Early Holocene thermal maximum in
western North America: new evidence from Castle Peak, British Columbia. *Geology*
17:277-280.

Clague, J.J., J.R. Harper, R.J. Hebda and D.E. Howes. 1982. Late quaternary sea
levels and crustal movements, coastal British Columbia. *Canadian Journal of Earth
Sciences* 19:597-618.

Hebda, R.J. 1982. Postglacial history of grasslands of southern British Columbia
and adjacent regions. In: *Grassland Ecology and Classification, Symposium
Proceedings*. June 1982. A.C. Nicholson, A. McLean. and T.E. Baker. (eds.). British
Columbia Ministry of Forests, Victoria, B.C. pp. 157-191.

Hebda, R.J. 1983. Lateglacial and postglacial vegetation history at Bear Cove Bog,
northeast Vancouver Island, B.C. *Canadian Journal of Botany* 61:3172-3192.

Hebda, R.J. in press. Late Quaternary paleoecology of Brooks Peninsula. In: *The Brooks Peninsula: an Ice-age Refuge.* R. Hebda and J.C. Haggarty (eds.). Royal British Columbia Museum, Victoria, B.C.

Hebda, R.J. and R.W. Mathewes. 1984. Holocene history of cedar and native cultures of the North American Pacific Coast. *Science* 225:711-713.

Heusser, L.E. 1983. Palynology and Paleoecology of postglacial sediments in an anoxic basin, Saanich Inlet, British Columbia. *Canadian Journal of Earth Sciences* 20:873-885.

Houghton, J.T., G.J. Jenkins and J.J. Ephraums. 1990. *Climate Change: the IPCC Scientific Assessment.* Cambridge University Press. Cambridge, U.K.

Mathewes, R.W. 1985. Paleobotanical evidence for climatic change in southern British Columbia during Late-glacial and Holocene time. In: *Climatic Change in Canada 5, Critical Periods in the Quaternary Climatic History of Northern North America.* C.R. Harrington (ed.). National Museums of Canada, National Museum of Natural Sciences, Syllogeus Series, no. 55, pp. 397-422.

Mathewes, R.W. and L.E. Heusser. 1981. A 12,000 year palynological record of temperature and precipitation trends in southwestern British Columbia. *Canadian Journal of Botany* 59:707-710.

Mathewes, R.W. and M. King. 1989. Holocene vegetation, climate and lake level changes in the Interior Douglas-fir Biogeoclimatic Zone, British Columbia. *Canadian Journal of Earth Sciences* 26:1811-1825.

Meidinger, D. and J. Pojar (eds.). 1991. *Ecosystems of British Columbia.* B.C. Ministry of Forests, Victoria, B.C.

National Wetlands Working Group. 1988. Wetlands of Canada. *Ecological Land Classification Series, No. 24.* Sustainable Development Branch, Environment Canada, Ottawa, Ontario and Polyscience Publications Inc., Montreal, Quebec.

Quickfall, G.S. 1987. *Paludification and Climate on the Queen Charlotte Islands During the Past 8,000 Years.* M.Sc. Thesis. Biology Department, Simon Fraser University. Burnaby, B.C.

Straley, G.B., R.L. Taylor and G.W. Douglas. 1985. *The Rare Vascular Plants of British Columbia.* National Museum of Canada, Syllogeus 59, Ottawa, Ontario.

Williams, H.F.L. and R.J. Hebda. 1991. Palynology of Holocene top-aggradational sediments of the Fraser River Delta, British Columbia. *Paleogeography, Palaeoclimatology and Paleoecology.* 86:287-311.

Chapter 26
Protected Areas in British Columbia: Maintaining Natural Diversity

British Columbia is the most ecologically diverse province or territory in Canada. Yet, that diversity is threatened by a range of activities, with habitat alteration and fragmentation posing perhaps the greatest threats, particularly in the remaining natural areas on Vancouver Island, in the Lower Mainland and in the southern Interior. Most watersheds in the coastal temperate rain forests have been disturbed by industrial activity (British Columbia Ministry of Environment, Lands and Parks and Environment Canada, 1993). Many of British Columbia's valley bottom, plateau and high elevation grasslands have been lost or significantly damaged. A significant number of the fresh water wetlands and estuaries have been modified or drained and are now considered at risk. Less than one percent of the original area of coastal old growth Douglas-fir forest now remains (Vold, 1992). There is growing concern, both in British Columbia and beyond its borders, about protecting this rich and diverse natural environment.

It is becoming increasingly accepted that a balance between natural resource protection and utilization can best be achieved by combining a system of integrated resource use and management with a comprehensive network of protected areas aimed at maintaining the functional integrity of ecosystems (McNamee, 1990; Task Force on Northern Conservation, 1984; World Resources Institute et al., 1992). If full native biodiversity is to be maintained, these two components must be combined in a conservation strategy for all land, fresh water and marine areas in the province. This approach recognizes that, as discussed in Chapter 1, protected areas are not closed, self-supporting ecosystems, but part of larger, interacting and interdependent systems. Properly integrated resource management practices on adjacent lands act as a buffer, helping to shield protected areas from the impacts of external developments, prevent the encroachment of inappropriate land uses, and provide habitat, dispersal and migration corridors. At the same time, protected areas offer unique opportunities for education, research and observation of long term natural changes, the products of which can be used to improve resource management.

British Columbia is a leading advocate of this two-pronged approach to conservation. The regional planning processes currently being put in place by the Commission on Resources and Environment and the province's evolving integrated resource planning program (of which Land and Resource Management Plans are an example) will facilitate a balanced and comprehensive approach to the stewardship of natural resources. Similarly, the commitment to develop a Protected Areas Strategy for the province acknowledges the need for a comprehensive network of specially protected areas.

Kenneth E. Morrison
Ministry of Environment,
Lands and Parks
800 Johnson Street
Victoria, B.C.
V8V 1X4

Anthony M. Turner
State of the Environment
Reporting
Environment Canada
1547 Merivale Road
Ottawa, Ontario
K2G 3J6

Protected areas are now recognized as an essential and irreplaceable component of the province's environmental and land use strategy. They provide a vital counterbalance to more intensively managed and developed lands; they protect the province's diverse and distinctive ecosystems; they maintain those essential ecological processes that depend on natural ecosystems; they preserve the diversity of species and the genetic variation within them, thereby preventing irreversible damage to our natural heritage; they maintain the productive capacities of ecosystems and safeguard habitats critical for the sustainable use of species; they provide opportunities for scientific research, education, training, recreation and tourism; they protect aesthetic and cultural resources; they are places of spiritual renewal and inspiration; and they help to strengthen our cultural identity (McNeely et al., 1990). Increasingly, protected areas are being viewed as a critical component of environmental stewardship and an important legacy.

There is growing concern, however, over whether protected areas achieve their environmental objectives. Designating a protected area does not automatically secure the preservation, in perpetuity, of the ecological integrity of the resources within its boundaries. Not only do protected areas suffer from internal pressures, such as recreational overuse, but they are increasingly surrounded by development on adjacent lands and subjected to external stresses (such as global warming, water pollution from upstream sources, and long-range atmospheric transport of toxic chemicals) for which legal boundaries are no match.

In this chapter, we try to assess the state of protected areas by examining two basic questions: Are there enough protected areas? and How effectively do protected areas preserve our natural heritage? We begin with an overview of protected area programs and an introduction to the World Conservation Union's (IUCN) universal classification system for protected areas, then we look closely at the number, size and representativeness of protected areas, and conclude with a discussion of the steps needed to complete the protected areas system in British Columbia.

Protected Area Programs

Among the many types of protected areas in British Columbia are national parks, ecological reserves, wilderness areas, provincial parks, migratory bird sanctuaries and wildlife management areas. These areas are administered under no fewer than 16 different protected area programs or activities, which vary considerably in legal securement, function, scope, size, objectives, management policies, and level of legislative protection afforded to ecosystems, species and natural features (Tables 26-1 and 26-2). Federal protected area holdings account for less than 9% of the total area protected in the province, provincial mechanisms for nearly 91%, and private interests and regional governments for the remaining fraction of a percent. Two agencies, BC Parks and Parks Canada, are the stewards of over 97% of the total protected area in the province (Table 26-1). BC Parks' programs are the foundation of the current protected areas system, as they encompass more than 88% of all protected lands in the province. The scope of the programs ranges

Legislation/Convention Program	Managing Agency	Designation	Program Goals, Objectives and/or Description
National Park Act	Environment Canada, Parks Canada	National Park/National Park Reserve (Statute)	. to protect for all time representative natural areas of Canadian significance and to encourage public understanding, appreciation and enjoyment of this natural heritage so as to leave it unimpaired for future generations
		National Marine Park/National Marine Park Reserve (Statute)	. to protect and conserve for all time representative marine natural areas of Canadian significance and to encourage public understanding, appreciation and enjoyment of Canada's marine heritage so as to leave it unimpaired for future generations
Migratory Birds Convention Act	Environment Canada, Canadian Wildlife Service	Migratory Bird Sanctuary (Statute)	. areas set aside for protection of special habitat and migratory birds using that habitat
Canada Wildlife Act	Environment Canada, Canadian Wildlife Service	National Wildlife Area (Statute)	. areas established for research, conservation and interpretation in respect to migrating birds and other wildlife
Park Act	Ministry of Environment, Lands and Parks, BC Parks	Provincial Park/Recreation Area (Statute or Order-in-Council)	. protect a system of representative and special landscapes and features that incorporate the greatest possible diversity of provincially significant biophysical resources . serve a variety of outdoor recreation functions including enhancing major tourism travel routes, providing attractions that serve as or enhance outdoor holiday destinations, providing backcountry adventure opportunities and ensuring access to local outdoor recreation opportunities for all residents of the province
		Nature Conservancy (Order-in-Council or Zoning Policy)	. established within provincial parks to give added protection to scenic and ecological values; roadless areas reserved absolutely for the preservation of representative ecosystems and landforms in their natural state
Ecological Reserve Act	Ministry of Environment, Lands and Parks, BC Parks	Order-in-Council	. protect a system of representative ecosystems . protect rare and endangered plants and animals in their natural habitats . preserve unique, outstanding or rare zoological, botanical or geological phenomena . serve as benchmarks for long-term scientific research and educational use . serve as examples of habitats recovering from modifications caused by human activities

Table 26-1: Overview of Existing and Potential Protected Area Programs.

357

Legislation/Convention Program	Managing Agency	Designation	Program Goals, Objectives and/or Description
Forest Act	Ministry of Forests	Wilderness Area (Order-in-Council)	. an area of land generally greater than 1000 ha that predominantly retains its natural character and on which human impact is transitory, minor and in the long term substantially unnoticeable . maintain and protect a wilderness resource representing BC's diverse natural environments and provide the opportunity for a wilderness experience
Wildlife Act	Ministry of Environment, Lands and Parks, Wildlife Branch	Wildlife Management Area (Order-in-Council) 3 types: 1) General Wildlife Management Areas 2) Wildlife Sanctuaries (no sites as yet designated) 3) Critical Wildlife Management Areas	. conservation and intensive management of areas of special importance to fish and wildlife species . areas which are particularly vulnerable to use or human disturbance of fish and wildlife populations . protect endangered or threatened species and their habitats
Creston Valley Wildlife Act	Cooperative endeavour between the Canadian Wildlife Service, Wildlife Branch and private interests	Creston Valley Wildlife Management Area (Statute)	. protection of major waterfowl breeding and nesting area
Environment and Land Use Act	Ministry of Environment, Lands and Parks, BC Parks	Purcell Wilderness Conservancy (Order-in-Council)	. area of natural environment containing special and representative features; roadless and uninfluenced by human activities
Park (Regional) Act	Regional Districts	Regional Park	. act permits Regional Districts to acquire, develop and administer regional parks and trails . serve a variety of functions including outdoor recreation and resource protection . although most are small, close to urban centres and receive substantial recreational pressures, some are of substantial size and have preservation of natural resources as their primary function

Table 26-1: Overview of Existing and Potential Protected Area Programs (Continued).

Legislation/Convention Program	Managing Agency	Designation	Program Goals, Objectives and/or Description
Canadian Heritage Rivers System	Cooperative program of the federal and provincial/territorial governments	Canadian Heritage River	. to give national recognition to the important rivers of Canada and to ensure long-term management that will conserve their natural, historical and recreational values for the benefit and enjoyment of Canadians today and in the future
Convention for the Protection of the World Cultural and Natural Heritage	Coordinated by Parks Canada	World Heritage Site	. to protect examples of natural and cultural heritage of global significance
Convention on the Conservation of Wetlands of International Importance	Coordinated by Environment Canada, Canadian Wildlife Service	Ramsar Site	. focus on wetlands of international significance and to acknowledge the significant value of wetlands as areas of high biological productivity
Man and the Biosphere Program	Various. Potential cooperative initiatives between federal, provincial and/or regional governments and private landowners. No biosphere reserve sites presently exist in the province.	Biosphere Reserve	. long-term goal of the program is to create an international network of biosphere reserves that will collectively represent the world's major ecological systems with different patterns of human use and adaptations to these systems . each biosphere reserve includes a core of relatively undisturbed land together with areas demonstrating ways in which such lands are being managed to meet human needs . objectives include conservation of representative features, long-term research and environmental monitoring
Western Hemisphere Shorebird Network Program	Coordinated by Environment Canada, Canadian Wildlife Service. No Western Hemisphere Shorebird sites presently exist in the province.	Western Hemisphere Shorebird Network Site	. cooperative program of government and private organizations which recognizes and protects essential staging areas for migratory birds

Table 26-1: Overview of Existing and Potential Protected Area Programs (Continued).

Program	# of Areas Designated	Hectares Designated	% of Provincial Land Base	% of Protected Areas	Notes
National Park/National Marine Park	6	630,200*	0.664	8.55	*Includes some marine waters, the extent of which are not currently available.
Migratory Bird Sanctuaries	7	3,091*	0.003	0.04	*Includes some marine waters, the extent of which are not currently available.
National Wildlife Areas	5	2,301	0.002	0.03	
Ecological Reserves	131	land: 111,106 marine waters: 47,647	0.117	2.15	
Provincial Parks/ Recreation Areas Class A Park	347*	land: 5,250,631 marine waters: 22,685	5.537	84.83	*Includes Khutzeymateen, although designation has not yet been determined.
Class B Park Class C Park Recreation Areas	2 22 36	3,778 728 land: 905,164 marine waters: 69,464	0.004 0.001 0.954		
Wilderness Conservancy	1	131523	0.139	1.78	
Wilderness Areas	4	130,000	0.137	1.76	
Wildlife Management Areas	12	20,172*	0.021	0.27	*Includes some marine waters, the extent of which are not currently available.
Creston Valley Wildlife Management Area	1	6,900	0.007	0.09	
Regional Parks	74	13,440	0.014	0.18	

Table 26-2: Number and Size of Existing Protected Area Programs in British Columbia.

Program	# of Areas Designated	Hectares Designated	% of Provincial Land Base	% of Protected Areas	Notes
NGO Lands e.g. BC Nature Trust Nature Conservancy of Canada	95 4	12,500 8,984	0.013 0.010	0.29	This list of NGO lands is incomplete. Some overlap exists between this category of lands and other protected areas in this table. Although the extent of overlap is not known at this time, it is expected to be minimal.
Canadian Heritage Rivers	1	67 km	N/A	N/A	The 67 km stretch of the Kicking Horse River designated under the Canadian Heritage Rivers program is in Yoho National Park.
World Heritage Sites	2	549,334	0.579	N/A	Includes Anthony Island and part of the Rocky Mountains designation. These sites are included in national and provincial park figures.
Ramsar Sites	1	300	0.000	N/A	The one site designated under this program is Alaksen National Wildlife Area.
Total	747*	7370314	7.621	N/A	Totals do not include new protected areas in Clayoquot Sound as these have not yet been designated or Canadian Heritage Rivers, World Heritage Sites or Ramsar Sites as these are included in other designations.

Table 26-2: Number and Size of Existing Protected Area Programs in British Columbia (Continued).

from protection of large representative ecosystems to that of areas less than a hectare in size and from multi-purpose to single purpose areas. At some sites, such as those ecological reserves established for the single purpose of protecting sea bird colonies, access is strictly controlled. Others encourage intensive recreation, permit certain commercial extractive activities, or allow habitat management or other similar manipulations of the environment. Despite such variations, all contribute to the protection of biological diversity.

World Conservation Union (IUCN) Categories of Protected Areas

The term, "protected area," is often widely interpreted and applied. Some groups see protected areas from a fairly narrow perspective, as ecological reserves where human activities are strictly controlled, for example, whereas others view them more openly as including areas where activities are largely unrestricted. To enable comparison between different types of protected areas, the World Conservation Union and its Commission on National Parks and Protected Areas has developed a universal classification system. This system is based on each protected area's management objectives and degree of human intervention. It features six categories which range from strictly protected scientific reserves to protected sites where the consumption of natural resources is permitted as long as the principle of sustainability is observed (Table 26-3). These categories are not hierarchical in the sense that one is more important than another; each is needed in a comprehensive network of protected areas.

British Columbia's protected areas cover the range of IUCN categories (Figure 26-1). Ecological reserves, nature conservancies, national parks and most provincial parks provide strict protection to natural areas, species and ecosystems, so they fall into IUCN categories I - III. Other protected areas, such as wildlife management areas, allow human intervention when it is compatible with their conservation objectives. In such areas, which fall under IUCN categories IV and V, extractive commercial activities may be permitted under conditions that do not disrupt the inherent natural values of the site. For example, in some areas managed for wildlife, hunting and

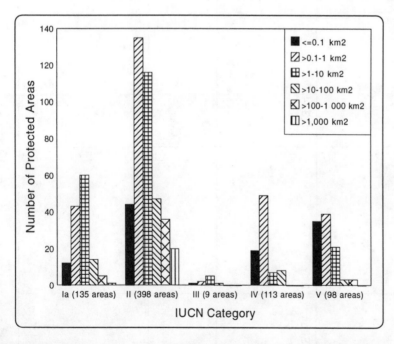

Figure 26-1. Number of Protected Areas by Size and IUCN Category. This six category protected areas framework was only recently adopted by IUCN. Protected areas have not yet been allocated to the new subcategory Ib in British Columbia, so the affected areas are still part of category II as in the old framework. In Canada, category VI is not considered protected. Source: State of the Environment Reporting Service, Environment Canada.

Category I – Strict Nature Reserves/Wilderness Areas
a) Areas managed mainly for science
These areas possess some outstanding or representative ecosystems, geological or physiological features and/or species of flora and fauna, and are reserved primarily for scientific research and/or environmental monitoring.
British Columbian example: ecological reserves
b) Areas managed mainly for wilderness protection
These large areas of unmodified or slightly modified land (or land and water) retain their natural character and influences and lack permanent or significant human habitation. They are protected and managed so as to preserve their natural condition.
British Columbian examples: wilderness conservation zones in provincial parks; legislated large nature conservancies

Category II – National Parks: Areas managed mainly for ecosystem conservation and recreation
These natural areas of land and/or sea are designated to: a) protect the ecological integrity of one or more ecosystems for this and future generations; b) exclude exploitation or intensive occupation; and c) provide a foundation for environmental and culturally compatible spiritual, scientific, educational, recreational and visitor opportunities. This category should perpetuate, in a natural state, representative samples of physiographic regions, biotic communities and genetic resources, and species to provide ecological stability and diversity.
British Columbian examples: national parks; Class A provincial parks; Purcell Wilderness Conservancy; wilderness areas where mining is not permitted

Category III – Natural Monuments: Areas managed mainly for conservation of specific natural features
These areas contain one or more special natural or natural/cultural features of outstanding or unique value because of inherent rarity, representative or aesthetic qualities or cultural significance.
British Columbian example: special feature-oriented Class A provincial parks

Category IV – Habitat/Species Management: Areas managed mainly for conservation through management intervention
These areas of land and/or sea subject to active intervention for management purposes so as to ensure the maintenance of habitats and/or to meet the requirements of specific species. Maintaining sustainable wildlife populations, as well as protecting rare and threatened species, is an integral function of these areas.
British Columbian examples: wildlife management areas; migratory bird sanctuaries; national wildlife areas

Category V – Protected Landscape/Seascape: Areas managed mainly for landscape/seascape conservation and recreation
The objective of this category is to maintain significant areas that characterize the harmonious interaction between nature and culture. These areas provide opportunities for public involvement through recreation and tourism, while supporting normal lifestyles and economic activities.
British Columbian examples: Class B and Class C provincial parks; most regional district parks.

Category VI – Managed Resource: Areas managed mainly for the sustainable use of natural ecosystems
These areas contain predominantly unmodified natural ecosystems which are managed to ensure long term protection and maintenance of biological diversity, while at the same time providing a sustainable flow of natural products and services to meet community needs.

Table 26-3: Categories of Protected Areas

some agricultural practices are allowed, but activities that destroy wildlife habitat are prohibited.

Number, Size and Rate of Establishment of Protected Areas

There are a minimum of 753 protected areas, encompassing 7.37 million hectares and covering about 7.68% of the total area of land and water in the province (Table 26-2). The number of protected areas given here differs from the number in Table 26-2 due to some overlaps [such as non-governmental organization lands managed by provincial agencies] and the exclusion of new protected areas in Clayoquot Sound from Table 26-2.). Area figures include some marine waters, the extent of which is not readily available but is minimal in terms of the total protected area. Strictly protected areas (IUCN categories I - III) encompass over 99% of the protected area of the province—about 7.55% of the total land and freshwater.

At present, British Columbia substantially exceeds the national figure for percentage of jurisdiction protected (4.8%), and ranks second among the provinces and territories in both total area protected and percentage of jurisdiction protected (World Wildlife Fund Canada, 1993). However, organizations, such as the World Commission on Environment and Development (1987) and the World Conservation Union (1991) at the international level, and World Wildlife Fund Canada (Hummel, 1989) and the federal government (Government of Canada, 1990) at the national level, have promoted conservation objectives which are now being extended to the provincial level. These objectives require that a minimum of 10-12% of the province's land and marine areas be given some form of protected status. Through its Protected Areas Strategy, the province has risen to the challenge by committing to protect 12% of its area.

One measure of the sustained viability of a protected area is size. Large protected areas generally have more ecological value and are better able to maintain biodiversity than small ones, though small ones also play an important role in preserving our natural heritage by affording protection to some rare and endangered species and their habitats, as well as linking reserves (Environment Canada, 1991a). A quick perusal of Figure 26-1 shows that most protected areas in the province are found in the smaller size categories: 50.3% are smaller than 1 square kilometre; 82.0% are smaller than 10 square kilometres; and only 8.2% are larger than 100 square kilometres. Though 72% of these areas fall into IUCN categories I - III, over 77% of them are smaller than 10 square kilometres in size. The small size of many protected areas in the province raises concerns over their long term viability.

The rate of protected area creation, in terms of both the number of sites established and the total area protected, is a relative measure of the progress in protected area initiatives (Turner et al., 1992). Identifying this rate also assists in determining what needs to be done to meet provincial objectives (Rubec et al., 1992). The protected area system in the province has grown sporadically throughout this century (Figure 26-2). At times, it has even shrunk because a protected area has been reduced in size, as was Hamber Provincial Park, or removed altogether, as was Liard River Provincial Park. Though the rate of establishment of and the amount of land dedi-

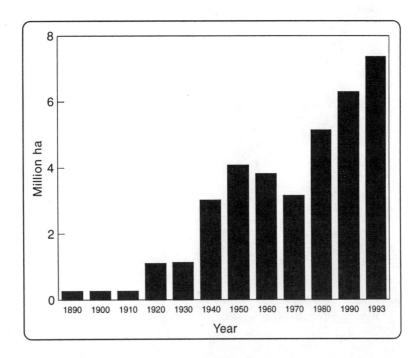

Figure 26-2. Trends in Growth of Protected Areas in British Columbia.
Source: B.C. Ministry of Environment, Lands and Parks.

cated to protected areas have been increasing in recent decades, they will need to increase still more if the province is to meet its goals of doubling the size of the protected areas system and filling in the gaps in ecosystem representation by the year 2000.

Ecosystem Representation

A fundamental goal of the protected areas system is to secure representative examples of the full environmental diversity of the province. The provincial government has committed itself to this goal through the Protected Areas Strategy (Province of British Columbia, 1993) and as a signatory to the Tri-Council Statement of Commitment to Complete Canada's Networks of Protected Areas (World Wildlife Fund, 1993). Figure 26-3 and Table 26-4 show the percentage of each ecosection which is protected. Currently, 21 ecosections have greater than 12% of their area protected. In contrast, 19 ecosections have no protected areas whatsoever and another 38 have less than 1% of their area protected. There are clearly significant gaps in the protected areas network, as the majority of ecosections are seriously under-represented. Areas containing ecosystems typical of these ecosections urgently need to be protected.

From a regional distribution perspective, some definite patterns emerge. A northwest-southeast axis running from the northern to southern provincial boundary in the interior is, for the most part, poorly represented. Similarly, with few exceptions, the environmental diversity of the northern half of the province is poorly represented. Marine ecosystems are also poorly represented (see Chapter 28 for a discussion of marine protected areas).

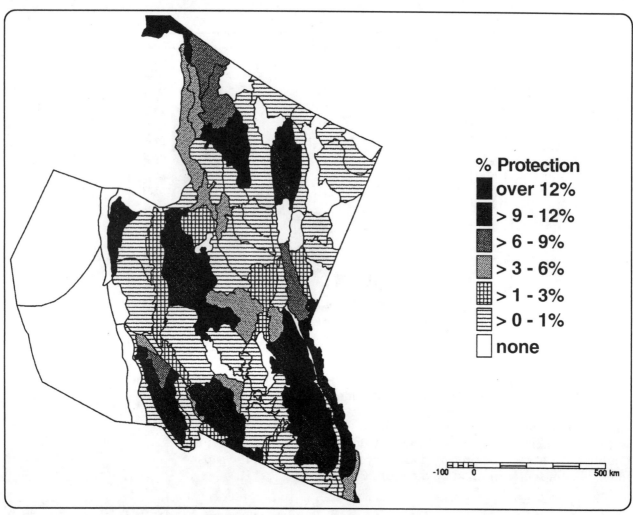

Figure 26-3. Percentage of Each Ecosection Which is Protected in British Columbia. Source: State of the Environment Reporting Service, Environment Canada, December, 1993.

Many of the ecosections are represented by protected areas which are likely too small (Figure 26-1), on their own, to preserve a reasonably complete sample of the ecological diversity of an ecosection. Since the National Conservation Areas Data Base does not contain detailed information on the ecological characteristics of each protected area, one cannot determine whether the protected areas are indeed representative of their ecosections or whether they include unique or uncommon elements. Therefore, the assumption is made that, the greater the number of areas and the greater the total area protected within an ecosection, the greater the likelihood that representative characteristics of the ecosystem have, to some degree, been protected.

More detailed frameworks for determining representativeness will be needed. For example, the considerable diversity within each ecosection is reflected in the number of biogeoclimatic units found therein, so the possibility of representing these units is currently being assessed through the Protected Areas Strategy. Gap analysis, a technique used to identify the gaps in representativeness, reveals that, although significant proportions of some ecosections are protected, important gaps remain. On Vancouver Island, for exam-

% of Ecosection Protected	Number of Ecosections	% of All Ecosections	% of Provincial Area (Total area is 948,596 square kilometres., excluding marine waters)
0	19	17.3	11.5
>0 to 1	38	34.5	36.2
>1 to 3	8	7.3	7
>3 to 6	11	10	8.9
>6 to 9	4	3.6	5.7
>9 to 12	9	8.2	11.9
>12	21	19.1	18.8

Table 26-4: Status of Ecoregion Protection Based on Amount of Protected Area.

ple, even in those ecosections with substantial representation (greater than 12%), wetlands and some mid- and low-elevation forested ecosystems are, to a large extent, poorly represented in existing protected areas (Earthlife Canada Foundation, 1993; Eng, 1992; Jones, 1992). Similarly, studies elsewhere in the province reveal that the figures on the percentage of ecosection protected may mask great unevenness in ecosystem representation (Canadian Parks and Wilderness Society, 1992 and 1993). High altitude habitats have received a disproportionate share of protective efforts, while other habitats of greater biological significance, such as forested valley bottoms, grasslands and aquatic ecosystems, have been neglected.

Completing the Protected Areas System

Many current public and private initiatives will affect the development of British Columbia's protected areas system. At the provincial level, there is the Protected Areas Strategy and the Commission on Resources and Environment (CORE). Federal initiatives include the Green Plan and ongoing Parks Canada and Environment Canada (Canadian Wildlife Service) programs. Among the private initiatives are cooperative agreements and various purchase plans.

In June, 1993, the province released the Protected Areas Strategy (Province of British Columbia, 1993), thereby committing itself to designating and managing a system of terrestrial and marine protected areas which aims to cover 12% of the province by the year 2000. The Strategy defines protected areas as land, fresh water and marine areas designated explicitly for the conservation of natural or cultural heritage and/or for recreation purposes that preclude commercial resource harvesting activities. The two goals for the protected areas system are:

○ to protect viable, representative examples of the natural diversity in the province representative of the major terrestrial, marine and freshwater ecosystems, the characteristic habitats, hydrology forms and the characteristic recreational and cultural heritage values; and

to protect the special natural, cultural heritage and recreational features of the province, including rare and endangered species and critical habitats, outstanding or unique botanical, zoological, geological and paleontological features, outstanding or fragile cultural heritage features, and outstanding outdoor recreational features such as trails.

The Strategy identifies processes and policies for identifying study areas; implementing interim management guidelines; and managing transitional issues, such as land and resource use tenures in candidate protected areas. Gap analyses, for example, are underway to determine to what extent the current system of protected areas represents the province's major ecosystems and to assist in identifying and selecting study areas possessing representative, rare, scarce and/or special elements which require protection. The Strategy also defines the relationship of the Protected Areas Strategy to other land use planning processes and presents an opportunity for the public to review and comment on protected areas legislation and management. In this way, the Strategy marks a new direction for the province—an integrated, coordinated and comprehensive approach to protected area planning which includes federal, provincial and regional initiatives.

One example of the way in which the Protected Areas Strategy integrates with other initiatives is its relationship to the Old Growth Strategy, which was completed in 1992 after two years of study and public involvement. The Old Growth Strategy identified the needs to include important old growth sites in a network of protected areas and to change forest management practices to ensure the perpetuation of old growth attributes in non-protected areas (British Columbia Ministry of Forests, 1992). Fourteen areas were recommended for protected area designation. These areas are now being considered in the context of the overall Protected Areas Strategy, through which still other important old growth sites will be identified.

Also in 1992, the provincial government established the Commission on Resources and Environment to: oversee land-use planning; conduct specific regional planning tasks; and move the province toward sustainable development (Commission On Resources and Environment, 1992). One of the Commission's principal tasks will be deciding upon, implementing and facilitating the protection of areas recommended through the Protected Areas Strategy.

In 1990, the federal government released Canada's Green Plan, a blueprint for a healthy environment (Government of Canada, 1990). This Plan commits the federal government to facilitating and contributing to setting aside 12% of Canada's total territory as protected space, in accordance with the goals of World Wildlife Fund Canada's Endangered Spaces campaign. To this end, the federal government plans to complete the terrestrial component of its national parks system and establish six new national marine parks by the year 2000. This work will be carried out by Parks Canada and the Canadian Wildlife Service in cooperation with the provinces and territories.

All or parts of eight of Parks Canada's 39 terrestrial natural regions and five of its 29 marine regions occur in British Columbia. At present, all but three of the terrestrial regions and one of the marine regions are represented by national parks or national park reserves. Potential candidates for a national park and a national marine park in the Strait of Georgia were identified through regional analyses between 1991 and 1993. Parks Canada has identified a site, Churn Creek, in the Chilcotin area of the Interior Dry Plateau natural region, which will be considered in the Commission on Resources and Environment's regional planning process for the Cariboo-Chilcotin. Both Parks Canada and the Canadian Wildlife Service are involved in the Protected Areas Strategy. The Canadian Wildlife Service recently completed a list of candidate sites for National Wildlife Areas, Migratory Bird Sanctuaries, Western Hemisphere Shorebird Network sites and Ramsar sites in the province.

Other means of acquiring and/or protecting significant natural sites are now either being considered or implemented. For example, a co-operative management agreement covering the Gwaii Haanas/ South Moresby National Park Reserve has been reached between Parks Canada and the Haida. Similarly, in 1992, a cooperative effort between BC Parks and the Nisga'a led to the establishment of the Nisga'a Memorial Lava Bed Provincial Park and Recreation Area. In 1993, a Memorandum of Understanding was signed between the Klahoose First Nation and the province to enable establishment of Von Donop Inlet Marine Park. Various provincial and federal government agencies and the Tsimshians are currently developing a cooperative approach to protecting and managing the Khutzeymateen as a grizzly bear sanctuary. As native land claim settlements and land use planning processes proceed, further cooperative and/or complementary initiatives between federal, provincial and aboriginal governments are anticipated.

The acquisition of properties by private groups or non-governmental organizations is gaining increasing momentum. Two of the more notable groups currently active in the province are the Nature Trust of British Columbia and the Nature Conservancy of Canada. Established in 1971, the Nature Trust acquires lands through gift or purchase and usually leases the land to a government or private agency for management. The Nature Trust has been involved in nearly 200 projects to date, conserving 12,500 hectares of land and water. The Nature Conservancy of Canada is a national organization similar to the Nature Trust. It recently received from Shell Canada Ltd. an area of 8,900 hectares in the Kootenays, the largest donation of its kind in Canadian history.

Ecological Integrity of Protected Areas

Setting aside an area does not in itself ensure its long-term protection or ecological integrity—that is, its capacity to withstand internal and external stresses. Maintenance of ecological integrity depends upon a variety of factors, including area viability, management policies and the compatibility of adjacent land uses.

In the absence of any comprehensive analysis of the ecological health of protected areas in the province, we just do not know how well their ecological integrity is being maintained, even after more than a century of managing them. We do know that protected areas face many threats, such as: oil spills; introduction of exotic species; incompatible land uses; global climate change; and recreation and tourism (Environment Canada, 1991a). Incompatible use of lands adjacent to protected areas, for instance, can lead to the loss of buffers, increased edge effects, fragmented or lost habitat, spill-over of these uses into protected areas, the introduction of exotic species, and the loss of wide-ranging wildlife species such as grizzly bears and caribou (Environment Canada, 1991a). Global climatic change may cause species extinction and/or migration and, conceivably, alter habitat to the point that protected areas can no longer sustain the ecosystems they were established to protect. This realization has led some to conclude that continuous intact ecosystems along north-south and elevational gradients will have to be protected to facilitate adaptation of ecosystems to new climatic regimes (Chapter 25, this volume).

Many protected area ecosystems have already been markedly changed by human activity. Commercial resource extraction activities, such as mining in Strathcona Park, have affected the integrity of some sites. As discussed in Chapter 17, accidental and intentional introductions of exotic flora and fauna have occurred. These exotic species often compete for suitable habitat with native species or have unexpected negative impacts on ecosystems. Non-native plant species have been used for landscaping and "beautifying" projects in protected areas. Coarse fish (commercially or recreationally unattractive species) have been poisoned and non-native species introduced to enhance fishing opportunities. Raccoons introduced to the Queen Charlottes are now threatening sea bird colonies. Knapweed has invaded some grassland ecosystems in protected areas and scotch broom now threatens endangered Garry oak ecosystems.

Some protected areas, most notably national and provincial parks, have dual and sometimes conflicting mandates: to protect ecosystems and to provide recreational opportunities. Recreation and tourism pressures in terms of visitation levels, new recreation opportunities and requests for expanded facility provision, are increasing. In 1992, BC Parks recorded more than 23 million visitors at its parks and recreation areas. Nearly two million visitors were recorded at the six national parks in the province that year. In some instances, these visitation levels have had negative impacts, such as site disturbances, species displacement and habitat destruction, on the ecological integrity of the protected area.

Intensive facility developments, such as campgrounds and day use areas occur in the majority of sites, whereas motels, golf courses and ski resorts occur in others. A townsite occurs within Yoho National Park, and transcontinental highway and railway corridors are found in some parks, such as Yoho and Revelstoke national parks and Mount Robson Provincial Park. Although these facilities occupy a very small percentage of the total area protected, they have an impact on larger areas and are sometimes located in important wildlife habitats.

Conflicts between recreationists and wildlife occur in a number of protected areas. There is increasing concern that the rising use of heli-hiking and heli-skiing, if not properly managed, may place increased stress on mountain ungulates. Similarly, possible conflicts between recreationists, such as snowmobilers and cross-country skiers, and mountain goats in the Babines Recreation Area are prompting BC Parks to initiate research into the possible impacts.

Various steps have been taken to mitigate recreation and tourism pressure. Use limits needed to be imposed at O'Hara Lake, the West Coast Trail and Bowron Lakes to protect both the natural environment and the quality of recreational experiences. In other sites, such as sea bird colonies and Robson Bight (Michael Bigg) Ecological Reserve, access restrictions needed to be imposed. Limiting or restricting access has been only one of the solutions adopted. Site hardening through, for example, provision of facilities like trails and boardwalks, has occurred in numerous areas. In others, facilities have been relocated away from sensitive features or major rehabilitation programs have been implemented (for example, trail redevelopment in Mount Assiniboine Provincial Park and alpine rehabilitation in Manning Provincial Park).

Many protected areas in the province are simply too small to maintain their ecological integrity and, therefore, biological diversity in today's rapidly changing world. None of them is large enough to include all the critical habitat requirements of wide-ranging mammal species, such as grizzly bears, caribou and wolves.

In effect, protected areas are becoming "islands in a sea of development," their ecological integrity threatened by fragmentation and encroachment of consumptive land uses, such as hydroelectric development, logging, mineral development, and oil and gas exploration and development. Their boundaries are frequently adjusted to accommodate surrounding land uses and their viewscapes negatively impacted, habitat fragmented or natural processes affected.

The importance of maintaining ecological integrity in protected areas is increasingly being acknowledged. Steps being taken to protect ecological integrity include the development of zoning frameworks, legislated boundaries, management policy changes, and increased integration with adjacent areas. Several agencies, such as BC Parks and Parks Canada, now prepare comprehensive management or master plans which are designed to protect the natural environments of protected areas. These plans include zoning frameworks that define levels of visitor use which are consistent with resource protection. Boundaries of protected areas are increasingly being established by legislation rather than by Order-in-Council, making boundary amendments more difficult. Management policies have also been strengthened in many regards. BC Parks has taken steps, for example, to prohibit all commercial resource extraction activities in all Class A parks. At the federal level, the amendments to the National Parks Act in 1988 improved the capability of Parks Canada to protect national park resources by placing priority on the maintenance of ecological integrity (Government of Canada, 1990).

The need to minimize the impact of adjacent land uses by more thoroughly integrating protected area planning with regional and local community planning was highlighted in the federal Green Plan (Government of Canada, 1990). Federal steps in this direction include the adoption of an ecosystem management approach by Parks Canada, the release of the first State of the Parks report (Environment Canada, 1991b), and more in-depth study of the threats posed by external land-use to the ecological integrity of Pacific Rim National Park Reserve. Some park managers, such as those at Revelstoke, Glacier and Pacific Rim national parks, have also begun to discuss ecosystem integrity with managers of adjacent lands and resources. Provincial initiatives, such as the Protected Areas Strategy, the Commission on Resources and Environment, Land and Resource Management Plans, and management planning exercises for individual protected areas, are also considering the relationships between protected areas and adjacent land uses.

Conclusions

Protected areas are an indispensable tool for preserving examples of much of British Columbia's natural diversity. Protected areas alone, however, are not enough. They will succeed in attaining their conservation objectives only if management of the surrounding land is compatible with those objectives. Not only must protected areas be sufficiently large to preserve ecological integrity in the long term, but good planning and management are needed within and sound stewardship and monitoring outside them. Other requirements are: comprehensive analyses of their ecological health; more active long-term management programs; regional strategies and greater cooperation with land managers beyond protected area boundaries; and the establishment of linkages between protected areas and buffer zones between protected areas and adjacent lands (Environment Canada, 1991a).

The time remaining for making the critical land allocation decisions required to ensure that British Columbia's ecosystems are adequately represented in protected area programs is shrinking as the pace of human activity and resulting landscape change quickens and intensifies. We need to ensure that the province's protected area system is soon completed and the full range of natural diversity protected.

Acknowledgements

Wayne McCrory, Valhalla Society, reviewed the manuscript.

References

British Columbia Ministry of Environment, Lands and Parks and Environment Canada. 1993. *State of the Environment Report for British Columbia.* Victoria, B.C.

British Columbia Ministry of Forests. 1992. *An Old Growth Strategy for British Columbia.* Victoria, B.C.

Canadian Parks and Wilderness Society, 1992. *B.C. Wildlands—Thompson-Okanagan Region.* Endangered Spaces Campaign Initiative No. 1. Vancouver, B.C.

Canadian Parks and Wilderness Society, 1993. *B.C. Wildlands—Southeast British Columbia Region.* Endangered Spaces Campaign Initiative No. 2. Vancouver, B.C.

Commission on Resources and Environment. 1992. *Report on a Land Use Strategy for British Columbia.* Victoria, B.C.

Earthlife Canada Foundation. 1993. *Gap Analysis: A Demonstration of Conservation Analysis Methods on Vancouver Island in the Northern Island Mountains Ecosection.* Earthlife Canada Foundation. Vancouver, B.C.

Eng, M. 1992. *Vancouver Island Gap Analysis.* B.C. Ministry of Forests, Research Branch, Victoria. Unpublished document.

Environment Canada. 1991a. *The State of Canada's Environment.* Ottawa, Ontario.

Environment Canada. 1991b. *State of the Parks—1990 Report.* Ottawa, Ontario.

Government of Canada. 1990. *The Green Plan.* Ottawa, Ontario.

Hummel, M. (ed.). 1989. *Endangered Spaces—the Future for Canada's Wilderness.* Key Porter Books. Toronto, Ontario.

Jones, T. 1992. *Preliminary "Coarse Filter" Gap Analysis for Vancouver Island.* Unpublished data prepared for BC Parks.

McNamee, K.A. 1990. Sustaining efforts to preserve Canada's wilderness heritage. In: *Heritage Conservation and Sustainable Development.* Heritage Resources Centre, University of Waterloo Occasional Paper 16. pp. 73-82.

McNeely, J.A., K.R. Miller, W.A. Reid, R.A. Mittermeier and T.B. Werner. 1990. *Conserving the World's Biological Diversity.* Gland, Switzerland and Washington, U.S.A. International Union for Conservation of Nature and Natural Resources, World Resources Institute, Conservation International, World Wildlife Fund-US and the World Bank.

Province of British Columbia. 1992. *Towards a Protected Areas Strategy.* Victoria, B.C.

Province of British Columbia. 1993. *A Protected Areas Strategy for British Columbia.* Victoria, B.C.

Rubec, C.D.A., A.M. Turner and E.B. Wiken. 1992. *Integrated Modelling for Protected Areas and Biodiversity Assessment in Canada.* Paper presented at Canadian Society of Land Ecology and Management, National Workshop '92, June 17-19, 1992, Edmonton, Alberta.

Task Force on Northern Conservation. 1984. *Report of the Task Force on Northern Conservation.* Ottawa: Department of Indian Affairs and Northern Development.

Turner, A.M., C.D.A. Rubec and E.B. Wiken. 1992. Canadian ecosystems: a systems approach to conservation. In: *Science and the Management of Protected Areas.* J.H.M. Williston et al. (eds.). Elsevier Science Publishers, Amsterdam, Netherlands. pp. 117-127.

World Commission on Environment and Development. 1987. *Our Common Future.* Oxford University Press, Oxford, U.K.

World Conservation Union. 1991. *Caring for the World: A Strategy for Sustainability.* Morges, Switzerland.

World Resources Institute, The World Conservation Union and United Nations Environment Programme. 1992. *Global Biodiversity Strategy—Guidelines for Action to Save, Study, and Use Earth's Biotic Wealth Sustainably and Equitably.*

Vold, T. 1992. *The Status of Wilderness in British Columbia: A Gap Analysis.* Appendix C: Representation by biogeoclimatic classification (Zone, Subzone, Variant). Ministry of Forests.

World Wildlife Fund Canada. 1993. *Endangered Spaces Progress Report Number 4.* Toronto, Ontario.

Editor's Note: On January 13, 1994, the provincial government announced the establishment of Ts'yl-os Provincial Park. This park, encompassing 233,240 hectares, is the sixth largest provincial park in British Columbia and will significantly increase representation (to over 12%) of the Central Chilcotin Ranges ecosection.

Chapter 27
The British Columbia Ecological Reserves Program[1]

The <u>Ecological Reserve Act of 1971</u> (British Columbia Statutes, 1971) established the Ecological Reserve Program, whose basic purpose is to "permanently preserve natural ecosystems, species and phenomena" (British Columbia Ecological Reserves Program, 1992). Its objectives are to: keep options open for research into natural ecological processes; preserve banks or reservoirs of genetic material; provide benchmark areas against which human modification of the rest of the province can be measured; and provide a variety of outdoor classrooms.

The British Columbia Ecological Reserve Program is administered by the Ministry of Environment, Lands and Parks. Staff at headquarters and throughout the province are aided by some 90 volunteer wardens.

At present (February 1993), 131 ecological reserves have been established in British Columbia (Figure 27-1), making this program the most successful of its kind in Canada. These reserves vary in size from 0.6 to 48,560 hectares, with an average size of 1,212 hectares. The total area encompassed by these reserves is 158,753 hectares, nearly one third of which is marine and the remaining two thirds of which cover 0.1% of the land area of the province. Reserves are located in all regions of the province and contain representative samples of many ecological zones and subzones, as well as of unique or special species and ecosystems. Table 27-1 provides a list of the reserves in numerical order, together with information on their location, main features, and size. Three of the reserves (numbers 44, 95, and 96) established under the British Columbia Ecological Reserves Program have since become part of the Gwaii/Haanas/South Moresby National Park Reserve.

Only provincial Crown lands may be made into ecological reserves. Occurrences of rare species, habitats or natural phenomena on private lands can only be protected after these lands are acquired by or for the Crown. Crown lands amount to 92% of the area of the province. However, the percentage of Crown land near centres of settlement and at low elevation in the south is considerably lower, and it is there that the majority of threatened and endangered species and habitats are found.

The kinds of lands sought for ecological reserves are:

1. representative samples of the great variety of natural biogeoclimatic zones and subzones which occur in British Columbia, and a range of significant habitats within them;

B.C. Parks
Ministry of Environment,
Lands and Parks
800 Johnson Street
Victoria, B.C.
V8V 1X4

---

[1] Reprinted from *Guide to Ecological Reserves in British Columbia* by permission of the British Columbia Ministry of Environment, Lands and Parks. Updated (to November, 1992) by ministry staff for this publication.

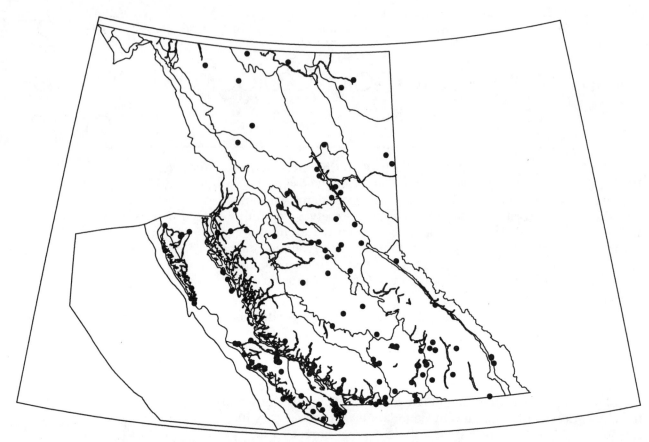

Figure 27-1. Locations of British Columbia's Ecological Reserves, With Ecoregion Boundaries. Source: B.C. Ministry of Environment, Lands and Parks.

2. areas where rare, endangered or sensitive plants, animals and habitats require protection from disturbance;

3. lands or waters that contain unique or special examples of botanical, zoological or geological phenomena;

4. areas needed for scientific research on and demonstration of biological productivity and diversity, and related aspects of the natural environment; and

5. examples of ecosystems that have been modified by humans and provide an opportunity to study the recovery of the natural system from modification.

It is this focus on preserving representative and unique samples of ecosystems and rare species, habitats and natural phenomena that makes the Ecological Reserves Program unique among the other protected area programs described in Chapter 24. Its objectives and approach offer the greatest opportunity to conserve biodiversity.

Number	Name	Location	Main Feature(s)	Area
1	Cleland Island	Clayoquot Sound	Seabird colony	7.7
2	East Redonda Island	N end of Georgia Strait	Two biogeoclimatic zones with many habitats	6,212
3	Soap Lake	S of Spences Bridge	Alkaline lake and Douglas fir forest	884
4	Lasqueti Island	Georgia Strait	Shoreline forest with Rocky Mountain juniper and cactus	201
5	Lily Pad Lake	S of Lumby	Undisturbed boggy lake on Interior Plateau	101
6	Buck Hills Road	S of Lumby	Western larch stand	16
7	Trout Creek	SSW of Summerland	Ponderosa pine parkland and several snake species	75
8	Clayhurst	S of Clayhurst	Peace River parklands	316
9	Tow Hill	Graham Island, Queen Charlotte Islands	Sandy beach, dunes, swamp and peat bogs	514
10	Rose Spit	Graham Island, Queen Charlotte Islands	Sand dunes and shoreline meadows	170
11	Sartine Island	One of Scott Islands	Seabird colony	13
12	Beresford Island	One of Scott Islands	Seabird colony	7.7
13	Anne Vallee (Triangle Island)	Outermost of Scott Islands	Largest seabird and sea-lion colonies in province	85
14	Solander Island	Off Cape Cook, Vancouver Island	Seabird colony	7.7
15	Saturna Island	Strait of Georgia	Coastal Douglas fir forest	131
16	Mount Tuam	Saltspring Island	Arbutus-Douglas fir forest	254
17	Canoe Islets	Near Valdes Island	Seabird colony	0.6
18	Rose Islets	Trincomali Channel, N of Reid Island	Cormorant and gull colony	0.8

Table 27-1: Location and Description of the Ecological Reserves of British Columbia.

Number	Name	Location	Main Feature(s)	Area
19	Mount Sabine	N of Canal Flats	Montane spruce forest	7.9
20	Columbia Lake	E side of Lake	Rare plants on limestone cliffs and along a calcareous stream	32
21	Skagit River Forest	S of Hope	Douglas fir forest	73
22	Ross Lake	S of Hope	Ponderosa pine in a semi-coastal environment	61
23	Moore/ McKenney/ Whitmore Islands	Hecate Strait	Seabird colony and Sitka spruce forest	73
24	Baeria Rocks	Barkley Sound	Gull colony and subtidal marine life	53
25	Dewdney and Glide Islands	Hecate Strait	Variety of maritime bog, pond and scrub forest communities	3,845
26	Ram Creek	East Kootenays	Hotsprings and associated plants; burned forest	121
27	Whipsaw Creek	SW of Princeton	Interior Douglas fir and ponderosa pine stands	32
28	Ambrose Lake	Sechelt Peninsula	Coastal bog lake	228
29	Tranquille	W of Kamloops	Sagebrush, ponderosa pine, Douglas fir plant communities	235
30	Vance Creek	N of Lumby	Forest stands transitional between Interior Douglas Fir and Interior Cedar-Hemlock Zones	49
31	Lew Creek	E of Upper Arrow Lake	A complete drainage basin including three biogeoclimatic zones	815
32	Evans Lake	Valhalla Park, W of New Denver	Subalpine forests including disjunct stand of yellow cedar	185
33	Field's Lease	W of Osoyoos Lake	Semi-arid shrub grassland communities	4.2

Table 27-1: Location and Description of the Ecological Reserves of B.C. (Continued).

Number	Name	Location	Main Feature(s)	Area
34	Big White Mountain	E of Kelowna	Subalpine and alpine plant communities	951
35	Westwick Lake	S of Williams Lake	Aspen parkland vegetation in Cariboo area	27
36	Mackinnon Esker	NW of Prince George	Long compound esker with lichen-woodland communities	583
37	Mount Maxwell	Saltspring Island	Garry oak stand	65
38	Takla Lake	E of Hazelton	Most northerly known occurrence of Douglas-fir	263
39	Sunbeam Creek	Near McBride	Alpine vegetation in Rockies	511
40	Kingcome/ Atlatzi River	Head of Kingcome Inlet	Rich alluvial swamps, bogs and forest	414
41	Tacheeda Lakes	N of Prince George	Representative sub boreal spruce forest	526
42	Mara Meadows	E of Salmon Arm	Unique calcareous fen; rare orchids	189
43	Mount Griffin	N of Mabel Lake	Interior cedar-hemlock and subalpine forest over wide elevational range	1,376
44	East Copper/ Jeffrey/Rankine Islands	East coast of Moresby Island	Seabird colonies; now part of South Moresby Park Reserve	121
45	Vladimir J. Krajina (Port Chanal)	West coast of Graham Island	Virgin littoral environment, lowland Sitka spruce forest, rare mosses, seabird colony, spawning salmon, endemic birds and mammals	9,834

Table 27-1: Location and Description of the Ecological Reserves of B.C. (Continued).

Number	Name	Location	Main Feature(s)	Area
46	Sikanni Chief River	NW of Hudson's Hope, headwaters of Sikanni Chief River	Engelmann spruce at norther extremity of range. Alpine flora and fauna	2,401
47	Parker Lake	W of Fort Nelson	Boreal bog habitat. Preservation of pitcher plant (Sarracenia purpurea) and other rare species	259
48	Bowen Island	W of Apodaca Provincial Park	Forest of Douglas-fir and red cedar. Dry subzone of Coastal Western Hemlock Zone	397
49	Kingfisher Creek	Hunters Range, ESE of Sicamous	Subalpine parkland vegetation in Monashee Mountains	1,441
50	Cecil Lake	NE of Fort St. John	Fen, black spruce bog, and aspen communities	129
51	Browne Lake	N of McCulloch	Meadow and forest rich in wild flowers	124
52	Drizzle Lake	SE of Masset	Lake and bogland; unique species of stickleback; nesting loons	837
53	Narcosli Lake	W of Quesnel	Waterfowl breeding grounds; well developed aquatic communities	1,098
54	Nitinat Lake	Vancouver Island SE of Bamfield	Population of Douglas fir in Coastal Western Hemlock Zone	79
55	Cardiff Mountain	SW of Hanceville	Example of lava plateau, basalt columns and crater lake	65
56	Goosegrass Creek	W of Mica Reservoir	An elevational sequence of the biogeoclimatic zones and a complete watershed	2,185
57	Chickens Neck Mountain	N of Dease Lake	Climax stand of white spruce and subalpine fir	680
58	Blue/Dease Rivers	W of Lower Post	Terrestrial and aquatic communities of the Boreal White and Black Spruce Zone	777

Table 27-1: Location and Description of the Ecological Reserves of B.C. (Continued).

Number	Name	Location	Main Feature(s)	Area
59	Ningunsaw River	SE of Bob Quinn Lake	Interior Cedar-Hemlock Zone near its northern limit; Engelmann Spruce-Subalpine Fir and Alpine Tundra Zones	2,046
60	Drywilliam Lake	S of Fraser Lake	An isolated old growth stand of Douglas fir in the Sub-Boreal Spruce Zone	95
61	Upper Shuswap River	E of Mabel Lake	Excellent alluvial stands of western redcedar	70
62	Fort Nelson River	NE of Fort Nelson	White spruce developing within alluvial stands of balsam poplar	121
63	Skeena River	Near confluence of Exchamsiks River	Mature cottonwood stands on floodplain islands	91
64	Ilgachuz Range	N of Anahim Lake	Isolated mountain pass with alpine-subalpine flora and fauna of biogeographic interest	2,914
65	Chasm	N of Clinton	Ponderosa pine near its northern limits	197
66	Ten Mile Point	Victoria	Inter and subtidal marine life	11
67	Satellite Channel	Between Saltspring Island and Saanich Peninsula	Rich subtidal marine life, particularly benthic infauna	343
68	Gladys Lake	Spatsizi Wilderness Park	Stone sheep, mountain goats and their environment	48,56
69	Baynes Island	Squamish River	Alluvial black cottonwood forest on undisturbed island	71
70	Mount Tinsdale	ESE of Barkerville	Extensive transition between alpine and subalpine zones	419
71	Blackwater Creek	NW of Mackenzie	Boreal forest and portion of extensive low moor area	234

Table 27-1: Location and Description of the Ecological Reserves of B.C. (Continued).

Number	Name	Location	Main Feature(s)	Area
72	Nechako River	W of Prince George	Best stand of tamarack west of Rocky Mountains	133
73	Torkelsen Lake	W of Babine Lake	Low moor wetlands with cloudberry and rich moss flora	182
74	U.B.C. Endowment Lands	Vancouver	Second-growth forest of Puget Sound Lowlands	90
75	Clanninick Creek	Near Kyuquot	Alluvial Sitka spruce	37
76	Fraser River	W of Chilliwack	Alluvial cottonwood forest on islands	76
77	Campbell-Brown (Kalamalka Lake)	W side of lake near Rattlesnake Point	Ponderosa pine-bunch-grass site; rattlesnake den	107
78	Meridian Road (Vanderhoof)	S of Vanderhoof	Engelmann spruce-subalpine fir-lodgepole pine forest	262
79	Chilako River	S of Vanderhoof, N of Batnuni Lake	Tamarack at its southern limit in BC; swamp, fen, and bog mosaic	64
80	Smith River	Near junction with Liard River	Complete drainage basin with boreal white and black spruce forest	1,326
81	Morice River	SW of Houston	Burned sub-boreal spruce forest	358
82	Cinema Bog	NNE of Quesnel	Lowland black spruce-sphagnum bog	68
83	San Juan Ridge	E of Port Renfrew	Protection of rare white avalanche lily (Erythronium montanum)	98
84	Aleza Lake	"Big Bend" of Fraser River	Sub-boreal spruce forest, lakes, and wetland ecosystems	242
85	Patsuk Creek	NW of Mackenzie	Paper birch and other seral forest species	554
86	Bednesti Lake	W of Prince George	Kettle lake wetland succession	139
87	Heather Lake	NW of Mackenzie	Productive aspen forest	235

Table 27-1: Location and Description of the Ecological Reserves of B.C. (Con't).

Number	Name	Location	Main Feature(s)	Area
88	Skwaha Lake	N of Lytton	Montane and sub-alpine forest with superb flower meadows	850
89	Skagit River Cottonwoods	Skagit River Valley	Excellent cottonwood stands	69
90	Sutton Pass	W of Port Alberni	Rare Alder's tongue fern (Ophioglossum vulgatum)	3.4
91	Raspberry Harbour	W side of Williston Lake	Tall lodgepole pines on excellent growth site	143
92	Skihist	NE of Lytton	Ungrazed ponderosa pine - bunchgrass site	36
93	Lepas Bay	Northern Graham Island, Queen Charlotte Islands	Storm-petrel colony on island	3.6
94	Oak Bay Islands	E of Victoria	Spring flowers, rare plants, a seabird colony, and marine life	211
95	Anthony Island	W of Moresby Island, Queen Charlotte Islands	Small islets with nine species of nesting seabirds and rich marine life; now part of South Moresby National Park Reserve	324
96	Kerouard Islands	S of Moresby Island, Queen Charlotte Islands	Major sea-lion rookery, seabird colony and rich marine life; now part of South Moresby National Park Reserve	130
97	Race Rocks	Metchosin, Vancouver Island	High-current marine community; winter sea-lion haul-out; nesting seabirds	220
98	Chilliwack River	U.S. border, SE of Chilliwack	Mature alluvial forest with large western redcedars	86
99	Pitt Polder	Near Maple Ridge	Two forested hills surrounded by swamp and bog	88
100	Haynes' Lease	N end of Osoyoos Lake	Most arid ecosystem in Canada, including rare plants and animals	101

Table 27-1: Location and Description of the Ecological Reserves of B.C. (Con't).

Number	Name	Location	Main Feature(s)	Area
101	Doc English Bluff	SE of Williams Lake	Limestone cliff with several species of rare flowers and ferns, colony of white-throated swifts, and golden eagle nest	52
102	Charlie Cole Creek	S of Teslin; ESE of Atlin	Cold-water springs used by ungulates as mineral licks	162
103	Byers/Conroy/ Harvey/Sinnett Islands	Hecate Strait	Eight species of nesting seabirds; tree nesting peregrine falcons; extensive subtidal area	12,205
104	Gilnockie Creek	E of Kingsgate in East Kootenays	Mature western larch	58
105	Megin River	NE of Estevan Point, W coast of Vancouver Island	Typical west coast alluvial forests	50
106	Skagit River Rhododendrons	SE of Hope	Two stands of Pacific Rhododendrons	70
107	Chunamon Creek	NE of Germansen Landing, Williston Lake	Two small drainages; Engelmann and white spruce forest on sites of varying productivity	344
108	Cougar Canyon	S of Vernon, E side of Kalamalka Lake	Mosaic of dry forest communities; chain of small lakes with associated wetlands	550
109	Checleset Bay	NW of Kyuquot, Vancouver Island	BC's major sea otter population; extensive subtidal environments	34,650
110	McQueen Creek	N of Kamloops	Native grassland with many wildflowers	35
111	Robson Bight (Michael Bigg)	Johnstone Strait, SE of Telegraph Cove	Heavily used killer whale habitat; undisturbed estuary; forested shoreline	1,753
112	Mount Tzuhalem	Duncan, Vancouver Island	Garry oak, spring wildflower ecosystems	18
113	Honeymoon Bay	Cowichan Lake, Vancouver Island	Outstanding population of pink fawn lily (Erythronium revolutum)	7.5

Table 27-1: Location and Description of the Ecological Reserves of B.C. (Continued).

384

Number	Name	Location	Main Feature(s)	Area
114	Williams Creek	SE of Terrace	Representative coastal western hemlock forest and outstanding terraced bogs	700
115	Gingietl Creek	30km upstream of mouth of Nass River	Undisturbed watershed containing coastal western hemlock, mountain hemlock, and alpine vegetation	2,873
116	Katherine Tye (Vedder Crossing)	SE of Chilliwack	Rare phantom orchid (Cephalanthera austinae)	3.1
117	Haley Lake	W of Nanaimo	Vancouver Island marmot (Marmota vancouverensis)	120
118	Nimpkish River	85 km W of Campbell River, Vancouver Island	Sample of Canada's tallest Douglas firs	18
119	Tahsish River	50 km S of Port McNeill, Vancouver Island	Pristine westcoast estuary	70
120	Duke of Edinburgh (Pine/Storm/ Tree Islands)	Off Vancouver Island, 35 km NW of Port Hardy	Largest seabird nesting colony in Queen Charlotte Strait	660
121	Brackman Island	N of Sidney, Vancouver Island, near Portland Island	Pristine, ungrazed Gulf Island vegetation and marine buffer	35
122	Tsitika Mountain	Tsitika Valley, NE Vancouver Island	Alpine communities, wet subalpine forest, unusual terraced fen, and small lake	554
123	Mount Derby	Tsitika Valley, NE Vancouver Island	Alpine peak and precipitous, partly forested slopes	557
124	Tsitika River	Tsitika Valley, NE Vancouver Island	Low-elevation swamp/fen/bog complex	110
125	Mount Elliott	Tsitika Valley, NE Vancouver Island	Representative subalpine subdrainage surrounding Cirque Lake	324
126	Claud Elliott Creek	Tsitika Valley, NE Vancouver Island	Representative hemlock, amabilis fir and red cedar forest	231

Table 27-1: Location and Description of the Ecological Reserves of B.C. (Con't).

Number	Name	Location	Main Feature(s)	Area
127	Big Creek	SW of Williams Lake	Representative, ungrazed grasslands in the Chilcotin region	257
128	Galiano Island	N end of Galiano Island	Rare undisturbed peat bog ecosystem in dry Coastal Douglas-Fir zone	30
129	Klaskish River	SW of Port Alice, Vancouver Island	Estuary and alluvial forest in Coastal Western Hemlock zone; native oysters	132
130	Mahoney Lake	S of Okanagan Falls	Saline lake with unique limnological features of international significance	29
131	Stoyama Creek	Near Boston Bar and the Fraser River Canyon	Meeting of three biogeoclimatic zones, with seven coniferous tree species; to conserve special seed provenances	76
132	Trial Islands	S of Oak Bay, Vancouver Island	The most outstanding assemblage of rare and endangered plant species in British Columbia	23
133	Gamble Creek	E of Prince Rupert	North coastal forest/bog complex and occurrence of pacific silver fir near the northern limit of its range	1,026
134	Ellis Island	W of Vanderhoof on Fraser Lake	Inland breeding colony of herring and ring-billed gulls	1

**Revised Total Area   158,753 hectares
** At least 50,000 ha of this total area cover marine water

Table 27-1: Location and Description of the Ecological Reserves of British Columbia (Continued).

Ecological reserves have been established in all of the 14 biogeoclimatic zones (Figure 18-1) and many of the sub-zones in British Columbia (Table 27-2), though some are better represented than others. Some mountainous reserves span two or even three zones due to their elevation differences. Although several reserves are located in each zone, not all have vegetation representative of the zone because they cover specialized habitats such as bogs, wetlands, lakes, alluvial deciduous forest, or wildflower stands. In the following discussion, we consider the degree to which the biogeoclimatic zones, as well as the marine environment, are represented in the ecological reserves system.

Thirteen reserves contain significant areas of the Alpine Tundra Zone. Most are in the southern interior and contain other zones as well. Four reserves, two in the northern and two in the southern interior, consist almost entirely of alpine habitat. Alpine representation in the southern coastal area is needed.

Biogeoclimatic Zone[1]	Ecological Reserve Number[2]
1. Alpine Tundra	(2), 31, 32, 34, 39, (43), (45), 46, 56, (57), 59, 64, 68, 70, 114, 115, (122), (123)
2. Spruce - Willow - Birch	46, 57, 68, 80
3. Boreal White and Black Spruce	8, 47*, 50, 57, 58, 62, 80, 102
4. Sub-Boreal Spruce	36, 38, 41, 60, 71, 72, 73, 79, 81, 82*, 84, 85, 86, 87, 91, 107
5. Sub-boreal Pine - Spruce Zone	53, 55
6. Mountain Hemlock	2, 45, 83, 114, 115, 117, 122, 123, 125, (131), 133
7. Engelmann Spruce - Subalpine Fir	31, 32, 34, (36), (38), 41, 43, 49, 56, 59, 64, 70, 78, 85, 88, 107, (131)
8. Montane Spruce	5, 6, 19, 26, 51, (88), 104
9. Ponderosa Pine	7, 29, 92, 130
10. Bunchgrass	33, 100, 110
11. Interior Douglas Fir	3, (7), 20, 21, 22, 27, 29, 30, 35, (51), 65, 77, 88, 101, (106), 108
12. Coastal Douglas Fir	4, 15, 16, 17*, 18*, 37, 94*, 97*, 112, 121, 128, 132*
13. Interior Cedar - Hemlock	(30), 31, 42*, 43, 56, 59, 61
14. Coastal Western Hemlock	1*, 2, 9, 10, 11*, 12, 13*, 14*, 23, 24*, 25, 28, 40, 44, 45, 48, 52, 54, 63, 69, 74, 75, 76, 89, 90, 93, 95, 96*, 98, 99*, 103, 105, 106, 109, 111, 113, 114, 115, 116, 118, 119, 120*, 122, 123, 124, 126, 129, 131, 133

Table 27-2: Representation of Biogeoclimatic Zones in British Columbia's Ecological Reserves
Key: 1. Follows classification used by the B.C. Ministry of Forests
    2. Reserves consisting entirely of intertidal/subtidal environments (66, 67, 111)
    () Indicates that zonal vegetation is only marginally represented
    * Reserves occurring in zones defined by their forest vegetation but having non forest cover over most of their land

Although only four reserves contain Spruce-Willow-Birch vegetation, some fairly extensive and representative tracts are included.

Seven reserves contain vegetation typical of the Boreal White and Black Spruce Zone, and two of them consist almost entirely of this type of vegetation. Four reserves in this zone contain vegetation, other than forests of white and black spruce (*Picea glauca* and *Picea mariana*), such as bogs, or alluvial deciduous stands.

The Sub-Boreal Pine-Spruce Zone, which occurs on the high plateau of the west central interior, is not well represented at present. Only two reserves currently exist in this zone.

Sixteen reserves fall partly or entirely within the extensive Sub-boreal Spruce Zone of the central interior. While typical, several are not true to the zonal name in that they are dominated by lakes, bogs, or stands of birch, aspen (*Populus tremuloides*), or tamarack (*Larix laricina*).

The Mountain Hemlock Zone, a coastal subalpine type, is represented in eleven reserves. Nearly half of these reserves occur on Vancouver Island. The southern mainland coast and mountains at the extreme northern extent of the zone are still poorly represented.

The Engelmann Spruce-Subalpine Fir Zone is quite well represented. Relatively typical forest stands predominate in at least ten reserves and cover parts of several others. Engelmann spruce-subalpine fir forests are marginally represented in two other reserves.

Seven ecological reserves contain elements of the Montane Spruce Zone, a zone of fairly limited extent in the southern interior. Trees other than spruce, such as lodgepole pine (*Pinus contorta*) and western larch (*Larix occidentalis*), are common in several of those reserves.

The Ponderosa Pine Zone is represented in four reserves, although some are transitional to the Interior Douglas Fir or Bunchgrass zones. These reserves are mostly in settled valleys and their average size tends to be small. Much the same is true of the closely associated Bunchgrass Zone. Additional grassland reserves are needed.

The Interior Douglas Fir Zone is quite well represented in the present system. At least seven reserves consist predominantly of Douglas-fir (*Pseudotsuga menziesii*) stands; 11 others have a mosaic of Douglas-fir, other tree species, and non-forest vegetation, or contain marginal examples.

The Coastal Douglas Fir Zone is not well represented because most lands within this zone are privately owned. Although 12 reserves are established within this zone, only about four may be considered to contain typical forest stands, and none of these are on southeastern Vancouver Island where the bulk of the zone occurs. The remainder contain atypical forest stands or were established mainly to preserve wildflowers or seabird nesting sites.

Considering its extent, the Interior Cedar Hemlock Zone is not well represented in the existing seven reserves. Five of these reserves contain representative forest stands, the sixth is marginally representative, and the seventh is predominantly bogland.

The Coastal Western Hemlock zone is better represented than any other. About 49 reserves fall entirely or partly within this zone. Many of these reserves were established to preserve special features, like seabird nesting islands, marine environment, or rare or unique plants. Many others contain atypical forest stands (such as alluvial sitka spruce [*Picea sitchensis*] or cottonwood [*Populus trichocarpa*]) forests or non-forest vegetation (grassy or shrubby islands, bogs, or lakes). Western hemlock (*Tsuga heterophylla*) stands are well represented, even if not in ideal distribution throughout the zone.

Intertidal and subtidal marine environments are contained within eleven reserves; an additional reserve is completely subtidal and one has intertidal but no subtidal habitat. High-energy marine environments of the west coasts of Vancouver Island and the Queen Charlottes are included in ecological reserves, as are somewhat more sheltered sites in Hecate Strait, Johnstone Strait, the Gulf of Georgia and the Strait of Juan de Fuca.

Representation of Unique and Rare Features

Much progress has been made toward the protection of unique, rare or sensitive ecosystems, habitats, species, and natural phenomena within ecological reserves in British Columbia. Quite a few reserves were created to preserve unique features, while many others contain both unique features and representative ecosystems. Although only a few reserves have been set aside primarily to preserve physical features, such as basalt columns or eskers, the ecological reserves system contains many interesting terrain features, like alpine glaciers, moraines, patterned ground, caves, sand dunes, sea cliffs, and islands. Sites suitable for the study of processes like formation and erosion of river bars, littoral drift, and slope instability are present. Freshwater features include complete watersheds, alpine tarns, bog lakes, alkaline lakes, hot and cold mineral springs, and a variety of stream types.

British Columbia's Ecological Reserve Act calls specifically for protection of rare, threatened and endangered species "in their natural habitat." Several reserves have as their major purpose the protection of rare or unique plant species or associations, such as outstanding wildflower stands (mostly on Vancouver Island and the Gulf Islands), or economically important tree stands on exceptional sites or beyond their normal distribution range, such as stands of Douglas-fir, ponderosa pine (*Pinus ponderosa*), and Engelmann spruce (*Picea engelmannii*) at or beyond their usual northern limits, tamarack beyond its usual southern limits, the only known yellow cedars (*Chamaecyparis nootkatensis*) in interior British Columbia, and exceptional stands of Garry oak *(Quercus garryana)*, arbutus *(Arbutus menziesii)*, paper birch *(Betula paperyphera)*, aspen, black cottonwood *(Populus tricocarpa)*, lodgepole pine, western red cedar (*Thuja plicata*), and sitka spruce within their usual zones of occurrence. Wetlands (fens, bogs, swamps, and estuaries) are major components of several reserves, while others feature dune ecosystems or limestone vegetation. Several reserves are important because their vegetation has been disturbed and provides opportunities to study long-term recovery.

389

Seabird colonies are an important and easily disturbed wildlife resource on the British Columbian coast. Most large and/or sensitive colonies are now preserved within 20 ecological reserves, some of which also contain other unique features, like sea lion breeding rookeries or haul-outs (nine reserves), sea otters (*Enhydra lutris*; one reserve), subtidal marine environments, or unique island vegetation. One reserve has been set aside primarily for Killer Whales (*Orcinus orca*). Other wildlife features preserved in ecological reserves are eagle, falcon, or Sandhill Crane (*Grus canadensis*) nesting habitat, endemic (native to the region) insular races of birds and mammals, and snake populations. Big game animals have not yet received much attention, mostly because they require very large reserves and because other forms of protection are available. However, one reserve was established to protect mineral springs used as a game lick, and another mainly to protect large mammals and their habitat in the Spatsizi area.

Freshwater organisms have not been a major focus of reserve establishment. However, one reserve was designated largely to protect unique stickleback populations, four contain spawning runs of salmon, and a third is a site of interesting aquatic insects including several "type" specimens.

Special features of marine reserves include protection of: rare invertebrates; communities adapted to heavy surf and strong currents; and intertidal or benthic habitats of exceptional diversity or productivity.

Research and Education

Research is encouraged in ecological reserves, provided it is not detrimental to the values of the ecological reserve.

Research and inventory projects have been carried out in about half of the reserves, and several involve ongoing, long-term studies (British Columbia Ecological Reserves Program, 1991). A few reserves have been the subject of considerable research, often on high-profile species such as seabirds (Cleland Island and Anne Vallee [Triangle Island] ecological reserves), the rare Vancouver Island marmot (*Marmota vancouverensis*), sea otters (*Enhydra lutris*) (Checleset Bay Ecological Reserve) and killer whales (*Orcinus orca*) (Robson Bight Ecological Reserve), but also involving less spectacular but equally important features like vegetation (several reserves), or sticklebacks (Drizzle Lake Ecological Reserve).

Both federal and provincial government agencies are presently using ecological reserves for long-term monitoring of such diverse biotic features as benthic marine invertebrates, insects, and forest stands.

Ecological reserves offer opportunities for educational activities ranging from simple observation to the teaching of complex ecological processes. Due to their proximity, several reserves, like Baeria Rocks near Bamfield Marine Station and Race Rocks near Lester Pearson College of the Pacific, are valuable to research institutions because they presently serve as important teaching aids for school and university staff, and as undisturbed locations for graduate students to carry out thesis research.

Conclusions

In terms of the number of and area covered by reserves, British Columbia's ecological reserve system is advanced relative to other provinces' systems. This is not to say, however, that the system is complete or without shortcomings at the present.

There are significant gaps and imbalances in representation, for instance, both of the geographical areas of the province and of the biogeoclimatic zones within those areas. For instance, neither the Sub-boreal Pine-Spruce nor the Coastal Douglas Fir Biogeoclimatic Zone is well represented. It is anticipated that the use of both ecosections (Demarchi et al., 1990) and biogeoclimatic zones, subzones and variants (Meidinger and Pojar, 1991) to complete the B.C. Protected Areas Strategy will alleviate these imbalances. There are further imbalances at the community type or vegetation cover level. Representation of vegetation types that have high productivity or large standing crops, such as old-growth forests or rich grasslands, is still poor.

Many reserves, especially those in the south, may not be ecologically viable in the long term because they are too small and, therefore, too subject to influences from surrounding land uses. Some reserves whose ecosystem evolution was influenced by fires will change where management through prescribed burning cannot be carried out in the foreseeable future.

The Ecological Reserves Act regulations are enforceable only through a complex process and are currently being updated to improve its enforcement capabilities. Some known infractions, such as trespass, grazing or vehicle use, are difficult to deal with in the absence of a clear boundary demarcation, such as fencing.

Finally, research is still very uneven, with the lion's share occurring in a few "popular" reserves. Not only do research opportunities need to be better advertised, but also budgets—or at least "seed money"—need to be provided.

Much work remains to be done. British Columbia still has the opportunity to develop an outstanding ecological reserve system. However, particularly in the south and at low elevations, opportunities to fill the gaps in the existing reserve system disappear fast, so they must be acted upon now or it will be too late.

Acknowledgements

George Douglas, British Columbia Conservation Data Centre, reviewed the manuscript.

References

British Columbia Ecological Reserves Program. 1991. *List of Reports and Publications for Ecological Reserves in British Columbia.* Ministry of Environment, Lands and Parks, Victoria, B.C.

British Columbia Ecological Reserves Program. 1992. *Guide to Ecological Reserves in British Columbia.* British Columbia Ministry of Environment, Lands and Parks, Victoria, B.C.

British Columbia Statutes. 1971. *Ecological Reserves Act.* Queens Printer, Victoria, B.C. pp. 81-83.

Demarchi, D.A., R.D. Marsh, A.P. Harcombe and E.C. Lea. 1990. The environment. In: *The Birds of British Columbia. Vol. 1: Nonpasserines, introduction and loons through waterfowl.* R.W. Campbell, N.K. Dawe, I. McTaggert-Cowan, J.M. Cooper, G.W. Kaiser, and M.C.F. McNall. (eds.). Royal British Columbia Museum in association with Environment Canada, Canadian Wildlife Service, Victoria, B.C. pp. 55-144.

Meidinger, D. and J. Pojar. 1991. *Ecosystems of British Columbia.* British Columbia Ministry of Forests.

Chapter 28
Conserving Marine Ecosystems: Are British Columbia's Marine Protected Areas Adequate?[1]

Despite the fact that the marine environment encompasses two-thirds of the Earth's surface, little attention has been paid to marine conservation issues until relatively recently (Eiswerth, 1990; Hawkes, 1990 and 1992; Earle, 1991; Thorne-Miller and Catena, 1991; MacInnis, 1992; Norse, 1993). Of the 27,000 kilometres of coastline in B.C., there is not one kilometre within which all marine organisms, both commercial and non-commercial, are completely protected from harvesting. Marine protected areas are essential components of any coastal conservation strategy (Lien and Graham, 1985; Bohnsack et al., 1989; Graham, 1990; Beatley, 1991). Dethier et al. (1989) note that,

> There should be a push for the establishment of large marine preserves. There is a rapidly increasing need for areas protected from all human disturbance that can serve as natural habitats, nurseries for young, sources of spores and larvae of harvested species, and 'controls' for research into the impacts that we are having on our nearshore ecosystems.

However, the establishment of marine protected areas in Canada has been impeded by a multiplicity of jurisdictions, agencies, and legislation. Four levels of government (federal, provincial, municipal, and first nations) are involved in governing the marine environment (Table 28-1). Different agencies and legislation control the water surface, the water body, the ocean bed, and the tidal strip and adjacent coastal land (Dorcey, 1983). More than thirty agencies and public groups were involved in the B.C. offshore hydrocarbon panel hearings (Langford et al., 1988). As well, federal, provincial, and interjurisdictional interests are not well coordinated on issues involving the coastal zone (Day and Gamble, 1990). The split between federal and provincial ownership and jurisdiction results in several areas of overlapping responsibility. "Nowhere in the coastal zone--upland, foreshore, subtidal land or water—is there a clear undisputed basis for one government to be the sole authority" (Dorcey, 1986).

Existing legislation requires that the relevant government "own" the affected seabed before a protected area can be established (Westwater Research Centre, 1992a). Regardless of which level of government owns the land/seabed upon which a marine protected area is to be located, the federal government retains exclusive constitutional jurisdiction over: issues transcending international boundaries, navigation (Transport Canada), marine pollution (Environment Canada), migratory birds (Environment Canada) and the

Michael W. Hawkes
Department of Botany
University of British Columbia
#3529-6270 University Blvd.
Vancouver, B.C.
V6T 1Z4

---

[1] Portions of this paper have been adapted from Hawkes (1990) and Hawkes (1992).

Federal Government

1. Fisheries, finfish and shellfish (includes all invertebrates), and marine mammals, *Department of Fisheries and Oceans*
2. Shipping and navigation, *Transport Canada*
3. National Marine Parks, *Canadian Heritage (Parks Service)*
4. Migratory birds, *Environment Canada (Canadian Wildlife Service)*
5. Marine pollution, *Environment Canada (Environmental Protection Service)*
6. Energy and mineral resources, *Energy, Mines and Resources Canada*

Provincial Government

1. Marine plants, *Ministry of Agriculture and Fisheries*
2. Sea bed, jurisdiction sometimes disputed between the federal and First Nations governments. Ownership of the seabed in territorial waters is disputed between federal and provincial governments, as are exploitation rights to the sea bed on the continental shelf.
3. Environmental protection, *Ministry of Environment, Lands and Parks*
4. Provincial marine parks and provincial marine ecological reserves, *Ministry of Environment, Lands and Parks*

Municipal Government

First Nations Government

Table 28-1: Some Major Jurisdictions Responsible for the Coastal Marine Environment.

conservation and management of all organisms in the water column (Department of Fisheries and Oceans). Depending on the location of the protected area, First Nations may also have rights to resources and land.

To review the current status of marine protected areas in B.C., let us look at national marine parks, provincial marine parks and provincial ecological reserves.

National Marine Park Reserves

The National Marine Parks Policy allows for multiple use, unlike terrestrial national parks, which are strict preserves. Multiple use is regulated under a zoning system as follows (Environment Canada, 1986): Zone I - Preservation, Zone II - Natural environment, Zone III - Recreation, Zone IV - General use, and Zone V - Park services. Only Zone I is given complete preserve status. A draft of proposed revisions to Canadian Parks Service policies (dated June 15, 1993) proposes a three zone system and a name change from national marine parks to national marine conservation areas. However, concern remains centred around the fact that,

> Fisheries will continue in marine parks, subject to protecting the ecosystem, to maintaining viable fish stocks and to attaining the purpose and objectives of the park....Jurisdiction over fisheries will, therefore, remain with the Minister of Fisheries and Oceans (Environment Canada, 1986).

To date, B.C. does not have any full-fledged national marine parks; rather, there is one national marine park reserve and a national park reserve with a marine component. Substantial areas of three of the five marine regions in British Columbia are partially protected (Lien, 1989) by the marine component of Pacific Rim National Park Reserve on the outer coast of Vancouver Island (Figure 28-1), and Gwaii Haanas/South Moresby National Marine Park Reserve (Figure 28-2) on the southern portion of Haida Gwaii (Queen Charlotte Islands).

Conserving Marine Ecosystems: Are British Columbia's Marine Protected Areas Adequate?

Pacific Rim National Park Reserve

On the exposed west coast of Vancouver Island, Pacific Rim National Park Reserve has a marine component that is 155.4 square kilometres in extent and is divided into three segments: Long Beach, the Broken Group Islands, and the West Coast Trail. An initial agreement to establish Pacific Rim National Park was signed by the federal and provincial governments on April 21, 1970. This agreement was subsequently amended on March 27, 1973, October 21, 1977, and February 19, 1987, (personal communication on Sept 6, 1990 with Claude Mondor, Environment Canada, Parks Service).

Under amendments to the National Park Act made in 1988, it became necessary to proclaim Pacific Rim as a National Park Reserve until resolution of the Nuu-chah-nulth Tribal Council native land claim, which includes all the land and marine environment within Pacific Rim. To date (January, 1993), however, Pacific Rim still has not been proclaimed in Parliament (personal communication on January 14, 1993 with Bill Henwood, Environment Canada, Parks Service). Once it is proclaimed, the National Park Act will apply to Pacific Rim National Park Reserve as if it were a national park. The terrestrial portion of the Park is being managed under the National Park Policy whereas the marine component is being managed under the National Marine Parks Policy (Environment Canada, 1991).

Current status of the fisheries in Pacific Rim National Park Reserve is as follows. The only commercial fin fisheries that are open are salmon and herring roe-on-kelp. All commercial invertebrate fisheries are closed except for Dungeness crab (*Cancer magister*) and squid (*Loligo opalescens*). The recreational fin fishery and invertebrate fishery are both open, but an application was made by Parks

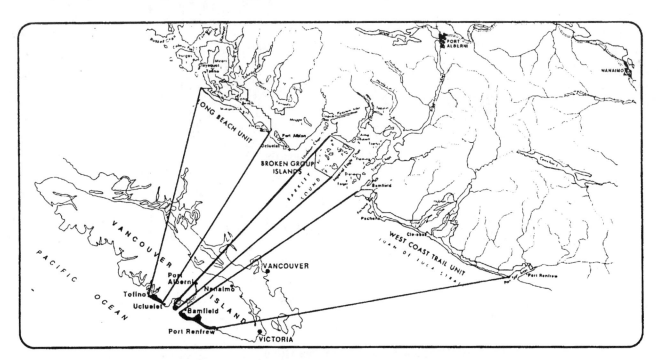

Figure 28-1. Location of Pacific Rim National Park Reserve. Source: Environment Canada 1991, p.3.

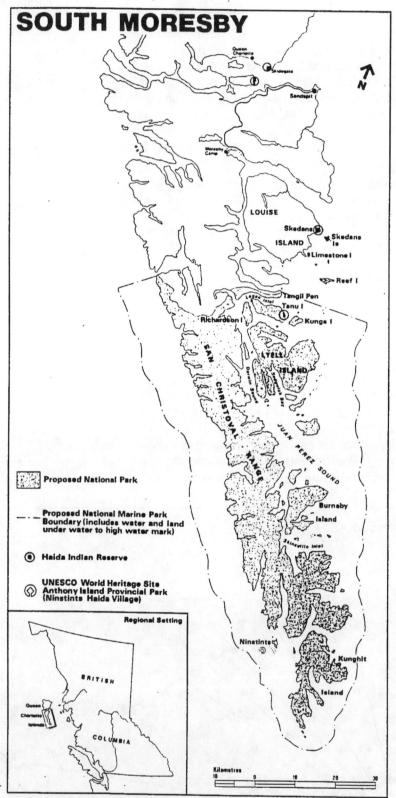

NOTE: Inclusion of Limestone, Reef and Skedans Islands in the national park is subject to further Federal-Provincial discussion.

Figure 28-2. Proposed Boundaries for Gwaii Haanas/South Moresby National Marine Park Reserve. Source: Claude Mondor, Parks Canada.

Canada (September, 1991) to the Department of Fisheries and Oceans requesting closure of all recreational invertebrate fisheries and the recreational rockfish fishery. As of January 1993, no closures have been made (personal communication on January 14, 1993 with Bill Henwood, Environment Canada, Parks Service).

Gwaii-Haanas/South Moresby National Marine Park Reserve

A Federal-Provincial Agreement (Canada and British Columbia 1988) established the Gwaii Haanas/South Moresby National Park Reserve and National Marine Park Reserve on July 12, 1988. The Canada/Haida agreement for joint management of the park was signed in 1992. The proposed marine park boundaries extend between six and ten kilometres from the coast and cover an area of 3,180 square kilometres, thereby encompassing a diversity of west coast marine environments. These boundaries are tentative, pending completion of offshore energy and mineral resource assessments (Lien, 1989). Petroleum and mineral resource assessments were finished in 1992, and final marine park boundaries were to be set by December 31, 1992 (personal communication on Sept 6, 1991 with Claude Mondor, Environment Canada, Parks Service). However, discussions between Parks Canada, the Haida Nation, and the Ministry of Energy, Mines, and Resources are still ongoing and it is expected that final park boundaries will be established sometime in 1993 (personal communication on January 14, 1993, with Bill Henwood, Environment Canada, Parks Service).

British Columbian Provincial Parks With Marine Components

Provincial marine parks, ecological reserves, and recreation areas that have marine components are subject to all of the jurisdictional complexities mentioned previously. Of key importance is the fact that the province lacks complete legislative and jurisdictional authority over the marine organisms and the marine environment in its parks. Pursuant to The Constitution Act (Canada, 1982), as amended, jurisdiction over harvesting of organisms (except for marine plants) in such marine protected areas rests with the federal Department of Fisheries and Oceans. The Department of Fisheries and Oceans has not invoked complete closures on the commercial and recreational harvesting of all fish and invertebrates in provincial marine protected areas. Several marine protected areas have no closures at all. Youds (1985) noted, "the marine conservation value of foreshore and subtidal provincial parkland [is] more symbolic than substantive." Recent initiatives are, however, improving this situation.

Thirty-one coastal provincial parks are listed on the *Coastal Marine Parks of British Columbia* map (BC Parks, 1989a). Forty-two percent of these parks are exclusively terrestrial, with no foreshore or subtidal component. Cape Scott, which has a large foreshore and subtidal component, is not listed because it is treated as a wilderness park (Youds, 1985). The primary focus of these parks is the recreational boater (Deardon, 1985; Chettleburgh, 1985) and not the preservation of representative marine ecosystems or biodiversity.

Parks Plan 90 (BC Parks, 1990a) offered little hope for improvement in the representation and protection of marine environments in the provincial parks system. In *Landscapes for B.C. Parks* (BC Parks, 1990a) it was stated that,

> It is a reasonable argument that park visitors, being largely land-based (even if they are boating or diving), are not inclined to seek recreational or appreciative experiences in these offshore, oceanic areas. These are areas of industrial interest (commercial fishing, shipping) but of marginal direct interest to the general public. Furthermore, the provincial Park Act has significant jurisdictional limitations in such areas. Therefore, it has been concluded that at this time offshore marine environments should be excluded from the Landscapes system and from consideration within the provincial parks system.

At present (December, 1993), there are 50 Class A provincial parks and 3 recreational areas that have an intertidal and/or subtidal component, totalling 92,399 hectares of marine waters (Table 28-2). Since Parks Plan 90 was written, BC Parks has recently shown renewed interest in provincial marine protected areas. The recent establishment of Broughton Archipelago Marine Park, which encompasses a large area of marine waters, constitutes a significant step toward more adequately protecting marine ecosystems and biodiversity.

British Columbian Ecological Reserves With Marine Components

In 1971, the Ecological Reserves Act was passed, formalizing the B.C. Ecological Reserves Program. Setting aside important, unique, or representative ecosystems and species is the main objective, with conservation and research being the primary function of the Ecological Reserves Program (BC Parks, 1989b). Most reserves are open to the public for non-destructive observational use but, unlike provincial parks, they are not created for outdoor recreation.

Of the 134 ecological reserves established in B.C. to date, 15 have intertidal and/or subtidal areas (Table 28-3). Only 13 of these are presently included in the ecological reserves system because reserves #95 and #96 have been transferred to federal jurisdiction to be included in Gwaii Haanas/S. Moresby National Park Reserve and National Marine Park Reserve. The 13 marine ecological reserves include estuarine, semi-protected, and exposed sites, as well as south and north coast localities. Most of these reserves are small (50-350 hectares), which brings up the whole question of the minimum reserve size necessary to maintain viable populations, a topic much discussed in the conservation biology literature (Gilpin and Soulé, 1986; Newmark, 1987; Usher, 1987). However, some of the smaller reserves seem appropriate for protecting local, unique, or especially diverse habitats, such as Race Rocks (reserve #97), Kerouard Islands, (reserve #96), and Pine, Storm, Tree Islands (reserve #120).

The 13 ecological reserves include 47,697 hectares of marine waters. Two of them are noteworthy for their size and extensive subtidal areas: Byers, Conroy, Harvey, and Sinnett Islands (reserve # 103) and Checleset Bay (reserve #109). The Checleset Bay Ecological Reserve is over 98% marine, runs 30 kilometres from east to west, and covers a total area of 346.5 square kilometres. It includes diverse habitats, both intertidal and subtidal. This showpiece reserve was set aside December 10, 1981 to provide sufficient high-quality marine habitat for re-introduced sea otters (*Enhydra lutris*), an endangered species.

Race Rocks (Ecological Reserve #97) in the Strait of Juan de Fuca has the most protected status of any marine protected area in the province. It is closed (by the Department of Fisheries and Oceans) to the commercial and recreational harvesting of all marine life except for recreational (sport) fishing of salmon and halibut. The reasoning behind this decision is that salmon and halibut are migratory finfish and, therefore, transient in the reserve, so closing these fisheries in the reserve will do nothing to conserve these species. However, accidental catch of resident fish in the reserve, especially rockfish, is a matter of concern.

An additional 12 ecological reserves which do not contain marine waters do, nevertheless, protect important marine features and organisms, such as seabird colonies and seal or sea lion haul-out sites. Collectively, provincial marine parks and ecological reserves cover 0.06% of B.C.'s marine environment (out to the 12 mile limit).

Name of Park or Recreational Area	Area (ha) of Marine Component
Botanical Beach	120.0
Brooks Peninsula (R.A.)	5,832.0
Broughton Archipelago Marine Park	10,034.0
Cape Scott Park	1,450.0
Codville Lagoon Marine Park	315.0
Copeland Islands Marine Park	257.0
Cormorant Channel Provincial Park	505.0
Desolation Sound Marine Park	2,550.0
Echo Bay	0.5
Fiordland (R.A.)	6,645.0
Gabriola Sands	4.5
Garden Bay	0.5
Green Inlet Marine Park	17.9
Hakai (R.A.)	56,987.0
Halkett Bay	19.0
Harmony Islands Marine Park	26.0
Helliwell Provincial Park	2,803.0
Jackson Narrows Marine Park	30.0
Kitson Island	24.7
Klewnuggit Inlet Marine Park	304.0
Lowe Inlet Marine Park	212.0
Manson's Landing	53.0
Miracle Beach Park	27.5
Mitlenatch I. Nature Park	119.0
Montague Harbour Marine Park	3.0
Naikoon Park	216.0
Newcastle Island Marine Park	34.0
Octopus Islands	43.0
Oliver Cove Marine Park	26.0
Parkinson Creek	80.0
Penrose Island Marine Park	1,079.0
Pirate's Cove Marine Park	7.0
Plumper Cove Marine Park	33.5
Porteau Cove	45.5
Princess Louisa Marine Park	27.0
Princess Margaret Marine Park	340.0
Raft Cove	265.0
Rathtrevor Beach Park	240.0
Rebecca Spit Marine Park	156.0
Roscoe Bay	47.0
Rugged Point	150.0
Saltery Bay	30.0
Sandwell	9.0
Sidney Spit Marine Park	177.0
Smuggler Cove Marine Park	16.0
Teakerne Arm	10.7
Thurston Bay Marine Park	80.0
Tribune Bay Park	23.0
Union Passage Marine Park	395.0
Wallace Island	0.2
Walsh Cove	46.0
Whaleboat Island	7.0
Winter Cove	16.0

Table 28-2: Provincial Parks and Recreation Areas with Marine Components

Reserve Number	Reserve Name and Location	Description	Area (ha) of Marine Component
24.	Baeria Rocks, Barkley Sound	Seabird colony and subtidal marine life	48.00
45.	V.J. Krajina, W. coast Graham I.	Virgin marine shoreline	100.00
66.	Ten Mile Point, Victoria	Inter- and subtidal marine life	10.00
67.	Satellite Channel, between Saltspring Island and Saanich Peninsula	Subtidal marine life	343.30
94.	Oak Bay Islands, east of Victoria	Seabirds and marine life	163.00
*95	Anthony Island, south of Gwaii Haanas (South Moresby)	Seabird colony	
*96.	Kerouard Islands, south of Gwaii Haanas (S. Moresby)	Seabird colony	
97	Race Rocks, southwest of Victoria	Outstanding marine community, sea-lion haul-out, seabirds	200.00
103.	Byers, Conroy, Harvey & Sinnett Islands, Hecate Strait	Important seabird and marine mammal breeding areas	11,780.00
109.	Checleset Bay, northwest of Kyuquot, Vancouver Island	Extensive marine shoreline, reefs and islets provide habitat for BC's prime sea otter population; seabirds, marine life	33,150.00
111.	Robson Bight (Michael Bigg), Johnstone Strait	Killer whales and a crucial part of their habitat; pristine estuary	1,248.00
119.	Tahsish River, west coast of Vancouver Island, south of Port McNeill	Pristine westcoast estuary	50.00
120.	Duke of Edinburgh (Pine/Storm/Tree Islands), northwest of Port Hardy, Vancouver Island	Largest seabird nesting colony in BC; spectacular inter- and subtidal marine life	535.00
121.	Brackman Island, north of Sidney, Vancouver Island	Inter- and subtidal marine life	30.00
129.	Klaskish River, southwest of Port Alice, Vancouver Island	Estuary with native oysters	40.00

*now part of Gwaii Haanas/South Moresby National Park Reserve and National Marine Park Reserve

Table 28-3: B.C. Ecological Reserves with Marine Components.

Inadequacy of Existing Marine Protected Areas

Is marine biodiversity adequately protected and are marine ecosystems adequately represented by our present system of marine protected areas? The answer has to be an unequivocal no. Existing areas have not been established with uniform biological criteria or an overview of the region. Few, if any, protect entire ecosystems (especially the land/sea interface), and the organisms in existing areas are inadequately protected.

As an example of the management problems common to marine protected areas in B.C., I would like to briefly review three of the marine ecological reserves: Baeria Rocks, Checleset Bay, and Robson Bight (Michael Bigg). In these three cases, the management problems (identified in the Management Statements produced by the Ecological Reserves Program [BC Parks, 1990b]) arise from the jurisdictional complexities in the marine environment, and the failure of other agencies, in particular the Department of Fisheries and Oceans, to establish complete protection, through closures, for all marine organisms in these reserves.

Baeria Rocks

Recreational fishing under the jurisdiction of the Department of Fisheries and Oceans (federal) has wiped out the entire adult rockfish population. Rockfish populations are in trouble in other areas of the coast, too. These fish are long-lived and resident; many live 70-80 years and don't reach reproductive maturity until 20 years of age (personal communication on August 29, 1990 with Andy Lamb, Department of Fisheries and Oceans).

Checleset Bay

Clearcut logging under the jurisdiction of the provincial Ministry of Forests has negatively impacted the reserve through increased runoff and ocean turbidity. The infamous clearcut on Mt. Paxton, which was portrayed as a three page foldout in *National Geographic Magazine* (Findley, 1990), borders on Checleset Bay. Until recently, both finfish and shellfish harvesting, under the jurisdiction of the federal Department of Fisheries and Oceans, were occurring in the reserve.

Marine charts do not indicate the existence of this (or any other) marine reserve, and logging barge traffic, which is under the jurisdiction of Transport Canada (federal), traverses the reserve.

Robson Bight (Michael Bigg)

In the Tsitika Valley, clearcuts made under the jurisdiction of the provincial Ministry of Forests are the big concern, as they may lead to increased erosion and sedimentation in the estuary. On October 3, 1990, the Federal government (Department of Fisheries and Oceans) recommended a moratorium on all logging in the area. The federal/provincial Johnstone Strait Killer Whale Committee final report released in June 1992 (Canada and British Columbia,

1992) recommended an immediate five year moratorium on all forest harvesting in the lower Tsitika. Other highlights of the twenty-seven management recommendations in the report are: the Department of Fisheries and Oceans should immediately designate a Special Management Zone in the killer whale (*Orcinus orca*) core area of western Johnstone Strait; BC Parks should expand the land portion of the Ecological Reserve south and east to provide a better buffer for the rubbing beaches, and expand the marine portion of the Reserve one kilometre east; Department of Fisheries and Oceans should eliminate mooring and commercial fishing within the expanded boundaries of the marine portion of the Ecological Reserve; the Reserve should be closed to all vessels, except by permit; and Department of Fisheries and Oceans should manage salmon stocks in a conservative manner in this area.

The management issues in both Checleset Bay and Robson Bight (Michael Bigg) underscore the need to protect whole ecosystems, not just individual species or portions of their immediate habitat. Yet, the many different levels of government with jurisdictional control over different parts of a single ecosystem are managing on a species rather than on an ecosystem basis.

Nor will simply setting aside marine protected areas be enough to preserve the biota. Monitoring and further research into individual species biology and ecosystem dynamics are particularly needed. So also are much more knowledge about the basic stewardship of protected areas (Deblinger and Jenkings, 1991) and significantly enhanced cooperation between provincial, federal, municipal and First Nations governments, including a more formal mechanism for facilitating cooperation on a regular basis (Westwater Research Centre, 1992b).

Even if marine protected areas within British Columbia were given complete sanctuary status, which they should be, they still would be inadequate for protecting much of the marine life within their boundaries if the marine environment outside their boundaries is not being utilized in a sustainable fashion. International transboundary issues with Alaska and Washington State are also a concern (Across the Border, 1992).

Threats to Marine Biodiversity

We do not have a history of sustainably managing living marine resources. Eradication of the sea otter from our coast around the turn of the century and the consequent sea urchin population explosion has left us with highly modified, rocky nearshore ecosystems that are much lower in seaweed biomass and diversity, than they should be (Chapter 11, this volume). By the early 1900s, whaling had eliminated humpback whale (*Megaptera novaeangliae*) populations from the Strait of Georgia (Merilees, 1985). Before 1913, over 3,800 Steller's sea lions (*Eumetopias jubatus*) used the islands of The Sea Otter Group in the Scott Islands as a breeding rookery. Between 1913 and 1938, the Department of Fisheries and Oceans machine-gunned 29,800 animals, resulting in the permanent demise of The Sea Otter Group rookery (Obee and Ellis, 1992). The three remaining sea lion rookeries in the Scott Islands are the most important ones along the B.C. coast. The current population of Steller's sea lions is only about one third of its former size, and the population has not recovered appreciably since being protected in 1970 (Olesiuk and Bigg, 1988).

Due to poor recruitment of young abalone into the population, both the commercial and recreational fisheries for northern abalone (*Haliotis kamtschatkana*) are closed for five years (Department of Fisheries and Oceans, 1990a). Many other non-traditional invertebrate species such as the geoduck clam (*Panopea generosa*), red sea urchin (*Stongylocentrotus franciscanus*) and sea cucumber (*Stichopus californicus*) represent "gold-rush" fisheries that are being operated on an experimental basis because their biology is incompletely known (Jamieson and Francis, 1986). Many of these species are locally ecologically important because they affect the distribution and abundance of other organisms in the system. Giant geoduck clams have an average age of seventy years and may live as long as one hundred and forty years (Jamieson and Francis, 1986). These clams are being harvested by divers using a technique that is the underwater equivalent of strip-mining.

Due to declining stocks, the commercial lingcod (*Ophiodon elongatus*) fishery in the Strait of Georgia is closed and strict size limits have been instituted for the sport fishery (Department of Fisheries and Oceans, 1990b). Rockfish populations throughout the coast have shown drastic declines, over the past twenty years, due to overfishing (personal communication on October 29, 1992 with Bernie Hanby, Sport Fishing Advisory Board).

Sources of pollutants in the B.C. marine environment include municipal effluent (point-source and non point-source), dumping of dredged material, pulp and paper mill discharges, mine tailings, oil spills and related environmental mishaps. Despite government assurances (Langford et al., 1988) that marine environmental quality has declined along only a small percentage of the B.C. coastline, there is cause for serious concern because the chronic, sublethal toxic and synergistic effects of pollutants in B.C.'s marine ecosystems are poorly understood (Kay, 1989).

In the most recent synopsis of marine environmental quality on the Pacific Coast, Wells and Rolston (1991) state, "Signs of widespread ecosystem stress have been detected and degradation is pronounced in many inshore waters along the coast." Dioxins/ furans and other organochlorine discharges from pulp mills are recent examples of land-based marine pollution that have had a significant local impact on marine organisms and associated fisheries in B.C. Chlorinated organic chemicals are also becoming more widespread through bioaccumulation in migratory seabirds and marine mammals (Wells and Rolston, 1991).

Some of the competing and potentially incompatible uses of the marine environment are summarized in Table 28-4. At present, all these uses are under the jurisdiction and control of various levels of government, different agencies, and legislation. Because the land and sea portions of the coastal zone have been managed separately, often with an ad-hoc approach, the ecological integrity of the zone has not always been adequately protected (Ray, 1988 and 1991). This fragmented approach has left us without an adequate overview of coastal zone planning, use, and management (Hildebrand, 1989).

1) Preservation/Conservation of wilderness, biodiversity, & unique or representative marine ecosystems
2) Marine environmental education areas
3) Scientific research areas
4) Preservation of the cultural heritage of first nations people
5) Historic sites
6) Recreation areas/tourism
7) Fisheries (finfish, invertebrates, marine plants)
8) Mariculture sites
9) Port sites
10) Oil & gas exploration/extraction
11) Mineral extraction
12) Ocean dumping
13) Discharge of pollutants (land-based, from point source & non-point source)
14) Marine transportation
15) Log handling
16) Undersea pipelines & transmission lines
17) Military uses

Table 28-4: Competing and Potentially Incompatible Uses of the Marine Environment.

Day and Gamble (1990) made a number of innovative recommendations for improving the approach to coastal zone management in British Columbia, but few if any have been acted on. As Côté (1989) pointed out, our policies for addressing problems in the coastal zone are largely focused on symptoms rather than causes.

Coastal Zone Stewardship

Our dealings with the marine environment have tended to be exploitative, with emphasis on short-term gain rather than long-term sustainability. We need to transform ourselves from "gold-rush fishers" to responsible stewards of our marine environment and its diverse biological treasures.

Integrated coastal planning, to date, has tended to focus on a few major river estuaries, such as the Fraser (Dorcey, 1991), the Squamish, and the Cowichan (Morgan et al., 1988). There is now an urgent need to extend this approach to the entire B.C. coast. Four key steps are needed to do so.

First, coastal zone planning, research, and management need to be coordinated and a coastal ecosystem classification scheme and inventory developed. Similar recommendations were made fifteen years ago by The Coastal Zone Resource Subcommittee (1977a) but, regrettably, these have been largely ignored. The marine and estuarine habitat classification system developed by Dethier (1990) for Washington State could be adapted for use in B.C.

Existing baseline information should be compiled and data gaps identified, as was done by the Coastal Zone Resource Subcommittee (1977b). Wherever possible, data should be incorporated into a central data base with GIS (Geographic Information System) capability, so that maps providing information specific to a particular location can be generated. The provincial Ministry of the Environment's Environmental Emergency Services Branch has developed such a marine data base for its Oil Spill Shoreline Sensitivity Model. At present, though, this data base is limited to the south coast and its main purpose is to identify shorelines sensitive to oil spills and cleanup operations (Howes and Wainwright, 1992). Only in this way will it be possible to determine the status of marine conservation and use in B.C.'s coastal zone.

The second step should be to ensure that the system of marine protected areas adequately represents genetic, species and ecosystem diversity. The Endangered Spaces Project's (Earthlife Canada Foundation, 1991) land-use planning proposal for the Province of B.C. would provide an excellent framework for such an undertaking for the marine environment.

It is hoped that the new provincial Protected Areas Strategy (BC Parks and Ministry of Forests, 1992) will address marine protected area issues. Since conservation proposals for terrestrial ecosystems recommend preservation of twelve percent of the land-base (Brundtland, 1987), we should also set a goal of preserving, in marine protected areas, twelve percent (by area and content) of all marine assets in the coastal zone of B.C. These areas should be one hundred percent protected from consumptive use of any marine organisms. Other jurisdictions are far ahead of us in marine conservation. New Zealand is targeting ten percent of its coastal marine environment to be set aside in marine reserves (Ballantine, 1991) and complete marine fishery reserves are being proposed for the southeast coast of the United States (Plan Development Team, 1990). Bohnsack (1992) has presented extensive evidence that such reserves are the best method for protecting biodiversity and natural community equilibrium.

The third step should be to allocate more money for research on the basic biology and population dynamics of fish, invertebrates, marine mammals, and marine plants. Improved stewardship of the coastal zone also depends on experimental research into ecosystem function (Dorcey, 1983).

Fisheries often have been managed to maximize economic returns, frequently with inadequate knowledge of the biology of harvested species or consideration of the effects of harvesting on the rest of the ecosystem. As noted by Jamieson and Caddy (1986) "Economic considerations form the bottom line of most management situations..." However, Jamieson (1986) cautions that "most invertebrate fisheries, then, will never warrant or command large investments in data collection, monitoring or research, and may never be supported by data bases and assessments adequate for active management." In view of this lack of information, experimental fisheries, such as the invertebrate fisheries and the newly established shark fishery (Pynn, 1991), should not be allowed. The shark fishery, especially, should be halted, since a recent report (Manire and Gruber, 1990) has revealed that many shark species in the United States and elsewhere may be headed toward extinction due to overfishing! There are four main problems with experimental fisheries (Dethier et al., 1989):

1. reliable population estimates are critical, but frequently difficult to obtain;

2. usually there are no controls (areas that are not fished and where populations are monitored);

3. by the time a problem is recognized (through, for example, drops in catch/unit effort or in local population size), it may be serious and potentially irreversible; and

4. it is politically difficult to close an established fishery.

The fourth and perhaps most important step is to recognize and respond to the need for increased public awareness and input on matters of coastal zone use and stewardship, in general, and marine protected areas, in particular (Kaza, 1988). There is an urgent need for conservation groups, and other non-governmental organizations, to extend their interests from their traditional terrestrial focus, or specific marine species bias, to the whole marine environment and its biota. Initiatives and groups such as the Vancouver Island Shorelines Workshop (Smiley et al., 1991; sponsored by The Federation of B.C. Naturalists and the Vancouver Island Natural History Clubs), the Marine Life Sanctuaries Society of B.C., the National Marine Conservation Forum, the Canadian Parks and Wilderness Society marine protected areas resolution (Ianson and Moore, 1992), and the Westwater Research Centre's Marine Protected Areas Workshops (Westwater Research Centre, 1992a, 1992b, and 1992c) are to be applauded. The only significant proposal for establishing more marine protected areas that I am aware of is The Valhalla Society's Endangered Wilderness map (Valhalla Society, 1988 and 1992).

Preservation and conservation of marine biodiversity from an ecosystem perspective should be the primary objectives for coastal zone stewardship. Socio-economic and political considerations are important, but should be secondary to achieving long-term viability of natural systems in the coastal zone.

If we want to keep our marine wilderness and biodiversity intact for future generations, we need to act now.

Acknowledgements

Canadian NSERC Operating Grant 580384 supported this work. Paul Gabrielson, William Jewell College made several helpful suggestions for improving the manuscript. Andy Lamb and the Marine Life Sanctuaries Society of B.C. provided information on the invertebrate fisheries. Doug Swanston brought to my attention the Marine Life Sanctuaries Society archives, which contain much early correspondence on the subject of marine parks (primarily the work of Betty Pratt-Johnson). The following individuals from the Ministry of Environment, Lands and Parks helped with information: Greg Chin, Kerry Joy, Kaaren Lewis, Mike Murta, Roger Norrish, and Hans Roemer. National parks people who were most helpful included: Bill Henwood and Claude Mondor. Larry Golden brought important references on marine invertebrate fisheries and marine environmental quality to my attention. Patricia Clay brought several important references on coastal zone management to my attention. Richard Kyle Paisley, Westwater Research Centre made many helpful suggestions regarding jurisdiction and legislation pertaining to the marine environment. Jim Bohnsack, US National Marine Fisheries Service, provided key literature on the ecological basis for using marine fishery reserves for reef resource management. Bill Ballantine, University of Auckland, Leigh Marine Laboratory provided inspiration and many helpful ideas regarding the establishment of marine protected areas. Denise Bonin suggested several improvements to the paper.

Conserving Marine Ecosystems: Are British Columbia's Marine Protected Areas Adequate?

References

Across the Border. A Scientific Meeting on Marine Environmental Conditions in Washington and British Columbia. 1992. Synopses of Presentations (with References and List of Contacts). Western Washington University, Bellingham, Washington.

BC Parks. 1989a. *Coastal Marine Parks of British Columbia.* Ministry of Environment, Lands and Parks, Victoria, B.C. Map.

BC Parks. 1989b. *Guide to Ecological Reserves in British Columbia.* Ministry of Environment, Lands and Parks, Victoria, B.C. Map.

BC Parks. 1990a. *Preserving Our Living Legacy. Parks Plan 90. Landscapes for B.C. Parks.* BC Parks, Planning and Conservation Services, Victoria, B.C.

BC Parks. 1990b. *Baeria Rocks, Robson Bight, and Checleset Bay Management Statements.* Ecological Reserves Program, Ministry of Environment, Lands and Parks, Victoria, B.C.

BC Parks and Ministry of Forests. 1992. *Towards a Protected Areas Strategy for B.C.* BC Parks and Ministry of Forests, Victoria, B.C. Map with text.

Ballantine, B. 1991. *Marine Reserves for New Zealand.* Leigh Marine Laboratory Bulletin No. 25. University of Auckland, Warkworth, New Zealand.

Beatley, T. 1991. Protecting biodiversity in coastal environments: Introduction and overview. *Coastal Management* 19:1-19.

Bohnsack, J.A. 1992. Reef resource habitat protection: The forgotten factor. In: *Stemming the Tide of Coastal Fish Habitat Loss.* R.H. Stroud (ed.). Marine Recreational Fisheries V. 14. pp. 117-129.

Bohnsack, J.A., H. Kumpf, E. Hobson, G. Huntsman, K.W. Able and S.V. Ralston. 1989. Report on the concept of marine wilderness. *Fisheries* 14:22-24.

Brundtland, G.H. (Chairperson). 1987. *Our Common Future. The World Commission on Environment and Development.* Oxford University Press, Oxford, England.

Canada. 1982. *The Constitution Act.*

Canada and British Columbia. 1988. *Memorandum of Agreement between the Government of Canada and the Government of the Province of British Columbia.* Government of Canada, Ottawa, Ontario.

Canada and British Columbia. 1992. *Johnstone Strait Killer Whale Committee Management Recommendations.* BC Parks and Department of Fisheries and Oceans, Victoria, B.C.

Chettleburgh, P. 1985. *An Explorer's Guide: Marine Parks of British Columbia.* Special Interest Publications (a Division of Maclean Hunter), Vancouver, B.C.

Coastal Zone Resource Subcommittee. 1977a. *The Management of Coastal Resources in British Columbia.* Vol. 1: 'State-of-the-Art.' Prepared for the B.C. Land Resources Steering Committee, Minister of Supply and Services, Vancouver, B.C.

Coastal Zone Resource Subcommittee. 1977b. *The Management of Coastal Resources in British Columbia.* Vol. 2:Review of Selected Information. Prepared for the B.C. Land Resources Steering Committee, Minister of Supply and Services, Vancouver, B.C.

Côté, R.P. 1989. *The Many Dimensions of Marine Environmental Quality.* Science Council of Canada, Ottawa, Ontario.

Day, J.C. and D.B. Gamble. 1990. Coastal zone management in British Columbia: An institutional comparison with Washington, Oregon, and California. *Coastal Management* 18:115-141.

Deardon, P. 1985. Desolation Sound Marine Park, British Columbia. In: *Marine Parks and Conservation. Challenge and Promise.* J. Lien and R. Graham (eds.). Vol. 2:1-211. NPPAC Henderson Park Book Series No. 10. National and Provincial Parks Association of Canada, Toronto, Ontario. pp. 155-165.

Deblinger, R.D. and R.E. Jenkings, Jr. 1991. Preserving coastal biodiversity: the private, nonprofit approach. *Coastal Management* 19:103-112.

Department of Fisheries and Oceans. 1990a. *Closure to Protect Abalone Stocks.* 1 November news release.

Department of Fisheries and Oceans. 1990b. *Lingcod Size Limit Introduced for Anglers.* 10 May news release.

Dethier, M.N. 1990. *A Marine and Estuarine Habitat Classification System for Washington State.* Washington State Department of Natural Resources, Olympia, Washington.

Dethier, M.N., D.O. Duggins and T.F. Mumford, Jr. 1989. Harvesting of non-traditional marine resources in Washington State: trends and concerns. *The Northwest Environmental Journal* 5:71-87.

Dorcey, A.H.J. 1983. Coastal zone management as a bargaining process. In: *Coastal Zone Management in British Columbia.* B. Sadler (ed.). Cornett Occasional Papers No. 3, Department of Geography, University of Victoria, Victoria, B.C. pp. 73-89.

Dorcey, A.H.J. 1986. *Bargaining in the Governance of Pacific Coastal Resources: Research and Reform.* Westwater Research Centre, University of British Columbia, Vancouver, B.C.

Dorcey, A.H.J. (ed.). 1991. *Water in Sustainable Development. Exploring Our Common Future in the Fraser River Basin.* Westwater Research Centre, University of British Columbia, Vancouver, B.C.

Earle, S.A. 1991. Sharks, squids, and horseshoe crabs--the significance of marine biodiversity. *BioScience* 41:506-509.

Earthlife Canada Foundation. 1991. Endangered Spaces. Project Synopsis. Earthlife Canada Foundation, Queen Charlotte City, B.C.

Eiswerth, M.E. 1990. *Marine Biological Diversity: Report of a Meeting of the Marine Biological Diversity Working Group.* Woods Hole Oceanographic Institution Technical Report 90-13.

Environment Canada. 1986. *National Marine Parks Policy.* Minister of the Environment, Ottawa, Ontario.

Environment Canada. 1991. *Background Information. Management Planning Program.* Pacific Rim National Park Reserve. Environment Canada, Parks Service, Ottawa, Ontario.

Findley, R. 1990. Endangered old-growth forests. Will we save our own? *National Geographic* 178(3):106-136.

Gilpin, M.E. and M.E. Soulé. 1986. Minimum viable populations: processes of species extinction. In: *Conservation Biology. The Science of Scarcity and Diversity.* M.E. Soulé (ed.). Sinauer Association, Sunderland, Mass. pp. 19-34.

Graham, R. (ed.). 1990. *Marine Ecological Areas in Canada. Perspectives of the Canadian Council on Ecological Areas Task Force on Marine Protected Areas.* Occasional Paper No. 9, The Canadian Council on Ecological Areas, Ottawa, Ontario.

Hawkes, M.W. 1990. Benthic marine algal flora (seaweeds) of B.C. Diversity and conservation status. *BioLine* 9:18-21.

Hawkes, M.W. 1992. Conservation status of marine plants and invertebrates in British Columbia: Are our marine parks and reserves adequate? In: *Community Action for Endangered Species. A public symposium on B.C.'s Threatened and Endangered Species and their Habitat, 28-29 September 1991.* S. Rautio (ed.). Federation of B.C. Naturalists and Northwest Wildlife Preservation Society, Vancouver, B.C. pp. 87-101.

Hildebrand, L.P. 1989. *Canada's Experience With Coastal Zone Management.* Oceans Institute, Halifax, Nova Scotia.

Howes, D and P. Wainwright. 1992. *Presentation to DFO Sensitivity Mapping Workshop.* Ministry of the Environment, Victoria, B.C.

Ianson, D and D. Moore. 1992. Resolution 16. Marine protected areas. *Parks and Wilderness Quarterly* 4(4):6-9.

Jamieson, G.S. 1986. A perspective on invertebrate fisheries management--the British Columbia experience. *Canadian Special Publication of Fisheries and Aquatic Sciences* 92:57-74.

Jamieson, G.S. and J.F. Caddy. 1986. Research advice and its application to management of invertebrate resources: an overview. *Canadian Special Publication of Fisheries and Aquatic Sciences* 92:416-424.

Jamieson, G.S. and K. Francis (eds.). 1986. Invertebrate and marine plant fishery resources of British Columbia. *Canadian Special Publication of Fisheries and Aquatic Sciences* 91, Department of Fisheries and Oceans, Ottawa, Ontario.

Kay, B.H. 1989. *Pollutants in British Columbia's Marine Environment. A Status Report.* SOE Report No. 89-1. Environment Canada, Ottawa, Ontario.

Kaza, S. 1988. Community involvement in marine protected areas. *Oceanus* 31:75-81.

Langford, R.W., S. Reid, and L. Harding. 1988. Marine environmental quality in Canada: status and issues. The Pacific coast of Canada. In: *Proceedings: Canadian Conference on Marine Environmental Quality.* P.G. Wells and J. Gratwick (eds.). International Institute for Transportion and Ocean Policy Studies, Halifax, Nova Scotia. pp. 209-228.

Lien, J. 1989. Eau Canada! A new marine-parks system. In: *Endangered Spaces. The Future For Canada's Wilderness.* M. Hummel (ed.). Key Porter Books, Toronto, Ontario. pp. 107-119.

Lien, J. and R. Graham (eds.). 1985. *Marine Parks and Conservation. Challenge and Promise.* Vol. 1:1-254, Vol. 2:1-211. NPPAC Henderson Park Book Series No. 10. National and Provincial Parks Association of Canada, Toronto, Ontario.

MacInnis, J. (ed.). 1992. *Saving the Oceans.* Key Porter Books, Toronto, Ontario.

Manire, C.A. and S.H. Gruber. 1990. Many sharks may be headed toward extinction. *Conservation Biology* 4:10-11.

Merilees, B. 1985. The Humpback Whales of Georgia Strait. *Waters, Journal of the Vancouver Aquarium* 8:1-24.

Morgan, B., J. O'Riordan, R.W. Langford and L. Harding. 1988. Marine environmental quality management in British Columbia. In: *Proceedings: Canadian Conference on Marine Environmental Quality.* P.G. Wells and J. Gratwick (eds.). International Institute for Transportion and Ocean Policy Studies, Halifax, Nova Scotia. pp. 173-183.

Newmark, W.D. 1987. A land-bridge island perspective on mammalian extinctions in western North American parks. *Nature* 325:430-432.

Norse, E.A. 1993. *Global Marine Biological Diversity. Strategy for Building Conservation Into Decision Making.* Island Press: Washington, D.C.

Obee, B. and G. Ellis. 1992. *Guardians of the Whales. The Quest to Study Whales in the Wild.* Whitecap Books, Vancouver, B.C..

Olesiuk, P.F. and M.A. Bigg. 1988. *Seals and Sea Lions on the British Columbia Coast.* Department of Fisheries and Oceans, Nanaimo, B.C.

Plan Development Team. 1990. *The Potential of Marine Fishery Reserves for Reef Fish Management in the U.S. Southern Atlantic.* NOAA Technical Memorandum NMFS-SEFC-261.

Pynn, L. 1991. Shark Bait. *Vancouver Sun* 19 November 1991. p. B1 and B14.

Ray, C. 1988. Ecological diversity in coastal zones and oceans. In: *Biodiversity.* E.O Wilson (ed.). National Academy Press, Washington, D.C. pp. 36-50.

Ray, C. 1991. Coastal-zone biodiversity patterns. *BioScience* 41:490-498.

Smiley, B., N. Layard, D. Carsen and S. Love (eds.). 1991. *Identifying and Protecting Sensitive Shoreline and Adjacent Wetland Habitat on the East Coast of Vancouver Island.* Report of the Vancouver Island East Coast Shoreline and Adjacent Wetlands Workshop. The Federation of B.C. Naturalists, Vancouver, B.C.

Thorne-Miller, B. and J.G. Catena. 1991. *The Living Ocean. Understanding and Protecting Marine Biodiversity.* Island Press, Washington, D.C. and Covelo, California.

Usher, M.B. 1987. Effects of fragmentation on communities and populations: a review with applications to wildlife conservation. In: *Nature Conservation: The Role of Remnants of Native Vegetation.* D.A. Saunders et al. (eds.). Surrey Beatty and Sons, Norton, N.S.W., Australia. pp. 103-121.

Valhalla Society. 1988. *British Columbia's Endangered Wilderness. A Proposal for an Adequate System of Totally Protected Lands.* Valhalla Society, New Denver, B.C. 1:2,000,000 map with accompanying text.

Valhalla Society. 1992 (2nd revised edition). *British Columbia's Endangered Wilderness. A Comprehensive Proposal for Protection.* Valhalla Society, New Denver, B.C. 1:2,000,000 map with accompanying text.

Wells, P.G. and S.J Rolston (eds.). 1991. *Health of Our Oceans. A Status Report on Canadian Marine Environmental Quality.* Environment Canada, Ottawa, Ontario and Dartmouth, Nova Scotia.

Westwater Research Centre. 1992a. *Marine Protected Areas (MPA) in British Columbia. Options for the Future.* Westwater Research Centre, University of British Columbia, Vancouver, B.C.

Westwater Research Centre. 1992b. *Marine Protected Areas (MPA) in British Columbia. Final Report to the Law Foundation of British Columbia.* Westwater Research Centre, University of British Columbia, Vancouver, B.C.

Westwater Research Centre. 1992c. Workshop Agenda. *Marine Ecological Areas in British Columbia.* Westwater Research Centre, University of British Columbia, Vancouver, B.C.

Youds, K. 1985. Marine parks in British Columbia. In: *Marine Parks and Conservation. Challenge and Promise.* J. Lien and R. Graham (eds.). Vol. 2:1-211. NPPAC Henderson Park Book Series No. 10. National and Provincial Parks Association of Canada, Toronto, Ontario. pp. 149-156.

Chapter 29
Thoughts on an Earth Ethic

In his highly regarded book, *Soil and Civilization*, Edward Hyams gave the second part of his text the subtitle: Man As a Parasite on Soil. As a not altogether inaccurate description of our place in the kingdom of life, his subtitle should remind us of the biological axiom that a good parasite does not destroy its host.

The word "parasite" is derived from the Greek "parasitos" and refers to one who eats at the table of another. *Webster's New Twentieth Century Dictionary* states that in biology a parasite is "a plant or animal that lives on or within another organism, from which it derives sustenance or protection without making compensation."

It does not require a great leap in thought to realize that human civilization is so utterly dependent on the life-supporting conditions provided by the ecosphere (the interrelated physical and biological elements of our world) that humanity would be well advised to recognize that activities which threaten the health of its host bode ill for itself. If we suffer today from the delusion that the winning hand in the game of life is economic determinism, we should realize that earth, the player on the other side of the board, has ecological determinism as its hole card... and that's the wild card that has taken the pot from every faltering species for over three billion years.

As an intelligent species, we need to call upon our intellectual faculties to develop rules of self-limitation in order to eliminate the threat we pose to the organized interdependencies which form the warp and woof of the fabric of that greater life in which we live and move and have our being. It is not merely idealistic, but absolutely imperative that we take positive, unfaltering steps toward a code of behaviour which will change our effect on the ecosphere from catabolic to anabolic. This new code of behaviour might be referred to as an Earth Ethic. What we might call it, though, is less important than the immediate development of attitudes and understandings which will enable us to accept the fact that we are not lords and masters of the universe. Although we flatter ourselves by thinking we are qualified to become stewards of the environment, we are simply a species which must come of age and adopt long term values which may offer survival. This will require us to abandon those short term values which offer temporary gratification but are bringing us to that point referred to by T.S. Eliot, whereat things will end (for us) with a bang or a whimper. We will always be parasites upon the earth because we depend upon it, while it does not depend upon us. But we can adopt values that treasure our host biosphere and behaviours that are less damaging to it.

Robert F. Harrington
Box 2, R.R. 2
Nakusp, B.C.
V0G 1R0

The Wisdom of the Ancients

We do not exactly have to begin with a "tabula rasa" (clean slate) if we are to strive toward an ethic which limits our efforts to

411

turn the world into our vision of a materialistic Heaven. Thinkers and philosophers of the past have already tried to point us in the right direction.

In his essay "The Land Ethic," forester Aldo Leopold (1949) compared a land ethic with the ethics of the Mosaic Decalogue (the ten commandments) and the Golden Rule, "do unto others as you would have them do unto you," and commented that "individual thinkers since the days of Ezekiel and Isaiah have asserted that the despoliation of land is not only inexpedient but wrong." He felt that the extension of ethics to embrace the land was both "an evolutionary possibility and an ecological necessity."

Our claim to superiority as a species is also challenged by Ecclesiastes 3:19: "For that which befalleth the sons of men befalleth the beasts; even one thing befalleth them; as the one dieth, so dieth the other; yea, they have all one breath; so that a man hath no preeminence above a beast; for all is vanity."

In the 25th and 26th chapters of Leviticus, the third book of the *Old Testament*, are quite specific instructions on the treatment of agricultural land. More than the rudiments of an earth ethic may be found in these verses, which remind us that the land belongs to God and that we are but sojourners with Him.

Thoughts about the relationship between people and the earth also appeared in the *Tao te Ching* (The Book of the Way and Its Virtue), written in the 6th century B.C. In the Tao, the world is described as "a sacred vessel not made to be altered by man." We are cautioned that there will be those who choose to take the whole world and tinker with it to suit their own wishes and aspirations, but that the world will be spoiled in the process. A life of wisdom, the Tao suggests, is a life lived in accordance with the immutable laws of the universe (Tzu, 1955).

In my book, *To Heal the Earth: The Case for an Earth Ethic*, I chronicled historical writings about the relationship between humans and the environment. There have been a great many more eloquent writers on this subject than I have mentioned here, but one, Marcus Aurelius Antoninus (A.D. 121-180), captured, for me, the essence of the genre. Aware of the scorched earth policies implemented during the Peloponnesian War (431-404 B.C.), and of their use by the Roman armies of his day, he recorded his thoughts in his famed *Meditations*, while campaigning along the Danube (170-175 A.D.) His writings include ideas on holism, the web of life, and the sacred relationship between all things. Fragments of his thoughts might serve well as background for the formulation of the ethic so desperately needed today:

> I travel the roads of nature until the hour when I shall lie down and be at rest; yielding back my last breath into the air from which I have drawn it daily, and sinking down upon the earth from which my father derived the seed, my mother the blood, and my nurse the milk of my being—the earth which for so many years has furnished my daily meat and drink, and though so grievously abused, still suffers me to tread its surface.

> Always think of the universe, as one living organism, with a single substance and a single soul; and observe how all things are submitted to the single perceptivity of this one whole, all are moved by a single impulse, and all play their part in the causation of every event that happens. Remark the intricacy of the skein, the complexity of the web. (Aurelius, 1964).

Moving Into the Present

These thoughts from our past are pertinent to us in British Columbia today. The soil which provides the seed, the blood and milk of our being is subject to megamachine disturbance of a magnitude that Marcus Aurelius was unable to anticipate. The air from which we draw our breath becomes more toxic daily from the sustained flatulence of factories and private cars. The waters, which form the bulk of our bodies, are altered by increased soil disturbance and by their role as a diluter of compounds foreign to living organisms.

While British Columbia still has clean free-flowing rivers, forests filtering air and forest soils cleansing waters, only the remotest corners of the province are unaffected by ecosystem fragmentation, and some are seriously degraded. We stand at an axial point in the history of our province, capable of restoring health to our ecosystems or equally capable of unthinkingly accepting the cliché that the "business of Canada is business" and, thereby, willingly trading our irreplaceable inheritance for glittering goals of transient value. Fortunately, thoughts about our relationship to the earth are found not only in classical literature but also abound in more recent writings.

A good example is found in *Man and Nature or Physical Geography as Modified By Human Action*, by George Perkins Marsh (1864). This book was the product of 17 years of study of human impact on the land in Europe and was written while Marsh was U.S. Ambassador to Italy. "Even now," he wrote, "we are breaking up the floor and wainscoting and doors and window frames of our dwelling, for fuel to warm our bodies and seethe our pottage." He offered much evidence that wanton destruction and profligate waste were rendering earth an unfit home for human life. Marsh contended that continued irresponsible behaviour would lead to such impoverished productiveness, shattered land surface, and climatic excess, "as to threaten the depravation, barbarism, and perhaps even extinction of the species." He recounted and documented incidents of devastation by torrents, and river inundations caused by extensive destruction of forests, and cited eradication of whole villages following logging on surrounding steep slopes. He particularly urged protection and restoration of forests, claiming, "We have now felled forest enough everywhere, in many districts far too much. Let us devise means for maintaining the permanence of its relations to the fields, the meadows, and the pastures, to the rain and the dews of heaven, to the springs and rivulets with which it waters the earth."

Still another eminent figure, Liberty Hyde Bailey (1858-1954), reiterated the call for an ethic, both as a religious man and as a scientist. Professor, and eventually Dean of the College of Agriculture at Cornell University, Bailey stated flatly: "We are parts in a living sensitive creation....The living creation is not man-centred; it is biocentric....We can claim no gross superiority and no isolated self-importance. The creation and not man is the norm."(Bailey, 1915)

One who chose to place the decline of society squarely in the realm of ethics was Dr. Albert Schweitzer. Dr. Schweitzer, possess-

413

ing doctorates in theology, philosophy and medicine, received the Nobel Peace Prize in 1952 for founding and directing Lambaréné Hospital in western equatorial Africa. In *The Decay and the Restoration of Civilization*, Schweitzer (1923) charged that, "The history of our times is characterized by a lack of reason which has no parallel in the past." Although our society was proud of its achievements, he contended that it had rejected belief in the, "one thing which is all essential: the spiritual advancement of mankind." Nothing other than an ethical movement, "might rescue us from the slough of barbarism." He spoke of society as over-organized and too willing to believe in propaganda and felt that change would only come when people acted as independent personalities, exerting greater influence upon the manipulated mass society than it exerted upon them.

The philosophy Schweitzer offered the world, one of reverence for life, in which the only justification for terminating life is in accordance with a principle of necessity, seems to have been understood intuitively by many of the native peoples of North America. These indigenous North Americans also recognized, in the depths of their souls, the aesthetic splendour in which they lived. Ohiyesa, a Santee Dakota physician and author revealed the appreciation of the Great Mystery in which his people lived:

> Whenever, in the course of the daily hunt the red hunter comes upon a scene that is strikingly beautiful or sublime—a black thundercloud with the rainbow's glowing arch above the mountain, a white waterfall in the heart of a green gorge; a vast prairie tinged with the blood-red of sunset—he pauses for an instant in the attitude of worship. He sees no need for setting apart one day in seven as a holy day, since to him all days are God's (McLuhan, 1971).

Sojourners we are indeed and we see the world through the pinhole vision of our own personalities and our own limited experiences. Our short lives ended, we return to the elemental wealth of the world. Modern physicists' convictions about the indestructibility of matter and energy suggest that, at least at the atomic level, we will gleam in the sunsets, float in the clouds, and dance in the leaves. As wholes are greater than the sums of their parts, so the ecosphere may embrace a cumulative consciousness: a unity of all the multiplicities that have lived. The ecosphere, purely and simply, is the sum of all living things. Bodies that are cooperative assemblages of astounding numbers of individual living cells could well be merely tiny reflections of the myriad organizational levels that are possible in the ecosphere.

Realization, not so much of the spiritual dimension of our relationship to the earth as of its repercussions on the quality of our lives and environment, is beginning to be reflected in contemporary thought and policy recommendations. Documents like the 1987 report, *Our Common Future*, by the World Commission on Environment and Development (WCED—commonly known as the Brundtland Commission) acknowledge not only the physical causes and effects of environmental and social degradation, but the ethical dimension of any solution. Sustainable use of resources, the Commission observed in explaining its concept of sustainable development (to which Canada is officially committed), depends upon equitable access to them. For example, when the profitable resources of a region are controlled internationally, the local people may be driven to overexploit marginal resources (WCED, 1987). In the words of Sergio Dialetachi, a speaker from the floor at the WCED Public Hearing at São Paolo, 28-29 October, 1985:

> We know that the world lives through an international finance crisis, which increases the misery and the poverty in the Third World and we sacrifice even more of our environment, though we know that this situation can be reversed, if we can use correctly new technology and knowledge. But for this we have to find a new ethic that will include the relationship between man and nature above all.

Professor William E. Rees, a Planning and Resource Ecologist at the University of British Columbia School of Community and Regional Planning, described the expanded "ecological footprint" of Vancouver and the lower mainland, verifying that it is much greater than the actual physical size of the city. He found that the resources consumed by residents in the lower Fraser Valley require at least 8.3 million hectares of land for production, although the valley itself is only about 400,000 hectares in size. If the pollution and cumulative thermodynamic aspect (heat generated by cities) are also considered, an even larger third dimensional effect becomes apparent.

Rees observes:

> ...urbanization and trade have the effect of physically and psychologically distancing urban populations from the ecosystems that sustain them. Access to bioresources produced outside their home region both undermines peoples' sense of dependency on 'the land' and blinds them to the far-off social and ecological effects of imported consumption. (Rees, 1992)

He notes that, when residents of a region both consume and control resources from other regions, their conflict of interest in the allocation of land and resources becomes obvious.

All of these writers, from Aldo Leopold to William Rees, and many more besides, have tried to stem the tide of environmental mismanagement. But only recently have attitudes toward the environment begun to change—and many believe that, despite the dire warnings of scientists and philosophers alike, this change is still only superficial. Some wonder how long it will take for humankind to find that its carefully constructed monuments to materialism and hedonism are built upon the illusion that humankind occupies some central and privileged position in the universe.

The Problem

The stumbling block to human survival was succinctly identified by the cartoon character Pogo in the words, "I have met the enemy and he is us."

This problem was stated more academically by Ernest H. Haeckel, (1834-1919), professor of Zoology at Jena from 1865-1909, who publicized the term ecology in 1869. Haeckel, in defining the word anthropism, wrote: "I designate by this term that powerful and worldwide group of erroneous opinions which opposes the human organism to the whole of the rest of nature, and represents it to be the preordained end of the organic creation, an entity essentially distinct from it, a god-like being." (Haeckel, 1929).

The paradox, and even the presumptuousness of homocentricism (as anthropocentricism is commonly called today) is that, as Marcus Aurelius realized, we are a product of the streams

415

and hills, oceans and atmosphere, prairies and forests of the ecosphere. We are a restless, dissatisfied novice species, clamouring for rulership of a planet toward which we display not even a rudimentary form of allegiance; an apparently malfunctioning part, intent on dominating a whole which has endured the collision of asteroids, and the demise of dinosaurs and mastodons, in fact, as is commonly said, of ninety nine or more percent of all the species that have ever lived. Stephen Jay Gould contends that our view that we are capable of acting as stewards for the world is, "rooted in the old sin of pride." He asks what argument we could use, having arrived, "just a geological microsecond ago" to justify our competence to manage the affairs of a planet 4.5 billion years old, which teems with life that has been evolving and diversifying for an estimated 3.5 billion years. He notes that we are one species among millions—"stewards of nothing." (Gould, 1990).

This ecosphere, to which we refuse to give standing in our courts or in our thoughts, is perfectly capable of disposing of a species like ours, which is so rash as to believe that, in the "geological microsecond" of its existence, it can impose its insubstantial culture upon the cosmos. Will we really be astounded to learn that the ecosphere has full standing in a Higher Court than ours? Might we fancy that we have grounds to object to a ruling that we have run out of time, that we are a pretender species without justification, and that we are to be shuffled into the limbo of extinction?

Like the fabled "wee little worm who lived in a hickory nut," we think we, "live in the heart of the whole round world and it all belongs to us." As Aldo Leopold suggested in his 1949 essay on a land ethic, this illusion that we are conquerors of the land community must yield to the realization that we are plain members and citizens of that community.

There is a disturbing possibility that we have created an Economic God with its own trinity of industry, science and technology. We may be so devoted to the Materialistic Heaven of our imagination that we will storm ahead through the shelving waters of environmental resilience until we founder on the rocks and shoals that mark the limit of those conditions which permit the existence of our species. Toxic air and polluted water, soil deficient in organic matter and carrying its own burden of harmful chemicals, epidemic cancer, hazardous waste in land fills and oceanic waters: all these suggest that we must arrive at a new approach to life.

Our government recently focused our attention on the political benefits of a constitution for Canada. We would be infinitely better off, and Canada might last considerably longer, if as much expense and thought were given to developing an earth ethic to protect the physical and biological realm we call Canada, which is itself a group of vital and diverse ecosystems within the ecosphere.

An Earth Ethic—A Step in The Right Direction

An earth ethic is basically a new and very realistic way of looking at the world—a new world view, put into action. It is a set of moral principles or values that are based upon the understanding that we are beings whose health from birth to death is the product of a healthy ecosphere, and that, in the sequence of time, there first had to be an ecosphere in which physical, chemical and biotic conditions enabled our species to exist. From such a realization, it is neither difficult nor unreasonable to further understand that changes we impose upon those conditions which support life should not be of sufficient magnitude to initiate major perturbations that would threaten the health of the ecosphere, or of the various ecosystems which cumulatively affect the health of the whole. No ecosystem should be sacrificed for the advantage of another.

A new ethic cannot be forced upon a society, but it can be encouraged to grow through changes in the institutions by which society is governed: education, law, politics and, ultimately, religion. Developing an intelligent love of and sense of responsibility for the ecosphere should become a primary focus in school programs from kindergarten to post-graduate years. The role of an intelligent, conservative consumer who understands the difference between needs and fickle wants should be stressed. With the sophisticated forms of media now available, the ecosphere could justifiably rise to stardom, and managing that ascent would be a significant, purposeful challenge for programming. In law and politics, the first fumbling steps have already been taken with the Earth Summit in Rio de Janeiro in June, 1993, and the popular demand that our political and business leaders become environmentally accountable.

In her article for the United Nation's Environment Programme magazine, international environmental lawyer Prue Taylor (1992) identifies the need for an ethic or guiding principle which would heed the intrinsic values of nature and intergenerational equity, and recognize that humanity is part of, not superior to, nature. An earth ethic based on such fundamental concepts would not only provide a sound foundation for wise decision-making, but would also be a moral rallying point in a world of shifting, relative and often meaningless values.

The Brundtland Commission (1987) offered some practical solutions to the ethical problems of temporal and spatial equity in resource use and consumption:

> Communities or governments can compensate for this isolation [of local people from central economic and political power] through laws, education, taxes, subsidies and other methods. Well-enforced laws and strict liability legislation can control harmful side effects. Most important, effective participation in decision-making processes by local communities can help them articulate and effectively enforce their common interest.

Such ideas are percolating down to the local level. For example, British Columbia's Forest Resources Commission recommended significant decentralization of land and resource allocation decision-making (Forest Resources Commission, 1991). An example of such decentralization may be noted in the fact that 77% of the voters in Revelstoke at a referendum held on February 20th, 1993, registered approval for purchase by the Municipality of Revelstoke of a controlling interest in one half of Tree Farm Licence Number 55, formerly held by Westar Timber Ltd. The other half of this tree farm licence will be controlled by Evans Forest Products Ltd. of Golden. Such local control of resources provides an opportunity for ecologically sensitive resource management. Local communities then have the opportunity to tackle their thorny unemployment problems and environmental blight by choosing labour intensive over machine

417

intensive projects. Brush and weed clearing with manual methods rather than by the broad scale application of herbicides, with their attendant problems of drift and toxic runoff, is a case in point. Such a choice reduces stress on the land while providing income and self esteem for the unemployed—it is a doubly ethical choice, for the earth and for its people.

To arrive at an earth ethic it will be necessary to break with the linear concept of progress that has dulled our senses since its inception in the late 1700s. Each one of us needs to realize that moderate expectations are realistic, while immoderate expectations are neither socially desirable nor ethically proper on a finite planet whose interwoven life systems are the basis of strength and health for every living organism. We can live lower on the food chain, live with newspapers a fourth of their present size, live without junk mail, live with less travel than presently occurs, and live with fewer, but ecologically safer products. We could develop procedures whose immediate adoption would bend the world away from the suicidal military and industrial policies that are desecrating the planet and fostering barbarism. There are many "how-to's" once we decide to change, but none as long as we prefer only cosmetic change. We would be healthier, happier, and our lives would be vastly enriched through development of a conscientious respect for the ecosphere.

The September, 1992, message that the ozone hole over the South Pole has increased by 15% to 8.9 million square miles and now occupies an area nearly the size of the North American continent is sufficient warning that time is of the essence. (1992 Antarctic Ozone Hole, 1992). While many species of polar marine life are already producing substances which act like sunscreen (Hanson, 1991), it is not likely that our species will be able to react in similar fashion. The scientific concern that, "Without the protective ozone veil, ultraviolet B would scrub earth clean of life," (Hanson, 1991), must certainly lend impetus to the desire for remedial change. While not all scientists predict catastrophe, there is agreement that a large increase in ultraviolet radiation would induce widespread, substantial change.

It will be difficult to convince the business community that "business as usual" is no longer a suitable philosophy. Yet, we may well be in a radical situation that will not be served by the tortuously slow change characteristic of this philosophy.

Some idea of the magnitude of the change we need was given by Harvard University paleontologist Stephen Jay Gould (1990) in his column in *Natural History*. He argues that it would be *enlightened self interest* on our part to adopt the "Golden Rule" as the major principle in our relationship toward the planet. Pointing out that the planet holds all the cards, such a principle, he says, "would be a blessing for us, and an indulgence for her." He also contends that, "We had better sign the papers while she is still willing to make a deal." He suggests that if we make the effort, earth will uphold her end of the bargain. But, he continues, "If we scratch her, she will bleed, kick us out, bandage up, and go about her business at her planetary scale."

Civilizations grow through fresh insights and accommodations, but ours has stagnated in the mire of unbridled greed and materialism. Evidence of ecological stress on the planet mounts steadily.

While the aims of society do not shift overnight, an actual, rather than a merely oratorical shift toward an ecocentric view of life on earth would indicate maturation of our civilization. The fourth dimension, time, urges that we change before there is no longer a possibility of change. When we have learned to act as restorers, rather than as ravagers of the planet, we will have earned our scientific name, Homo sapiens, "wise person."

The unfamiliar word we now need introduce to our vocabularies and behaviour is *usufruct*. George Perkins Marsh (1864), stated something worth carving in stone at the entrances to the halls of Academia. "The earth was given to (humankind) for usufruct alone, not for consumption, still less for profligate waste."

As *Webster's New Twentieth Century Unabridged Dictionary* says: "Usufruct—in law, the right of enjoying a thing which belongs to another and of deriving from it all the profit or benefit it may produce, *provided it be without altering or damaging the substance of the thing.*" (emphasis added).

It is with this sort of awareness that an earth ethic needs to be approached.

Acknowledgements

Dr. William Rees, University of British Columbia, reviewed the manuscript.

References

1992 Antarctic ozone hole biggest on records. Anonymous. 1992. *Global Environmental Change Report.* Vol. IV, No. 19, Oct. 9th, 1992. Arlington, Ma. p. 4.

Aurelius, Marcus. 1964. *Meditations. trans. Maxwell Staniforth.* Penguin Books, Harmondsworth, England. pp. 73, 78.

Bailey, L.H. 1915. *The Holy Earth. Charles Scribner's Sons.* New York; reprinted edition, State University of New York, 1980, Ithaca, New York. p.13.

Forest Resources Commission. 1991. *The Future of our Forests.* Province of British Columbia.

Gould, S.J. 1990. This view of life, the golden rule—a proper scale for our environmental crisis. *Natural History.* September, 1990. pp. 24-30.

Haeckel, E. 1929. *The Riddle of the Universe.* Watts and Co. London. p. 9.

Harrington, R.F. 1990. *To Heal the Earth: The Case for an Earth Ethic.* Hancock House Publishers Ltd, Surrey, B.C. p. 223.

Hanson, B. 1991. The ultraviolet zone. *The Amicus Journal, Summer, 1991.* New York, New York. pp. 24-25.

Hyams, Edward. 1952. *Soil and Civilization.* Thames and Hudson, London and New York.

Leopold, A. 1949. *A Sand County Almanac.* Reprinted in 1987. Oxford University Press. New York and Oxford. pp. 201-226.

Marsh, G.P. 1864. *Man and Nature.* David Lowenthal (ed.). Reprinted in 1967. Harvard University Press, Belknap Press. Cambridge, Mass. pp. 36, 43-52, 279-281.

McLuhan, T.E. 1971. *Touch the Earth! A Self-portrait of Indian Existence.* Pocket Books, New York. p. 36.

Rees, W.E., 1992. Ecological footprints and appropriated carrying capacity: what urban economics leaves out. *Environment and Urbanization* 4(2):121-130.

Schweitzer, A. 1923. *The Decay and the Restoration of Civilization.* Unwin Books, London. pp. 59, 62, 68-70.

Taylor, P. 1992. The failure of international environmental law. *Our Planet*. United Nations Environment Program. Nairobi, Kenya. 4(3):14-15.

Tzu, L. 1955. *The Way of Life*. New American Library, Inc., Mentor Books. New York and Toronto. pp. 81, 117.

World Commission on Environment and Development (The Brundtland Commission). 1987. *Our Common Future*. Oxford University Press.

Chapter 30
Conclusions and Recommendations

Ecologically speaking, British Columbia is relatively healthy. Its forest, grassland, and marine ecosystems, though altered by human activity, are more diverse than those in any other part of Canada and there are still vast areas of relatively untouched mountain wilderness. These ecosystems support a tremendous diversity of species and subspecies, some of which are unique to Canada.

Lee E. Harding
Canadian Wildlife Service
P.O. Box 340
Delta, B.C.
V4K 3Y3

Yet, the chapters of this book are permeated with a sense of urgency. The authors have observed here, and in similar ecosystems elsewhere, the impacts of a rapidly expanding human population that, for the most part, still views the natural environment as infinitely exploitable. To a person, they have concluded there is little time left in which to develop the systems needed to protect the biodiversity of this province. While they highly approve of recent government initiatives, they stress the urgent need for further action and more profound changes in management systems. For example, the authors of Chapters 4-16 believe that at least 50 species of invertebrates, perhaps 15 species of benthic marine algae, five amphibians, 26 fish, 63 bryophytes, 124 vascular plants and several bird species are threatened or endangered, and many more (such as 634 vascular plants) may be rare or vulnerable. At least 23 species and subspecies (five mammals, four birds, one reptile, one fish, five bryophytes, four vascular plants, and three invertebrates) have already gone extinct or been extirpated from the province, although three have since been reintroduced. A further 26 plants have been recorded historically, but have not been seen in the province since 1950. Many more species, while still secure somewhere in the province, have disappeared from some regions. Similarly, 61 of the 124 classified biogeoclimatic zones, subzones and variants, comprising about 37% of the province's land area, have less than 1% of their area left in the "primitive" state. Within the four most fragmented ecosystems—the Coastal and Interior Douglas fir Zones, the Bunchgrass Zone and the Ponderosa Pine Zone—are many of the most endangered species. The authors' recommendations for conserving species and ecosystem diversity fall into five categories: philosophical change, resource management, research and monitoring, institutional arrangements, and legislation.

Philosophical Change

If we are to protect biodiversity, then we must change our philosophical approach to our natural environment. Recognizing the commitment of time and money that would be required to do detailed ecological studies of the few species that have been identified as rare and endangered (let alone those that have not yet been identified), many authors agree that a more habitat and ecosystem based approach to conservation is required. Specific recommendations are that we need to:

- shift from a homocentric to an ecocentric viewpoint of our natural environment (Chapter 1);

- change our focus from the study and conservation of components of ecosystems to that of ecosystems and the ecosphere as a whole (Chapters 1, 19, 25, 28 and 29); and

- adopt an earth ethic which places higher priority on long term conservation of natural systems than on unsustainable short term political and economic objectives (Chapters 1 and 29).

Resource Management

The need for philosophical change is strongly reflected in the conclusions and recommendations regarding resource management. Many authors point to habitat loss or fragmentation as a key cause of species decline and extirpation. They talk further of the many different impacts, from hunting to development to pollution, that can accumulate to place a species in jeopardy and recommend that only by conserving contiguous areas of habitat can a species be protected. Authors acknowledge that, at the ecosystem level, forest and grassland resource management have improved considerably, but conclude that conversion from natural to managed forests still represents an overall loss of natural ecosystem diversity, as does the large scale but subtle alteration of grasslands by grazing and cultivation. Though the marine environment of the Strait of Georgia remains diverse, there are early signs of stress from pollution (uptake of organic contaminants and prevalence of disease near pollution sources) and intensive fishing, which could potentially alter the trophic structure of the Strait. Specific recommendations include:

- shift from species based to ecosystem based protection and recovery programs for rare and endangered species (Chapters 8, 9, 12, 13, 14, 16, 17, 20, 25, and 28);

- allow for a greater density of protected areas in locations where there is an unusual density of rare flora and fauna, which may mean increasing habitat protection in the most populated areas of the province, particularly the southwest corner (Chapters 5, 6, 8, 10, 12, 13, 15 and 22);

- improve frameworks for determining ecosystem representativeness and establish more protected areas in those ecoregions that are poorly represented (Chapter 26);

- allow for buffers around and links between protected areas, particularly along climatic and hydrological gradients and in urban areas (Chapters 22, 25, 26 and 28);

- change forest harvest rates and some management practices to enable maintenance of natural physical and biological ecosystem structures, as well as important ecological functions such as nutrient cycling and mediation of hydrological and carbon cycles (Chapters 19 and 25);

○ identify the specific successional stages needed to achieve grassland biodiversity goals, and manage livestock to maintain or produce those stages (Chapter 20);

○ apply integrated coastal planning to the entire B.C. coast (Chapter 28);

○ discourage the importation and spread of alien species (Chapters 12, 13, 17 and 22);

○ ensure that resource exploitation is sustainable, both by the resource being exploited and by the ecosystem supporting the resource (Chapters 9, 11, 14, 21 and 29); and

○ plan development to minimize impacts on local biodiversity (Chapters 12, 13 and 14).

Research and Monitoring

The call for more research and monitoring is almost universal. Many authors conclude that the study of biodiversity is, to a large extent, dependent upon the disciplines of biosystematics and taxonomy and their infrastructure—zoological parks, herbaria, botanical gardens, museums and other specimen collections. Specific recommendations include:

○ increase funding for biosystematics and taxonomy and their infrastructure (Chapters 3, 4, 6, 12 and 15);

○ expand research and training in biosystematics and taxonomy so as to increase the conceptual and theoretical understanding of the elements of biodiversity (Chapter 3);

○ incorporate into biosystematics more powerful tools, such as multivariate statistical analyses of traditional morphological traits, chromosomes, and molecular genetics—used for delimiting the patterns of geographic variation and identifying unique races or populations in the province (Chapter 15);

○ survey biological diversity throughout the province and develop basic inventory and reference collections (Chapters 3-11, 14, 15, 20 and 25);

○ monitor changes in status, abundance or distribution of rare, threatened and endangered species and carry out further research into ecosystem dynamics (Chapter 28);

○ emphasize species/habitat relationships in grassland research and monitoring and document the short and long term impacts of grassland disturbances on biodiversity (Chapter 20);

○ study the extent and causes of the decline of neotropical migrant songbirds and determine whether there is any relationship between their decline and irruptions of the forest pests upon which many of them feed (Chapters 19 and 23); and

O complete paleoecological studies of major biogeoclimatic regions, gathering information on potential changes in species range, vegetation, hydrology and landscape process, to help develop a conservation strategy which ensures that native flora persist and new vegetation types arise from as broad a base of native species as possible (Chapter 25).

Institutional Arrangements

The recommendations for change in philosophical approach, resource management and research and monitoring lead most authors to the conclusion that institutional change is also needed. The detection of cumulative impacts, in particular, demands improved cooperation between institutions involved in tracking biodiversity. Specific recommendations include:

O enhance cooperation between provincial, federal, municipal and First Nations governments and develop a more formal mechanism for facilitating that cooperation on a regular basis (Chapter 28);

O encourage public, industry and government cooperation in biodiversity conservation (Chapters 20, 28, and 29);

O improve public education regarding the importance of biodiversity, particularly of the lesser known taxa and of ecosystems as a whole (Chapters 7, 12, 13, 20, 22 and 29);

O decentralize resource management decision-making to the community level, while maintaining sufficient provincial control to ensure sustainable resource use (Chapter 29);

O initiate a biodiversity evaluation program for privately owned lands so that development can be planned with biodiversity considerations in mind (Chapter 22); and

O consider developing a system for purchasing and placing particularly important sites in a land trust (Chapter 22).

Legislation and Regulation

The conclusion that improvements in legislation and regulation are needed is inescapable for many authors. Several point out that designation of a species or subspecies as endangered or threatened by the National Committee on the Status of Endangered Wildlife in Canada or placement on the province's Red List confers no legal protection. Others observed that there is no federal law in Canada pertaining specifically to the introduction of exotic marine organisms through shipping practises. Specific recommendations include:

○ expand the scope of endangered species legislation to include invertebrates, bryophytes, vascular plants and aquatic organisms (Chapters 4, 7, 10, 11, and 14);

○ identify and implement some means of maintaining air quality in forests intended to preserve lichen diversity—perhaps by granting British Columbian parks and ecoreserves some jurisdiction over air quality in their preserves (Chapter 8);

○ improve legal control over the trade in non-native species of plants and animals (Chapters 12 and 17);

○ continue control of the emissions of toxic chemicals and other pollutants into marine waters (Chapter 21); and

○ consider establishing, where practicable and not already in existence, municipal or district by-laws to protect trees and ecologically sensitive lands within urban environments (Chapter 22).

In Closing

If only God would again say, "Build an ark," we could do that and then march in the animals, two by two. But this time, it won't be that easy. The flood of humanity threatens to overrun the planet, and no ark of our construction could be big enough, nor do we have the time or the technology to even identify, let alone collect and preserve viable populations of every species. No, this time the ark is the Earth, and there is no safe haven. We have to stop the flood.

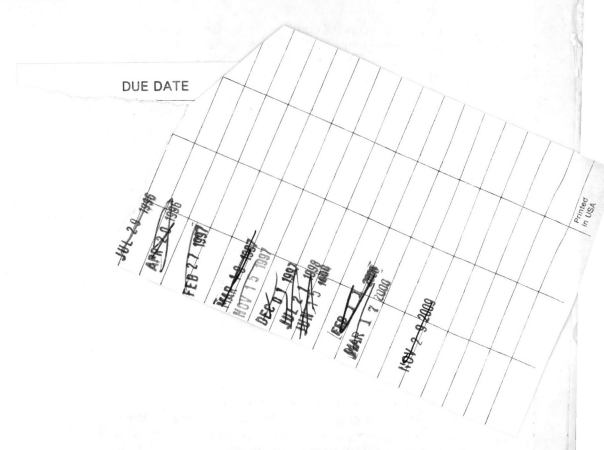